3D Stem Cell Culture

3D Stem Cell Culture

Editor

Joni H. Ylostalo

MDPI • Basel • Beijing • Wuhan • Barcelona • Belgrade • Manchester • Tokyo • Cluj • Tianjin

Editor
Joni H. Ylostalo
University of Mary Hardin-Baylor
USA

Editorial Office
MDPI
St. Alban-Anlage 66
4052 Basel, Switzerland

This is a reprint of articles from the Special Issue published online in the open access journal *Cells* (ISSN 2073-4409) (available at: https://www.mdpi.com/journal/cells/special_issues/3D_culture).

For citation purposes, cite each article independently as indicated on the article page online and as indicated below:

LastName, A.A.; LastName, B.B.; LastName, C.C. Article Title. *Journal Name* **Year**, *Volume Number*, Page Range.

ISBN 978-3-03943-803-7 (Hbk)
ISBN 978-3-03943-804-4 (PDF)

Contents

About the Editor

Joni H. Ylostalo is an Associate Professor at the University of Mary Hardin-Baylor (UMHB). Dr. Ylostalo obtained his BSc in Biochemistry and MSc in Biochemistry, with a focus in Biotechnology and Molecular Biology, from the University of Oulu. His MSc research involved biochemical studies of type IX collagen. Dr. Ylostalo obtained his PhD in Biomedical Sciences from Tulane University, where he worked with mesenchymal stem cells (MSCs). His research was focused on gene expression changes of MSCs during differentiation and culture expansion. Dr. Ylostalo completed his postdoctoral work at Texas A&M, studying the aggregation and 3D culture of MSCs to enhance their therapeutic potential. He continued his MSC research at Texas A&M as a faculty member focusing on the therapeutic applications of 3D culture-activated MSCs. Since transitioning to UMHB, he has continued his MSC research among other research and service opportunities. Dr. Ylostalo has published over 50 articles that span the fields of biochemistry, cell biology, bioinformatics, regenerative medicine, and biology education. He has served as an editor and reviewer for numerous journals while giving plenary talks at various conferences.

Editorial

3D Stem Cell Culture

Joni H. Ylostalo

Department of Biology, University of Mary Hardin-Baylor, 900 College Street, Box 8432, Belton, TX 76513, USA; jylostalo@umhb.edu; Tel.: +1-254-295-5534

Received: 16 September 2020; Accepted: 24 September 2020; Published: 27 September 2020

Abstract: Much interest has been directed towards stem cells, both in basic and translational research, to understand basic stem cell biology and to develop new therapies for many disorders. In general, stem cells can be cultured with relative ease, however, most common culture methods for stem cells employ 2D techniques using plastic. These cultures do not well represent the stem cell niches in the body, which are delicate microenvironments composed of not only stem cells, but also supporting stromal cells, extracellular matrix, and growth factors. Therefore, researchers and clinicians have been seeking optimal stem cell preparations for basic research and clinical applications, and these might be attainable through 3D culture of stem cells. The 3D cultures recapitulate the in vivo cell-to-cell and cell-to-matrix interactions more effectively, and the cells in 3D cultures exhibit many unique and desirable characteristics. The culture of stem cells in 3D may employ various matrices or scaffolds, in addition to the cells, to support the complex structures. The goal of this Special Issue is to bring together recent research on 3D cultures of various stem cells to increase the basic understanding of stem cells and culture techniques, and also highlight stem cell preparations for possible novel therapeutic applications.

Keywords: stem cell; 3D; culture condition; expansion; niche; regenerative medicine; scaffold; organoid

---

Stem cells are cells that demonstrate the abilities to self-renew and differentiate. Many types of stem cells can be isolated from embryonic or adult tissues, varying in their potency from pluripotent to unipotent depending on the stem cell type. Many cells can be found throughout the human body, often localized in niches that provide a nurturing microenvironment for the stem cells while directing their proliferation and differentiation. Stem cells are typically cultured in 2D tissue culture plastic for ease and maximal expansion, as large numbers are often needed for translational research and therapies. The 2D cultures do not, however, represent the natural environment of stem cells in the body well. Many beneficial properties of stem cells might be lowered or even lost in 2D cultures. Therefore, the use of 3D culture techniques has become more common in basic and translational research. The 3D cultures often provide more complete cell-to-cell and cell-to-matrix interactions, mimicking the natural environment in which the stem cells reside better than the traditional 2D cultures. Furthermore, many desired cellular characteristics are maintained or even promoted in 3D cultures, further supporting their use in basic and translational research. In this Special Issue, recent advances in the 3D cultures of stem cells are highlighted, bringing together a collection of articles presenting various stem cell types and their characteristics in 3D environments.

This Special Issue includes articles highlighting the effective use of induced pluripotent stem cells (iPSCs) to develop cardiac microtissue [1], neurospheres [2], and cortical progenitors [3]. In addition, this Special Issue contains work on scaffolds with mesenchymal stem cells (MSCs) [4] and embryonic stem cells (ESCs) [5]. Furthermore, research on generating intestinal organoids from stem cells [6,7] and nephrogenesis studies utilizing ESCs [8] are included in this Special Issue, along with a comprehensive review on the 3D culture of hematopoietic stem cells (HSCs) [9].

A very relevant problem in cardiology is cardiac fibrosis. In the work by Blyszczuk et al., a model of fibrotic cardiac microtissue was generated using iPSC-derived cardiomyocytes and cardiac fibroblasts by treatment with transforming growth factor β1 (TGF-β1) or by use of cardiac fibroblasts from heart failure patients [1]. The authors demonstrated that activated cardiac fibroblasts could, via direct stimulation of β-adrenoreceptor signaling, promote cardiac contraction rate. Furthermore, the generated model could be used as a high-throughput model for drug testing or general studies of cardiac fibrosis [1].

A 3D in vitro model utilizing iPSCs was generated by Kobolak et al. to study neurotoxicity [2]. In this work, iPSC-derived, 3D, free-floating neurospheres, exhibiting the various cells of the nervous system, were developed and tested in neurotoxicity studies. Furthermore, these neurospheres could open some further opportunities for the detection of developmental neurotoxicity, and hence support the existing animal models [2].

The work by Cutarelli et al. employed iPSC-derived cortical progenitors and scaffolds to mimic the radially oriented cortical radial glia fibers that play an important role in the development of the cerebral cortex [3]. The authors used silicon vertical micropillar arrays to promote the expansion and preservation of stemness of the cortical progenitors. Furthermore, the described model of iPSC-derived cortical progenitors and silicon micropillars could be used in cortical tissue engineering [3].

Another in vitro 3D model was generated by Ciardulli et al. for the study of MSCs and their tenogenic differentiation [4]. This work employed a hyoluronate/poly-lactic-co-glycolic acid (PLGA)/fibrin 3D scaffold with MSCs that were studied under static and cyclic-strain conditions. The research demonstrated that MSCs grown in this scaffold increased their tenogenic marker and pro-repair cytokine expression, supporting the notion of this model as a potential predictive system to be used in future studies employing the sustained release of biochemicals [4].

To circumvent some of the challenges with the 2D culture expansion of ESCs, McKee et al. developed a 3D scaffold to mimic in vivo stem cell niches [5]. The authors used polyethylene glycol (PEG) polymers with thiol and acrylate end-groups to guide the self-assembly of the scaffolds. ESCs grown in these scaffolds were able to maintain their viability, self-renewal, and differentiation potential. ESCs in the scaffolds also exhibited a high expression of pluripotency markers and some mechanosensitive genes, supporting the notion of these scaffolds as potentially helpful for many ESC studies [5].

To aid in the study of intestinal disorders, Kramer et al. developed intestinal organoids from stem cells obtained from various parts of the intestine [6]. These intestinal organoids showed long-term expansion of the cells while inhibiting cell differentiation when cultured in expansion media, while differentiation into goblet and enteroendocrine cells was promoted with culture in differentiation media. These organoids could possibly be employed in various in vitro models of functional intestinal disorders [6].

Another study utilizing intestinal stem cell organoids explored the cellular origin of carcinogenesis [7]. In this study, Yen et al. developed a tumorigenesis model based on carcinogenesis and genetically engineered mice. The authors were able to demonstrate the interplay between extrinsic carcinogen and intrinsic genetic modification and their contribution towards transformation while elucidating the involvement of molecular factors, such as protein phosphatase 2A (PP2A), in this process [7].

A study by Tan et al. utilized ESCs to generate ureteric bud progenitor cells [8]. The authors then used these progenitor cells to induce nephrogenesis in co-culture with primary metanephric mesenchyme. These kidney organoids exhibited nephron structures with collecting ducts connected to nephron tubules. This study demonstrated a relatively simple and reproducible way of generating ureteric bud progenitors [8].

This Special Issue includes articles highlighting recent discoveries in 3D stem cell cultures using MSCs [4], ESCs [5,8], iPSCs [1–3], and intestinal stem cells [6,7]. This Special Issue also includes a review article by Ribeiro-Filho et al. that discusses the 2D and 3D culture of hematopoietic stem cells (HSCs) as a relevant model to study both normal and abnormal hematopoiesis [9]. Furthermore, this Special

Issue includes articles that utilize various scaffolds together with stem cells from various sources to study 3D cultures [3–5].

Funding: This research received no external funding.

Conflicts of Interest: The author declares no conflict of interest.

References

1. Błyszczuk, P.; Zuppinger, C.; Costa, A.; Nurzynska, D.; Di Meglio, F.D.; Stellato, M.; Agarkova, I.; Smith, G.L.; Distler, O.; Kania, G. Activated Cardiac Fibroblasts Control Contraction of Human Fibrotic Cardiac Microtissues by a β-Adrenoreceptor-Dependent Mechanism. *Cells* **2020**, *9*, 1270. [CrossRef]

2. Kobolak, J.; Teglasi, A.; Bellak, T.; Janstova, Z.; Molnar, K.; Zana, M.; Bock, I.; Laszlo, L.; Dinnyes, A. Human Induced Pluripotent Stem Cell-Derived 3D-Neurospheres are Suitable for Neurotoxicity Screening. *Cells* **2020**, *9*, 1122. [CrossRef] [PubMed]

3. Cutarelli, A.; Ghio, S.; Zasso, J.; Speccher, A.; Scarduelli, G.; Roccuzzo, M.; Crivellari, M.; Maria Pugno, N.; Casarosa, S.; Boscardin, M.; et al. Vertically-Aligned Functionalized Silicon Micropillars for 3D Culture of Human Pluripotent Stem Cell-Derived Cortical Progenitors. *Cells* **2019**, *9*, 88. [CrossRef] [PubMed]

4. Ciardulli, M.C.; Marino, L.; Lovecchio, J.; Giordano, E.; Forsyth, N.R.; Selleri, C.; Maffulli, N.; Porta, G.D. Tendon and Cytokine Marker Expression by Human Bone Marrow Mesenchymal Stem Cells in a Hyaluronate/Poly-Lactic-Co-Glycolic Acid (PLGA)/Fibrin Three-Dimensional (3D) Scaffold. *Cells* **2020**, *9*, 1268. [CrossRef] [PubMed]

5. McKee, C.; Brown, C.; Chaudhry, G.R. Self-Assembling Scaffolds Supported Long-Term Growth of Human Primed Embryonic Stem Cells and Upregulated Core and Naïve Pluripotent Markers. *Cells* **2019**, *8*, 1650. [CrossRef] [PubMed]

6. Kramer, N.; Pratscher, B.; Meneses, A.M.C.; Tschulenk, W.; Walter, I.; Swoboda, A.; Kruitwagen, H.S.; Schneeberger, K.; Penning, L.C.; Spee, B.; et al. Generation of Differentiating and Long-Living Intestinal Organoids Reflecting the Cellular Diversity of Canine Intestine. *Cells* **2020**, *9*, 822. [CrossRef] [PubMed]

7. Yen, Y.T.; Chien, M.; Lai, Y.C.; Chen, D.P.; Chuong, C.M.; Hung, M.C.; Hung, S.C. PP2A Deficiency Enhances Carcinogenesis of Lgr5. *Cells* **2019**, *9*, 90. [CrossRef] [PubMed]

8. Tan, Z.; Rak-Raszewska, A.; Skovorodkin, I.; Vainio, S.J. Mouse Embryonic Stem Cell-Derived Ureteric Bud Progenitors Induce Nephrogenesis. *Cells* **2020**, *9*, 329. [CrossRef] [PubMed]

9. Ribeiro-Filho, A.C.; Levy, D.; Ruiz, J.L.M.; Mantovani, M.D.C.; Bydlowski, S.P. Traditional and Advanced Cell Cultures in Hematopoietic Stem Cell Studies. *Cells* **2019**, *8*, 1628. [CrossRef] [PubMed]

Article

Activated Cardiac Fibroblasts Control Contraction of Human Fibrotic Cardiac Microtissues by a β-Adrenoreceptor-Dependent Mechanism

Przemysław Błyszczuk [1,2,*], Christian Zuppinger [3], Ana Costa [4], Daria Nurzynska [5], Franca Di Meglio [5], Mara Stellato [1], Irina Agarkova [6], Godfrey L. Smith [4], Oliver Distler [1] and Gabriela Kania [1,*]

1   Center of Experimental Rheumatology, Department of Rheumatology, University Hospital Zurich, Wagistr. 14, 8952 Schlieren, Switzerland
2   Department of Clinical Immunology, Jagiellonian University Medical College, 30-663 Cracow, Poland
3   Department for BioMedical Research, Department of Cardiology, University Hospital Bern, 3008 Bern, Switzerland
4   Institute of Cardiovascular and Medical Sciences, University of Glasgow, Glasgow G12 8TA, UK
5   Department of Public Health, University of Naples "Federico II", 80131 Naples, Italy
6   InSphero, 8952 Schlieren, Switzerland
*   Correspondence: przemyslaw.blyszczuk@uzh.ch (P.B.); gabriela.kania@uzh.ch (G.K.)

Received: 13 April 2020; Accepted: 19 May 2020; Published: 20 May 2020

Abstract: Cardiac fibrosis represents a serious clinical problem. Development of novel treatment strategies is currently restricted by the lack of the relevant experimental models in a human genetic context. In this study, we fabricated self-aggregating, scaffold-free, 3D cardiac microtissues using human inducible pluripotent stem cell (iPSC)-derived cardiomyocytes and human cardiac fibroblasts. Fibrotic condition was obtained by treatment of cardiac microtissues with profibrotic cytokine transforming growth factor β1 (TGF-β1), preactivation of foetal cardiac fibroblasts with TGF-β1, or by the use of cardiac fibroblasts obtained from heart failure patients. In our model, TGF-β1 effectively induced profibrotic changes in cardiac fibroblasts and in cardiac microtissues. Fibrotic phenotype of cardiac microtissues was inhibited by treatment with TGF-β-receptor type 1 inhibitor SD208 in a dose-dependent manner. We observed that fibrotic cardiac microtissues substantially increased the spontaneous beating rate by shortening the relaxation phase and showed a lower contraction amplitude. Instead, no changes in action potential profile were detected. Furthermore, we demonstrated that contraction of human cardiac microtissues could be modulated by direct electrical stimulation or treatment with the β-adrenergic receptor agonist isoproterenol. However, in the absence of exogenous agonists, the β-adrenoreceptor blocker nadolol decreased beating rate of fibrotic cardiac microtissues by prolonging relaxation time. Thus, our data suggest that in fibrosis, activated cardiac fibroblasts could promote cardiac contraction rate by a direct stimulation of β-adrenoreceptor signalling. In conclusion, a model of fibrotic cardiac microtissues can be used as a high-throughput model for drug testing and to study cellular and molecular mechanisms of cardiac fibrosis.

Keywords: cardiac microtissues; iPSC-derived cardiomyocytes; cardiac fibroblasts; cardiac fibrosis; cardiac rhythm; TGF-β signalling; drug screening; in vitro model

---

1. Introduction

Cardiac fibrosis refers to an excessive accumulation of stromal cells and extracellular matrix (ECM) proteins in the myocardium and represents a common pathophysiological scenario in a broad range of cardiovascular conditions, including myocardial infarction, hypertension, myocarditis, and hypertrophic or dilated cardiomyopathy [1]. Cardiac fibroblasts and myofibroblasts represent the most extensively characterised stromal cell types involved in fibrotic processes in the heart [2]. In the traditional view, resident cardiac fibroblasts become activated and overproduce ECM proteins such as collagen type I and III and fibronectin. Fibrogenesis is regulated by multiple profibrotic inputs, which activate a complex signalling network. In cardiac fibrosis, transforming growth factor β (TGF-β) is recognised as the key profibrotic cytokine activating quiescent cardiac fibroblasts. Constitutive overexpression of TGF-β was shown to induce interstitial cardiac fibrosis and cardiac hypertrophy in transgenic mice [3], whereas targeting TGF-β or its downstream mediators successfully reduced or prevented cardiac fibrosis in various animal models [4–6]. Progressive fibrosis causes tissue stiffening and may affect cardiac rhythm. Clinical data associate fibrotic heart condition with increased resting heart rate [7] and with elevated risk of developing life-threating arrhythmias [8,9]. The underlaying mechanisms remain obscure, however.

In the body, heart rhythm established by pacemaker cells located in the sinoatrial node of the heart is mainly under control of the autonomic nervous system. Sympathetic stimulation increases the rate, whereas parasympathetic stimulation decreases it. Neurotransmitter noradrenaline released by the sympathetic neurons increases heart rate by activating β-adrenergic receptors on cardiomyocytes. Under stress condition, β-adrenoreceptors are further activated by adrenaline—a catecholamine produced by the adrenal glands and released into the circulation. Binding of agonists to β-adrenergic receptors causes the formation of cyclic adenosine monophosphate and activation of protein kinase A that phosphorylates a number of proteins including myosin light chains and thus increases the ability of cardiomyocytes to relax following excitation contraction coupling [10]. In the heart, the increased relaxation is associated with increased rate and strength of contraction.

Because extracardiac stimuli are important regulators of the heart rhythm, it is difficult to study intracardiac mechanisms of cardiac contractility using in vivo systems only. Data from ex vivo and in vitro cardiac models may shed more light on specific aspects of intracardiac rhythm regulation. Disease models using human cells offer a natural advantage over animal models by providing a proper genetic and cellular context. In cardiovascular research, difficulties in obtaining and culturing contractile primary human cardiomyocytes limited the development of in vitro human models for many years. The human induced pluripotent stem cell (iPSC) technology opened new perspectives and became a commonly used, unlimited source of spontaneously beating cardiomyocytes for regenerative medicine, drug discovery and toxicity screening [11].

A lack of the physiologically relevant experimental setting is a major drawback of many in vitro models. In general, traditional two-dimensional (2D) cell culture systems poorly mirror the biomechanical and biochemical microenvironment of the tissue. Three-dimensional (3D) cell culture technologies, instead, much better reflect the organ complexity and represent an attractive alternative to the commonly used 2D models. 3D cell culture systems, called microtissues or organoids, include cellular models on various bioscaffolds, but also scaffold-free spheroids comprising of one or more cell types. In recent years, various types of cardiac microtissues have been successfully used predominantly in regenerative medicine, but also as platforms for drug screening or as disease models [12]. Self-aggregating, scaffold-free cardiac microtissues consisting of human iPSC-derived cardiomyocytes and human primary cardiac fibroblasts cultured in nonadherent plates represent a high-throughput in vitro model of human cardiac tissue. These spontaneously beating cardiac microtissues show typical structures including well-developed myofibrils and recapitulate cardiac functionality including responsiveness to electrical stimulation [13]. In this work, we aimed to address how fibrotic changes affect contractility of human cardiac microtissues. We found that profibrotic condition and, in particular, activation of cardiac fibroblasts increase beating rate of cardiac microtissues, and this effect is at least partially mediated by endogenous activation of β-adrenergic receptors. Furthermore, we demonstrated that human cardiac microtissues represent an easy-to-use, reproducible high-throughput method for small molecule screening.

2. Material and Methods

2.1. Cellular Sources

Frozen human iPSC-derived cardiomyocytes (>99% troponin-positive) were purchased from Cellular Dynamics International, and were used directly upon thawing. Foetal human cardiac fibroblasts were purchased from Sigma (Cell Applications) and cells from passages 10–15 (>99% vimentin-positive, >99% collagen I-positive and <5% positive for filamentous form of α-SMA) were used. Adult cardiac fibroblasts were isolated from cardiac tissue samples obtained from the left atria of hearts from the patients with end-stage heart failure due to ischemic heart disease undergoing heart transplantation and from the waste fragments of donor normal hearts that were trimmed off while adjusting atrium size and shape during transplantation. Specimens were collected without patient identifiers according to the protocol revised by the Ethics Committee of the University of Naples Federico II (approval number 79/18) and in conformity with the principles outlined in the Declaration of Helsinki. Patients characteristic is available in Supplementary Table S1.

2.2. Isolation of Human Adult Cardiac Fibroblasts

Adult cardiac fibroblasts were isolated as described previously [14]. Briefly, cardiac tissue samples were minced and then enzymatically disaggregated by incubation in 0.25% trypsin and 0.1% (*w/v*) collagenase II (both Sigma-Aldrich, Basel, Switzerland) for 30 min at 37 °C. The digestion was stopped by adding Hanks' Balanced Salt solution supplemented with 10% foetal bovine serum (FBS, Gibco, Paislay, UK). The tissue was further disaggregated by pipetting; noncardiomyocyte cells were separated from debris and cardiomyocytes by sequential centrifugation and passage through 20 μm cell sieve. Fibroblasts were isolated from cell suspension by immunomagnetic cell sorting through positive selection with anti-fibroblast MicroBeads (MiltenyiBiotec, Bergisch Gladbach, Germany). Following selection, fibroblasts were cultured in Dulbecco's Modified Eagle's Medium (DMEM, Sigma, Basel, Switzerland) supplemented with 10% FBS, penicillin 10,000 U, and streptomycin 10 mg/mL (all from Sigma, Basel, Switzerland) at 37 °C in 5% CO_2. The aCFs from patients with heart failure and unaffected donors were used individually. Cells from passages 4–6 were used.

2.3. Fabrication and Culture of Microtissues

Three types of microtissues containing: (1) iPSC-derived cardiomyocytes and fibroblasts mixed in ratio 4:1, (2) fibroblasts only, and (3) iPSC-derived cardiomyocytes only were generated by self-assembling in 96-well GravityTRAP plates (InSphero, Schlieren, Switzerland), and cultured as one microtissue per well. Accordingly, for generation of one microtissue, 5000 cells suspended in 70 μL maintenance medium were placed in the GravityTRAP plates for three days in the slanted position. Maintenance medium contained 2% FBS (Gibco, Paislay, UK), 50 μM phenylephrine hydrochloride (Sigma), 0.3 μM L-ascorbic acid (Sigma), 50 μM 2-mercaptoethanol (Gibco, Paislay, UK), 50 U/mL penicillin/streptomycin (Gibco, Paislay, UK) in high glucose DMEM (Sigma). Upon self-assembly, microtissues were further cultured in the GravityTRAP plates in the maintenance medium for 10 days. Microtissues were cultured under standard culture conditions (at 37 °C, 5% CO_2 in a humidified incubator). Medium was changed at days 0, 2, 4, and 7. Fibrotic differentiation of cardiac microtissues was induced at day 0 with 10 ng/mL recombinant human TGF-β1 (Peprotech, London, UK) and TGF-βR1 was blocked with 10^{-8}–10^{-5} g/mL SD208 (Tocris, Zug, Switzerland). To address β-adrenoceptor-dependent mechanisms, microtissues were treated with 10 nM isoproterenol (Tocris) and/or 1 μM nadolol (Sigma) 15–90 min prior to video recording. Controls received solvents only.

2.4. Live Imaging and Video Analysis

Randomly chosen live microtissues were analysed using microscopes equipped with a humidified chamber at 37 °C and 5% CO_2. Pictures of microtissues were captured using the Live Cell Imaging System (Olympus, Shinjuku, Japan) and size of microtissues was measured with the Xcellence Pro software (Olympus) as area in pixels. Increase in size was calculated as the size of microtissue at day 10/size of microtissue at day 0. Videos of contracting microtissues were recorded using the AxioObserver Z1 (Zeiss, Hombrechtikon, Switzerland) and the ZEN software. Videos were processed using Fiji software and custom-made macro. Contraction analysis was performed using Fiji software and MUSCLEMOTION macro [15].

2.5. FluoVolt Measurements

Voltage experiments were conducted on day 11. The microtissues were washed in serum-free medium consisting of DMEM (Gibco, Paislay, UK) supplemented with 10 mM D-Galactose (Sigma) and 1 mM sodium pyruvate (Sigma). The microtissues were loaded with voltage-sensitive dye: 0.1% FluoVolt and 1% PowerLoad (ThermoFisher, Waltham, WA, USA) in the above listed serum-free medium for 25 min at 37 °C 5% CO_2. The voltage-sensitive dye was removed by washing in serum-free medium, and the multiwell plate was placed in an environmentally controlled stage incubator (37 °C, 5% CO_2, >75% humidity) in the CellOPTIQ® platform (Clyde Biosciences, BioCity Scotland) 30 min before experimentation. The fluorescent signal was recorded with an excitation at 470 nm, and emitted light was collected from the entire microtissue using a 10× Fluor objective and the intensity recorded by a photomultiplier tube at 510–560 nm at 10 kHz. A 15 s recording was taken of each microtissue. Offline analysis was performed using CellOPTIQ®.

2.6. Electrical Pacing

Prior to experiment, microtissues were transferred into a glass bottom cell culture dish. Electrical field pacing of microtissues was performed using a modified glass bottom culture dish with inserted platinum wires attached to a MyoPacer (IonOptix, Westwood, MA, USA) set to a frequency of 3 Hz. Microtissues were recorded using the Eclipse TE2000-U microscope (Nikon, Tokyo, Japan) equipped with heating chamber, temperature controller (Ibidi, Graefelfing, Germany) and HeroBlack6 GoPro camera (Back-Bone Gear, Ontario, Canada).

2.7. Quantitative RT-PCR

For one sample, 12 microtissues were pooled. Total RNA was isolated using the Quick RNA Micro Kit (Zymo Research, Irvine, CA, USA). cDNAs were amplified using the cDNA Synthesis Kit (Roche). Gene expression was detected using oligonucleotides complementary to transcripts of the analysed genes and the GoTaq qPCR Master Mix (Promega, Dubendorf, Switzerland) using the 7500 Fast Real-Time PCR System (Applied Biosystems, Waltham, MA, USA). Oligonucleotide sequences are available in the Supplementary Table S2. Transcript levels of human RPLP0 were used as endogenous reference, and relative gene expression was analysed using the $2^{-\Delta\Delta Ct}$ method.

2.8. Immunocytochemistry

In total, 25–30 microtissues were pooled, fixed overnight with 4% paraformaldehyde and embedded in 2% agarose and in paraffin. Microtissue sections of 3–5 μm were cut (Sophistolab, Muttenz, Switzerland) and deparaffinised following standard methods. Sections were blocked with 10% goat serum (Vector Laboratories, Burlingame, CA, USA) and stained with primary/secondary antibodies: rabbit anti-human α-SMA (Sigma, dilution 1:750)/goat anti-mouse AP (Dako, Glostrup, Denmark, dilution 1:50), rabbit anti-human vimentin (Abcam, Cambridge, UK, dilution 1:250)/goat anti-rabbit (Vector Laboratories, Burlingame, CA, USA, dilution 1:200), rabbit anti-human periostin (Abcam, dilution 1:250)/Polymer anti-rabbit (Histofine, Nichirei Biosciences Inc. Tokyo, Japan), rabbit anti-human fibronectin (Abcam, dilution 1:100)/Polymer anti-rabbit (Histofine), rabbit anti-human Ki67 (Abcam, dilution 1:100)/goat anti-rabbit (Vector Laboratories, dilution 1:200), rabbit anti-human connexin 43 (Abcam, dilution 1:200)/goat anti-rabbit (Vector Laboratories, dilution 1:500), and rabbit anti-human troponin T (Origene Technologies, Rockville, Maryland, USA, dilution 1:2000)/Polymer anti-rabbit (Histofine). Detection was performed with the Bond Polymer HRP Refine Detection kit (Leica, Heerbrugg, Switzerland) according to the manufacturer's guidelines for troponin T, and with DAB or Vector Red (Vector Laboratories) for all other antibodies. Nuclei were counterstained with haematoxylin (J.T.Backer, Gliwice, Poland). Direct Red Sirius Red (Sigma-Aldrich, Basel, Switzerland) staining was used to detect collagen deposition. Stained sections were analysed using the Olympus DP80 microscope and the cellSens Standard imaging software (Olympus, Shinjuku, Japan). Intensity of immunopositive signals and nuclei counterstaining for each section were quantified using the ImageJ software. Staining intensity is presented in arbitrary units (AU) representing immunopositive signal intensity corrected by the counterstained nuclei value.

2.9. Procollagen Type I and IL-6 ELISA

Supernatants from individual microtissues were collected at day 10 and stored at −80 °C. Undiluted supernatants were analysed with the human Procollagen I alpha 1 DuoSet ELISA (R&D Systems, Abingdon, UK) and the human IL-6 DuoSet ELISA (R&D Systems) following manufacturer's protocol. Optical density was measured with the Synergy microplate reader (Biotek, Winooski, Vermont, USA) and the Gen5 software.

2.10. Caspase 3/7 Activity

Supernatants containing one microtissue each were collected at day 10 and immediately measured for apoptosis with the Caspase-Glo 3/7 Assay (Promega, Dubendorf, Switzerland) following the manufacturer's protocol. Luminescence was measured with the Synergy microplate reader (Biotek, Winooski, Vermont, USA) and the Gen5 software.

2.11. Western Blotting

Cellular proteins of foetal human cardiac fibroblasts were extracted with RIPA buffer (Sigma-Aldrich) supplemented with protease inhibitor cocktail (Complete ULTRA Tablets, Roche) and phosphatase inhibitors (PhosphoStop, Roche) from cultivated cells. Protein concentration was quantified by colorimetric bicinchoninic acid assay according to the manufacturer's protocol (Thermo Fisher Scientific). SDS-PAGE electrophoresis and wet-transfer method were used to separate and transfer proteins on nitrocellulose membranes, followed by 45 min incubation in blocking solution (tris buffered saline, Tween-20 (TBST, Thermo Fisher Scientific) containing 5% skim milk powder (Becton Dickinson AG, Allschwil, Swizterland). Membranes were incubated overnight with the following primary antibodies: anti-αSMA (1:1000, A2547, clone 1A4, Sigma-Aldrich), anti-Fibronectin (1:2000, Abcam: ab2413) or GAPDH (1:10,000, clone 14C10, #2118, Cell Signalling). Horseradish peroxidase (HRP)-conjugated secondary antibodies (1:5000) were used for detection with ECL substrate (SuperSignal West Pico Plus, Thermo Fisher Scientific) and development on the Fusion Fx (Vilber, Collegien, France). Densitometric analyses were performed with ImageJ 1.47t. Fold changes were computed after normalization to GAPDH.

2.12. Cell Contraction Assay

To assess the contractile properties of foetal human cardiac fibroblasts, the Contraction Assay Kit (Cell Biolabs, San Diego, CA, USA) was used following the manufacturer's protocol. Cells were cultivated with or without 10 ng/mL TGF-β for 72 h and reseeded in collagen gels for a further 72 h. Each condition was analysed in triplicates or quadruplicates. Images were taken at time 0, 24, 48 and 72 h after reseeding in a collagen gel. Areas of the gels were measured by ImageJ. Percentage of contraction of all conditions was measured compared to the average of unstimulated cells at day 0.

2.13. Statistics

Normally distributed data were analysed by two-tailed unpaired or paired Student's *t*-test and unpaired two-tailed ANOVA followed by uncorrected Fisher's LSD post-hoc test. All analyses were computed using GraphPad Prism 8 software. Differences were considered as statistically significant for $p < 0.05$.

3. Results

3.1. Cardiac Fibroblasts Improve Integrity and Contractility of Human Cardiac Microtissues

Using human iPSC-derived cardiomyocytes (iCMs) and human foetal cardiac fibroblasts (fCFs), we developed a high throughput in vitro model of human cardiac tissue [16]. Cardiac microtissues consisting of 5000 cells each were generated by self-assembly (Figure S1). iCMs assembling without fibroblasts, however, formed loose cell aggregates (Figure 1A, Video S1). Addition of fCFs to iCMs in a ratio of 1:4 (fCFs:iCMs) allowed for formation of compact, spontaneously contracting cardiac microtissues (Figure 1A,B, Video S2) In the next step, we performed transcriptional profiling of microtissues by qPCR for genes characteristic for cardiomyocytes and fibroblasts. Seven genes characteristic for cardiomyocytes were detected in microtissues containing iCMs (iCM and iCM:fCF microtissues), but not in those generated with fCFs only. In contrast, nine out of twelve genes associated with fibroblasts and fibrosis were consistently detected in all types of microtissues (Figure 1C). However, expression levels of most of fibroblastic genes were substantially higher in fCFs than in iCMs microtissues. As expected, the expression profile of iCMs:fCFs microtissues showed high levels of all cardiac and fibroblastic genes (Figure 1C).

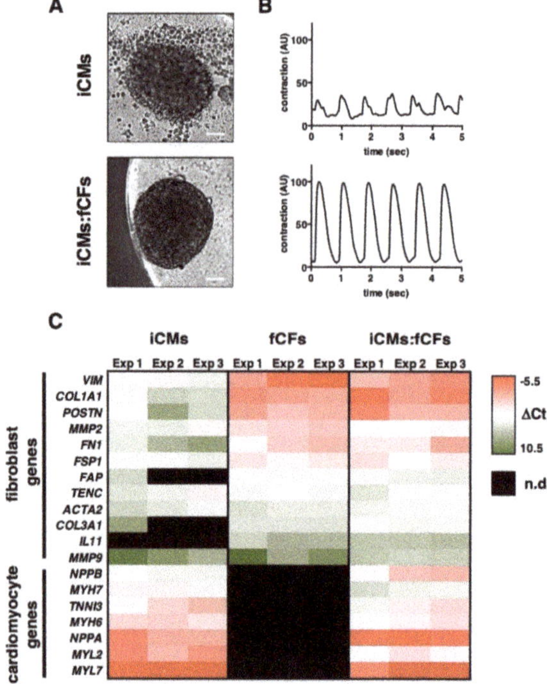

Figure 1. Characteristics of cardiac microtissues. Panel (**A**) illustrates typical morphologies (bar = 50 μm) and panel (**B**) shows typical contraction patterns of cardiac microtissues generated using iPSC-derived cardiomyocytes (iCMs) only (top) or iCMs mixed with foetal cardiac fibroblasts (fCFs) in a ratio of 4:1 (iCMs:fCFs, bottom). Representative contractions are available in the Supplementary Materials (Videos S1 and S2). A heat map in panel (**C**) indicates expression of cardiomyocyte and fibroblast genes in microtissues containing iCMs only (left), fCFs only (center) or iCMs:fCFs (right). Each segment indicates the average (n = 3–5) expression of one experiment. Lower $-\Delta$ C$_t$ values indicate higher relative expression. n.d.—not detected.

3.2. TGF-β1 Induces Fibrotic Changes in Human Cardiac Microtissues

TGF-β1 represents a potent inducer of profibrotic changes in cardiac fibroblasts. In the first step, we exposed microtissues to TGF-β1 for 10 days. In the presence of TGF-β1, both fCFs and iCMs:fCFs microtissues significantly increased their size (Figure 2A, Figure S2).

In the next step, we measured procollagen type I secretion, which is an important hallmark of the ongoing fibrotic processes. As expected, fCFs microtissues produced markedly more collagen I upon TGF-β1 stimulation. Under this stimulatory condition, significant increase in procollagen type I production was also observed for iCMs:fCFs microtissues (Figure 2B). Importantly, iCMs microtissues did not produce detectable levels of procollagen I (Figure 2B) pointing to fibroblasts as the main source of collagen I in cardiac microtissues. Increased collagen production in iCMs:fCFs microtissues treated with TGF-β1 was confirmed by picrosirius red staining (Figure 2C). To address whether treatment with TGF-β1 induced apoptosis in this model, we measured caspase 3/7 activity in iCMs:fCFs microtissues. We observed similar caspase 3/7 activity in iCMs:fCFs microtissues treated with and without TGF-β1 (Figure 2D). Additionally, iCMs:fCFs microtissues treated with TGF-β1 produced significantly less IL-6 (Figure 2E).

Cardiac fibrogenesis is a transcriptionally regulated process. Upon TGF-β1 stimulation, genes characteristic for cardiomyocytes were only slightly dysregulated in iCMs microtissues. In iCMs:fCFs microtissues, instead, cardiomyocyte-specific genes were downregulated following TGF-β1 treatment (Figure 2F). Analysis of fibrotic genes showed that most of them were strongly upregulated in fCFs and iCMs:fCFs microtissues in response to TGF-β1 (Figure 2F).

TGF-β1-induced changes in gene expression was followed by the analysis of the respective proteins in cardiac microtissues. To this aim, selected markers were analysed by immunohistochemistry in iCMs:fCFs microtissues cultured with or without TGF-β1. We found significantly less expression of the cardiac cell marker troponin T and gap junction protein connexin 43 in iCMs:fCFs microtissues exposed to TGF-β1. Instead, proteins produced by quiescent (vimentin, fibronectin) or activated (α-SMA, periostin) fibroblasts were more abundant in TGF-β1-treated cardiac microtissues. The number of proliferative (Ki67-positive) cells was also higher in iCMs:fCFs microtissues cultured with TGF-β1 (Figure 3, Figure S3). In conclusion, our results indicate that TGF-β1 effectively induced fibrogenesis in cardiac microtissues.

3.3. Pharmacological Targeting of TGF-βR1 Signalling Prevents from Fibrotic Changes in Human Cardiac Microtissues

In the next step, we asked whether the fibrotic cardiac microtissue model is relevant to test anti-fibrotic compounds. To this aim, we used the TGF-βR1 inhibitor SD208 and tested it in the iCMs:fCFs microtissues cultured in the presence or absence of TGF-β. We observed that SD208 inhibitor effectively blocked TGF-β1-mediated microtissue growth (Figure 4A).

Furthermore, we found that SD208 suppressed procollagen type I production in iCMs:fCFs microtissues cultured not only with, but also without TGF-β1 (Figure 4B). Given the effect of the inhibitor in the absence of exogenous TGF-β1, we analysed gene expression in iCMs:fCFs microtissues cultured with and without SD208. Expression of 12 out of 18 analysed genes were significantly changed in response to SD208 treatment (Figure 4C). Similar results were obtained for treatment of iCMs:fCFs microtissues with SD208 in the presence of TGF-β1 (Figure 4D). Next, we analysed dose-dependent responses of microtissues to SD208. In these experiments, iCMs:fCFs microtissues were cultured in the presence of TGF-β1 and serial dilutions of SD208. Our results showed dose-dependent responses of iCMs:fCFs microtissues to SD208 in terms of microtissue size (Figure 4E), collagen type I secretion (Figure 4F) and expression of profibrotic genes (Figure 4G). Taken together, these data suggest that cardiac microtissues represent a useful model for pharmacological studies.

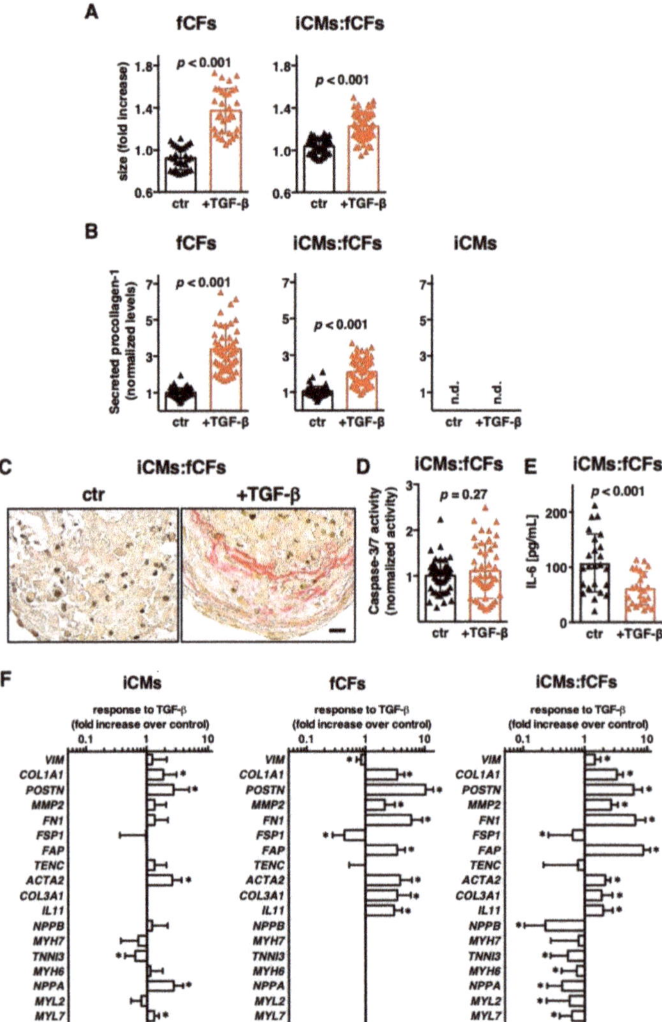

Figure 2. TGF-β1 activates foetal cardiac fibroblasts in microtissues. Panel (**A**) demonstrates changes in size of microtissues generated with fCFs only (fCFs, left) and iCMs mixed with fCFs in ratio 4:1 (iCMs:fCFs, right) cultured in the presence (red) or absence (black) of TGF-β1 (10 ng/mL) for 10 days. Panel (**B**) shows relative levels of procollagen I (measured by ELISA), at day 10 in supernatants of all three microtissue types: fCFs (left), iCMs:fCFs (middle) and iCMs (right). Graphs show cumulative data of 2–5 independent experiments. Each dot represents data of one microtissue. Panel (**C**) illustrates representative picrosirius red staining in iCMs:fCFs microtissues at day 10 (bar = 10 μm). Panel (**D**) shows caspase 3/7 activity measured at day 10 in iCMs:fCFs microtissues. Graphs show cumulative data of 3 independent experiments. Panel (**E**) shows IL-6 levels measured by ELISA, at day 10 in supernatants of iCMs:fCFs microtissues. Graphs show cumulative data of 3 independent experiments. Each triangle represents data of one microtissue. Panel (**F**) summarizes fold changes in gene expression in indicated microtissues in the presence of TGF-β1 (in relation to expression in the absence of TGF-β1). * $p < 0.05$. Graphs show cumulative data of 3–4 independent experiments, $n = 11$–20. For all graphs, p values were calculated with the Student's t-test. n.d.—not detected.

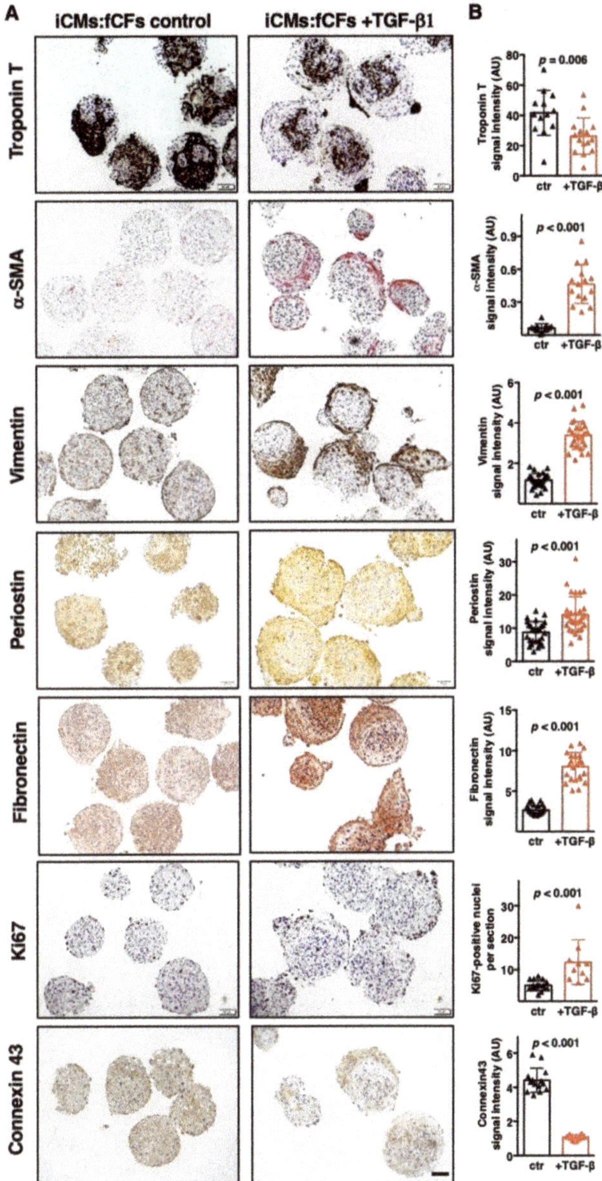

Figure 3. TGF-β1 induces fibrotic phenotype in cardiac microtissues. Immunohistochemistry of iCMs:fCFs microtissues cultured in the presence or absence of TGF-β1 (10 ng/mL) at day 10. Panel (**A**) illustrates representative staining for the indicated proteins at the indicated condition (bar = 50 μm). Higher magnification pictures are available in the Supplementary Materials (Figure S3). Panel (**B**) shows quantification of the respective staining for microtissues cultured in the presence (red) or absence (black) of TGF-β1. Graphs show cumulative data of 3 independent experiments. Each triangle represents data for one microtissue. *p* values were calculated with the Student's *t*-test.

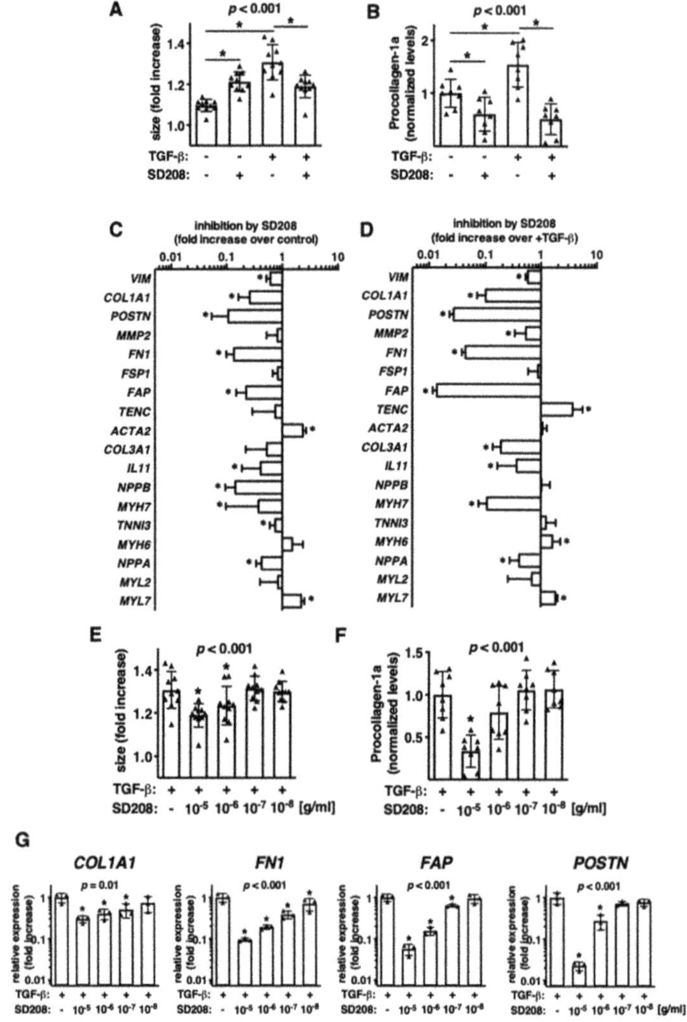

Figure 4. Pharmacological targeting of microtissues with TGF-βR1 inhibitor SD208. (**A–D**) iCMs:fCFs microtissues were cultured in the presence or absence of TGF-β1 (10 ng/mL) and SD208 (10 µg/mL). Panel (**A**) shows changes in size of microtissues and panel (**B**) normalized levels of secreted procollagen I (measured by ELISA in supernatants) at the indicated conditions at day 10. Each dot represents data for one microtissue. *p* values were calculated with ANOVA followed by uncorrected Fisher's LSD tests for selected groups, * *p* < 0.05. Panels (**C,D**) show changes in gene expression in response to treatment with SD208 of microtissues cultured in the absence (**C**) or presence (**D**) of TGF-β1. *n* = 6, *p* values were calculated with the Student's *t*-test, * *p* < 0.05. (**E–G**) iCMs:fCFs microtissues were cultured in the presence of TGF-β1 (10 ng/mL) and serial dilutions of SD208. Panel (**E**) shows changes in microtissue size, panel (**F**) normalized levels of secreted procollagen I and panel (**G**) relative expression of selected genes at the indicated conditions at day 10. Each dot represents data for one microtissue. Each triangle represents data for one microtissue. *p* values were calculated with ANOVA followed by uncorrected Fisher's LSD tests versus +TGF-β group, * *p* < 0.05.

3.4. Activated Cardiac Fibroblasts Increase the Contraction Rate of Cardiac Microtissues

Contractility represents the main functional feature of cardiomyocytes. Using high-speed movies and motion-tracking analysis, we examined contractile properties of iCMs:fCFs microtissues at day 10. Fibrotic changes triggered by TGF-β1 substantially deteriorated the contraction pattern of iCMs:fCFs microtissues. We consistently observed significantly increased contraction rate and reduced contraction amplitudes of fibrotic microtissues in all experiments. A more detailed analysis of contraction pointed mainly to the shortening of relaxation phase in iCMs:fCFs microtissues upon treatment with TGF-β1 (Figure 5A, Figures S4 and S5, Video S3).

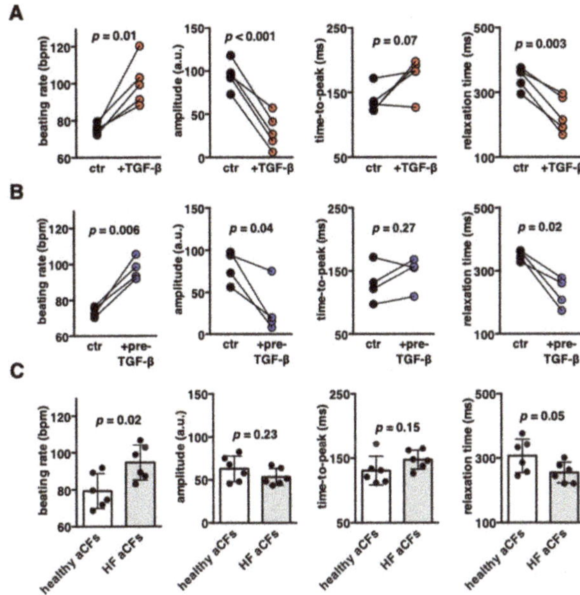

Figure 5. Contractile properties of cardiac microtissues containing foetal or adult cardiac fibroblasts. Panel (A) shows quantification of contraction parameters of iCMs:fCFs microtissues cultured in the presence (red) or absence (black) of TGF-β1 (10 ng/mL) at day 10. Quantification of contraction parameters of microtissues containing fCFs pretreated with TGF-β1 for 3 days prior microtissue formation (blue) or untreated fCFs (black) recorded at day 10 are shown in panel (B). Each dot represents average data of one experiment. Data of individual experiments are available in the Supplementary Materials (Figures S5 and S8). *p* values were calculated with the paired Student's *t*-test. Panel (C) shows quantification of contraction parameters of cardiac microtissues containing aCFs from unaffected hearts (white) or heart failure (HF) patients (grey). Each dot represents average data of one experiment (*n* = 14–18). *p* values were calculated with the Student's *t*-test. Representative contraction records are available in the Supplementary Materials (Figures S5, S10B and S11 and Videos S2–S6).

We hypothesized that TGF-β1 affected the contractility of iCMs:fCFs microtissues by activation of fCFs, rather than by a direct effect on iCMs. To validate this hypothesis, we generated cardiac microtissues using fCFs pretreated with TGF-β1 for three days (prior microtissue formation) and analysed their contractility in the absence of exogenous TGF-β1 at day 10. Pretreatment with TGF-β1 effectively activated fCFs as indicated by increased production of collagen, α-SMA and fibronectin in 2D cultures (Figure S6). We observed that iCMs:fCFs microtissues containing activated fCFs displayed increased size and secreted significantly more procollagen I (Figure S7), showed increased contraction rate, reduced contraction amplitudes and shortened relaxation phase (Figure 5B, Figures S4 and S8, Video S4), and successfully reproduced the contraction pattern observed in fibrotic (+TGF-β1) cardiac microtissues.

To further address the relevance of CFs to cardiac microtissue contractility, we compared cardiac microtissues containing adult CFs (aCFs) obtained from unaffected hearts of mostly young individuals (healthy aCFs) to aCFs from patients with heart failure (HF aCFs). Cardiac microtissues containing healthy aCFs and HF aCFs showed similar sizes at day 10 (Figures S9 and S10A). We found that iCMs:aCFs microtissues with HF aCFs showed increased contraction rate and shortened relaxation phase in comparison to cardiac microtissues containing healthy aCFs (Figure 5C, Figures S4 and S10B, Videos S5 and S6). Of note, microtissues containing healthy iCMs:aCFs more often responded to TGF-β1 stimulation by increasing contraction rate than those with HF aCFs (four of six vs. one of six, Figure S11). All these data indicate that contractile properties of cardiac microtissues depend on the activation status of CFs.

3.5. Unaffected Repolarization Phase in Fibrotic Cardiac Microtissues

In cardiomyocytes, mechanical contraction is triggered by electrical excitation. Next, we addressed the effect of external electrical stimulation of cardiac microtissues. Stimulation at 3 Hz induced a beating rate of 178–179 bpm in responsive microtissues, reduced contraction amplitude and substantially shortened contraction duration in both control and TGF-β1-treated iCMs:fCFs microtissues (Figure 6A, Videos S7 and S8), pointing to their full electrical responsiveness. Next, we analysed changes in membrane potential of control and fibrotic iCMs:fCFs microtissues using the relevant fluorescent probe (Figure 6B).

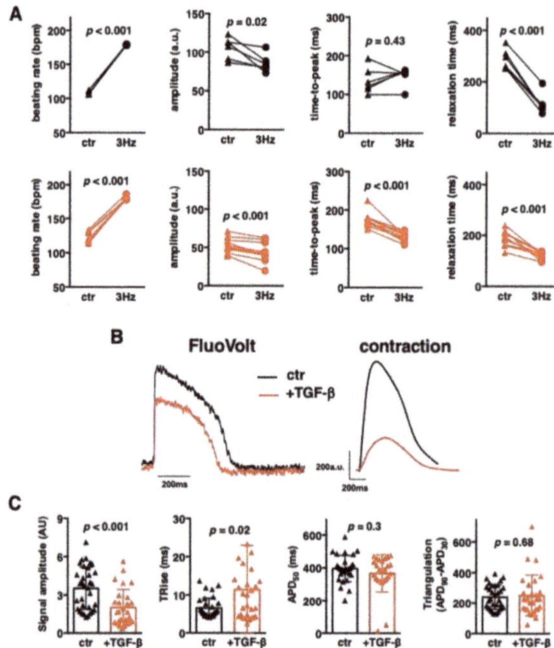

Figure 6. Cardiac microtissue electrophysiology. iCMs:fCFs microtissues were cultured in the presence (red) or absence (black) of TGF-β1 (10 ng/mL) for 10 days. Panel (**A**) shows quantifications of the contraction parameters recorded during spontaneous contractile activity (triangles, ctr) and then upon electrical stimulation with 3 Hz (circles, 3 Hz). Each dot represents data for one microtissue at the indicated condition, lines match data obtained from the same microtissue. *p* values were calculated with the paired Student's *t*-test. Panel (**B**) illustrates representative action potentials recorded with FluoVolt probe (left) and the respective contractions (right). Quantifications of action potential parameters are shown in panel (**C**). Graphs show cumulative data of 3 independent experiments. Each dot represents data for one microtissue. *p* values were calculated with the Student's *t*-test. APD—action potential duration, TRise—depolarisation phase.

We found significantly lower action potential amplitudes in TGF-β1-treated iCMs:fCFs microtissues in comparison to controls (Figure 6C), suggesting reduced number of electrically active cells in fibrotic microtissues. Instead, there was no difference in the action potential duration, although treatment with TGF-β1 slightly prolonged the depolarization phase (Figure 6C). It seems that electrical activity of iCMs in cardiac microtissues was not significantly affected by profibrotic changes.

3.6. Endogenous β-Adrenergic Receptor Signalling Controls Increased Contraction Rate in Fibrotic Cardiac Microtissues

Action potential triggers cardiomyocyte contraction, but β-adrenoceptor-coupled mechanisms can modulate it. First, we used iCMs:fCFs and iCMs:aCFs microtissues and treated them with β-adrenoceptor agonist isoproterenol. As expected, stimulated microtissues showed increased contraction rate (Figure 7A, Figure S12). Isoproterenol significantly shortened both the contraction and relaxation phases of spontaneously contracting microtissues. In contrast, contraction amplitude remained unaffected. Addition of the β-adrenergic receptor blocker nadolol effectively prevented isoproterenol-induced changes in microtissue contraction (Figure 7A). These data confirmed functional β-adrenoceptors in our cardiac microtissues.

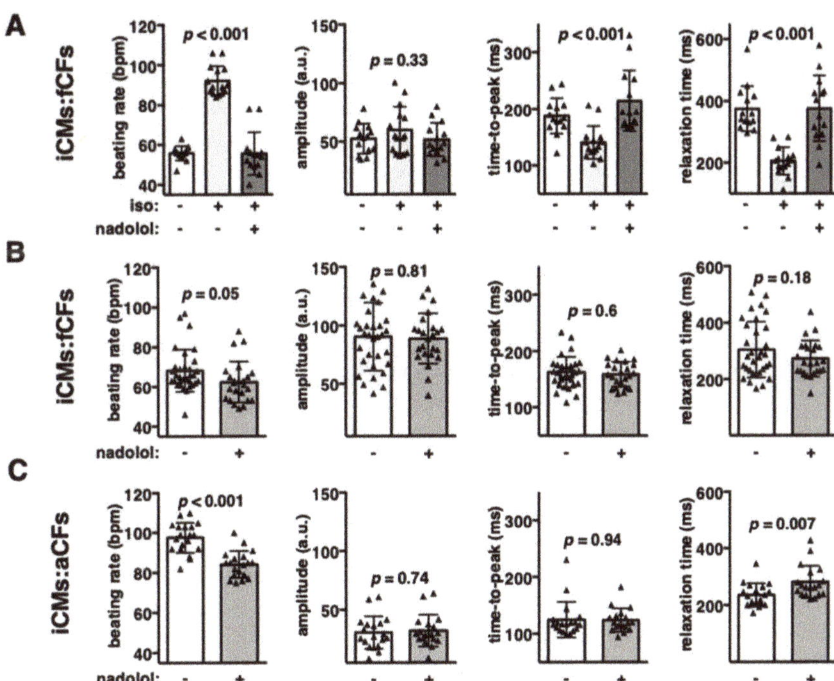

Figure 7. β-adrenergic receptor signalling in cardiac microtissues. Panel (**A**) shows quantification of contraction parameters of iCMs:fCFs microtissues stimulated with β-adrenoceptor agonist isoproterenol (iso, 10 nM) in the presence or absence of β-adrenoceptor blocker nadolol (1 µM). Unstimulated microtissues were used as controls. Graphs in panels (**B**,**C**) show contraction parameters of iCMs:fCFs (**B**) and iCMs:aCFs (**C**) microtissues treated with vehicle or 1 µM nadolol in the absence of isoproterenol. Each triangle represents data for one microtissue. *p* values were calculated with ANOVA for (**A**) and with the Student's *t*-test for (**B**,**C**).

Next, we addressed the relevance of endogenous β-adrenoceptor signalling in cardiac microtissues. To this aim, we analysed the contractility of iCMs:fCFs and iCMs:aCFs microtissues in the presence and absence of nadolol. Treatment with nadolol only slightly reduced contraction of iCMs:fCFs microtissues

(Figure 7B). In the next experiment, we used iCMs:aCFs microtissues showing high spontaneous contraction rate. In this model, nadolol significantly reduced contraction rate, whereas contraction amplitude remained unchanged (Figure 7C). We found that nadolol specifically affected relaxation, but not contraction phase. These results suggest activation of endogenous β-adrenoceptor signalling in iCMs:aCFs microtissues.

4. Discussion

In 3D microtissue models, fibroblasts have been recognized to improve microtissue architecture and biomechanics by providing proper ECM. We confirm that CFs significantly increased the integrity and synchronized contractility of human cardiac microtissues. As recently demonstrated, the use of a 4:1 ratio (iCMs:CFs) results in cardiac microtissues with stable phenotype and beating properties for up to one month [16]. Our data further demonstrated that contractility of these cardiac microtissues was substantially affected under profibrotic condition (i.e., in the presence of TGF-β1). TGF-β1 is a profibrotic cytokine that enhances fibroblast proliferation and production of ECM proteins. In our model, TGF-β1 induced fibrotic phenotype in cardiac microtissues, as indicated by elevated levels of fibrotic genes, secretion of procollagen I and fCFs proliferation. Previously, exogenous TGF-β1 has been shown to induce fibrosis in rat [17,18] or human [19] cardiac microtissue models. Furthermore, fibrotic phenotypes in cardiac microtissues or in biowire were also produced by collagen/fibroblast enrichment [20,21] or by constitutive activation of profibrotic pathways [22].

Our data demonstrated that pretreatment of fCFs with TGF-β1 prior to microtissue formation resulted in a similar effect on microtissue contractility as continuous stimulation with TGF-β1. This result suggested that activation of CFs, rather than a direct effect of TGF-β1 on iCMs, was a main mechanism responsible for deteriorated contraction of fibrotic cardiac microtissues. This idea was further strengthened by the observation that aCFs induced a contraction pattern (high contraction rate with low amplitude) of cardiac microtissues similar to those containing activated fCFs. CFs become activated following heart failure and with aging [23]. We observed that cardiac microtissues containing aCFs obtained from aged heart failure patients showed significantly higher beating rate than microtissues containing aCFs from young, healthy hearts. All these data suggested that activation status of CFs was a key factor determining contraction pattern of cardiac microtissues.

Fibrotic condition in the heart is associated with arrhythmias. Published data showed that activated rodent CFs could cause asynchronous contraction [18] and induce proarrhythmic changes in cardiomyocytes by affecting action potential profile [22,24]. In our model, fibrotic microtissues showed arrhythmic contractions only occasionally and their action potential profile remained unaffected. Instead, the fibrotic condition was associated with reduced action potential signal amplitude, suggesting a lower number of electrically active iCMs in these microtissues. Impaired conduction and contraction of these microtissues might be a consequence of iCMs decoupling and coupling of iCMs with CFs, reduced levels of gap junction protein connexin 43 and/or overproduction of cytokines and ECM by activated CFs. Secreted factors by activated CFs, biomechanical signalling through excessive ECM deposition and remodelling of mechanical junctions are known factors disturbing proper propagation of electrical impulses [25]. Signal conduction abnormalities, such as re-entry, can cause cardiac arrhythmias and may have serious clinical implications for patients with fibrotic hearts [9]. Clinical data suggest that ventricular fibrosis represents a strong predictor of life-threatening ventricular arrhythmia and sudden cardiac death in ischemic and nonischemic heart disorders [8].

In our model, profibrotic condition was consistently associated with increased beating rate of cardiac microtissues. Similar data were obtained in human cardiac microtissues containing mesenchymal stem cells instead of CFs following treatment with TGF-β1 [19]. Physiologically, increased beating rate of the heart is achieved by activation of β-adrenoreceptors. In normal adult cardiomyocytes, β1-adrenoreceptors are dominant, whereas iCMs contractility can be regulated by both β1- and β2-adrenoreceptors [26]. Previous data showed that the beating rate of human iCMs could be increased by treatment with the β-adrenergic agonist isoproterenol [26,27]. In line with these data, we could demonstrate the responsiveness of human cardiac microtissues containing iCMs to isoproterenol. In case of cardiac microtissues showing high beating rate (due to presence of activated aCFs), blockade of β-adrenergic receptors in the absence of exogenous agonists significantly decreased frequency. This finding indicated that activated CFs triggered endogenous β-adrenoreceptor signalling in cardiac microtissues. Mechanistically, stimulation of β-adrenoreceptors increases beating rate by enhancing cardiomyocyte relaxation [10]. Indeed, we observed that the reduced beating rate observed in the presence of β-adrenoreceptor blockers was associated with longer relaxation time. Accordingly, increased contraction rate of fibrotic cardiac microtissues correlated with shortened relaxation time. These results may suggest that activated CFs produce β-adrenergic agonists. Alternatively, fibrotic condition may sensitize β-adrenoreceptors. Cellular response to β-adrenergic stimulations is further modulated by multiple factors, such as cell surface receptor density, activity of the downstream signalling pathway, or by specific costimulations [28]. For example, fibrotic hearts of TGF-β-overexpressing mice showed increased density of β-adrenoreceptors on cardiomyocytes and altered β-adrenergic signalling [3]. Furthermore, collagen receptor $β_1$-integrin [29], connective tissue growth factor [30] or tissue stiffness [31] have also been shown to regulate β-adrenergic signalling.

A successful preclinical platform has to combine physiological-like relevance with the high-throughput processing. Our fibrotic cardiac microtissue model is established in a 96-well format and certain parameters, such as size, contractile properties, changes in membrane potential, procollagen-I production or apoptosis can be measured for individual microtissues. In particular, recently developed software measuring the contractile activity of the whole microtissue allows for quick and easy analysis of the key properties of beating cardiac microtissues [15]. In our experiments, transcriptomic data were obtained from 10–12 pooled microtissues, but we used a simple RT-PCR to analyse gene expression. Application of the next-generation sequencing technology would allow obtaining transcriptomic data from one microtissue. Thus, this model could be used as a high-throughput platform for pharmacological screening with a timeframe for up to at least 10 days. The proof-of-concept experiments with the TGF-βR1 inhibitor SD208 confirmed that fibrotic cardiac microtissues represent a robust model for drug screening.

Although our human cardiac microtissue-based fibrotic platform combines technical ease, high throughput and reproducibility with physiological-like relevance, it also needs to be discussed in the context of its limitations. iCMs are characterized by their foetal developmental status [32]. Currently, there is no satisfactory alternative to iCMs. The usefulness of adult primary human cardiomyocytes is limited, as these cells contract upon chemical or electrical stimulation only [33]. Despite recent advances in the maturation of iCMs through improved biomechanics and cocultures with other cell types, so far, their maturation status does not reflect adult cardiomyocytes [13,34–36]. Another limitation of our model is the heterogeneity of iCMs. Conventional differentiation of iPSCs into iCMs results in populations of pacemaker, ventricular-like, atrial-like cells as well as nonbeating progenitors. The use of defined types of differentiated cardiomyocytes could further improve the relevance of the cardiac microtissue model. Moreover, addition of other cell types, such as endothelial cells or macrophages, to iCMs and fCFs might result in formation of microtissues that better mimic the structure and function of human cardiac tissue.

Supplementary Materials: The following are available online at http://www.mdpi.com/2073-4409/9/5/1270/s1, Figure S1: Schematic presentation of experimental setup with time lines, Figure S2: Kinetics of microtissue growths, Figure S3: TGF-β1 induces fibrotic phenotypes in cardiac microtissues, Figure S4: Contraction patterns of cardiac microtissues containing foetal or adult cardiac fibroblasts, Figure S5: Effect of exogenous TGF-β1 on contractility of cardiac microtissues - individual experiments, Figure S6: Activation of foetal cardiac fibroblasts with TGF-β1, Figure S7: Effect of activated foetal cardiac fibroblasts on size and procollagen I secretion, Figure S8: Effect of activated foetal cardiac fibroblasts on contractility of cardiac microtissues - individual experiments, Figure S9: Size of cardiac microtissues containing adult cardiac fibroblasts, Figure S10: Size and contractile properties of cardiac microtissues containing adult cardiac fibroblasts, Figure S11: Effect of exogenous TGF-β1 on contractility of healthy iCMs:aCFs and HF iCMs:aCFs microtissues, Figure S12: b-adrenergic receptor stimulation of iCMs:aCFs microtissues; Table S1: Clinical characteristics of patients from which aCFs were obtained, Table S2: Primers used for quantitative RT-PCR; Video S1: Control iCMs only, Video S2: iCMs_fCFs control d10, Video S3: iCMSfCFs TGF-β1, Video S4: iCMsfCFs pre-TGF-β-3d_d10 D10, Video S5: iCMs_healthy aCFs control d10, Video S6: iCMS_HF aCMs control d10, Video S7: iCMs_fCFs control, Video S8: iCMs_fCFs 3Hz.

Author Contributions: Conceptualization: P.B., G.K.; Methodology: P.B., C.Z., A.C., D.N., F.D.M., I.A., M.S., G.K.; Data curation: P.B., C.Z., A.C., M.S., G.K.; Formal analysis: P.B., G.L.S., C.Z., G.K.; Funding acquisition: G.K.; Supervision: O.D., G.K.; Writing—original draft: P.B., G.K.; Writing—review & editing: C.Z., G.L.S., O.D. All authors have read and agreed to the published version of the manuscript.

Funding: This research was founded by the Swiss Heart Foundation and Vontobel Foundation.

Acknowledgments: The authors acknowledge financial support of the Swiss Heart Foundation and Vontobel Foundation as well as technical assistance of the Center for Microscopy and Image Analysis of the University of Zurich and Sophistolab AG.

Disclosures: Irina Agarkova is employed by InSphero company, which produces microtissues. Godfrey Smith is a founder, shareholder, executive and honorary Chief Scientific Officer of Clyde Biosciences Ltd. company, which provides an assay service to the pharmaceutical industry using commercial iPSC-derived cardiomyocytes.

Conflicts of Interest: The authors declare no conflict of interest.

References

1. Kong, P.; Christia, P.; Frangogiannis, N.G. The pathogenesis of cardiac fibrosis. *Cell. Mol. Life Sci.* **2014**, *71*, 549–574. [CrossRef]

2. Travers, J.G.; Kamal, F.A.; Robbins, J.; Yutzey, K.E.; Blaxall, B.C. Cardiac fibrosis: The fibroblast awakens. *Circ. Res.* **2016**, *118*, 1021–1040. [CrossRef]

3. Rosenkranz, S.; Flesch, M.; Amann, K.; Haeuseler, C.; Kilter, H.; Seeland, U.; Schluter, K.D.; Bohm, M. Alterations of beta-adrenergic signaling and cardiac hypertrophy in transgenic mice overexpressing TGF-beta(1). *Am. J. Physiol. Heart Circ. Physiol.* **2002**, *283*, H1253–H1262. [CrossRef] [PubMed]

4. Ikeuchi, M.; Tsutsui, H.; Shiomi, T.; Matsusaka, H.; Matsushima, S.; Wen, J.; Kubota, T.; Takeshita, A. Inhibition of TGF-beta signaling exacerbates early cardiac dysfunction but prevents late remodeling after infarction. *Cardiovasc. Res.* **2004**, *64*, 526–535. [CrossRef] [PubMed]

5. Kuwahara, F.; Kai, H.; Tokuda, K.; Kai, M.; Takeshita, A.; Egashira, K.; Imaizumi, T. Transforming growth factor-beta function blocking prevents myocardial fibrosis and diastolic dysfunction in pressure-overloaded rats. *Circulation* **2002**, *106*, 130–135. [CrossRef] [PubMed]

6. Kania, G.; Blyszczuk, P.; Stein, S.; Valaperti, A.; Germano, D.; Dirnhofer, S.; Hunziker, L.; Matter, C.M.; Eriksson, U. Heart-infiltrating prominin-1$^+$/CD133$^+$ progenitor cells represent the cellular source of transforming growth factor beta-mediated cardiac fibrosis in experimental autoimmune myocarditis. *Circ. Res.* **2009**, *105*, 462–470. [CrossRef]

7. Clements, I.P.; Miller, W.L.; Olson, L.J. Resting heart rate and cardiac function in dilated cardiomyopathy. *Int. J. Cardiol.* **1999**, *72*, 27–37. [CrossRef]

8. Disertori, M.; Mase, M.; Ravelli, F. Myocardial fibrosis predicts ventricular tachyarrhythmias. *Trends Cardiovasc. Med.* **2017**, *27*, 363–372. [CrossRef]

9. Nguyen, T.P.; Qu, Z.; Weiss, J.N. Cardiac fibrosis and arrhythmogenesis: The road to repair is paved with perils. *J. Mol. Cell. Cardiol.* **2014**, *70*, 83–91. [CrossRef]

10. Najafi, A.; Sequeira, V.; Kuster, D.W.; Van der Velden, J. Beta-adrenergic receptor signalling and its functional consequences in the diseased heart. *Eur. J. Clin. Investig.* **2016**, *46*, 362–374. [CrossRef]

11. Karakikes, I.; Ameen, M.; Termglinchan, V.; Wu, J.C. Human induced pluripotent stem cell-derived cardiomyocytes: Insights into molecular, cellular, and functional phenotypes. *Circ. Res.* **2015**, *117*, 80–88. [CrossRef] [PubMed]

12. Zuppinger, C. 3D culture for cardiac cells. *Biochim. Biophys. Acta* **2016**, *1863*, 1873–1881. [CrossRef] [PubMed]

13. Beauchamp, P.; Moritz, W.; Kelm, J.M.; Ullrich, N.D.; Agarkova, I.; Anson, B.D.; Suter, T.M.; Zuppinger, C. Development and characterization of a scaffold-free 3D spheroid model of induced pluripotent stem cell-derived human cardiomyocytes. *Tissue Eng. Part C Methods* **2015**, *21*, 852–861. [CrossRef] [PubMed]

14. Castaldo, C.; Di Meglio, F.; Miraglia, R.; Sacco, A.M.; Romano, V.; Bancone, C.; Della Corte, A.; Montagnani, S.; Nurzynska, D. Cardiac fibroblast-derived extracellular matrix (biomatrix) as a model for the studies of cardiac primitive cell biological properties in normal and pathological adult human heart. *Biomed. Res. Int.* **2013**, *2013*, 352370. [CrossRef] [PubMed]

15. Sala, L.; van Meer, B.J.; Tertoolen, L.G.J.; Bakkers, J.; Bellin, M.; Davis, R.P.; Denning, C.; Dieben, M.A.E.; Eschenhagen, T.; Giacomelli, E.; et al. MUSCLEMOTION: A versatile open software tool to quantify cardiomyocyte and cardiac muscle contraction in vitro and in vivo. *Circ. Res.* **2018**, *122*, e5–e16. [CrossRef]

16. Beauchamp, P.; Jackson, C.B.; Ozhathil, L.C.; Agarkova, I.; Galindo, C.L.; Sawyer, D.B.; Suter, T.M.; Zuppinger, C. 3D co-culture of hiPSC-derived cardiomyocytes with cardiac fibroblasts improves tissue-like features of cardiac spheroids. *Front. Mol. Biosci.* **2020**, *7*, 14. [CrossRef]

17. Figtree, G.A.; Bubb, K.J.; Tang, O.; Kizana, E.; Gentile, C. Vascularized cardiac spheroids as novel 3D in vitro models to study cardiac fibrosis. *Cells Tissues Organs* **2017**, *204*, 191–198. [CrossRef]

18. Sadeghi, A.H.; Shin, S.R.; Deddens, J.C.; Fratta, G.; Mandla, S.; Yazdi, I.K.; Prakash, G.; Antona, S.; Demarchi, D.; Buijsrogge, M.P.; et al. Engineered 3D cardiac fibrotic tissue to study fibrotic remodeling. *Adv. Healthc. Mater.* **2017**, *6*. [CrossRef]

19. Lee, M.O.; Jung, K.B.; Jo, S.J.; Hyun, S.A.; Moon, K.S.; Seo, J.W.; Kim, S.H.; Son, M.Y. Modelling cardiac fibrosis using three-dimensional cardiac microtissues derived from human embryonic stem cells. *J. Biol. Eng.* **2019**, *13*, 15. [CrossRef]

20. Van Spreeuwel, A.C.C.; Bax, N.A.M.; Van Nierop, B.J.; Aartsma-Rus, A.; Goumans, M.T.H.; Bouten, C.V.C. Mimicking cardiac fibrosis in a dish: Fibroblast density rather than collagen density weakens cardiomyocyte function. *J. Cardiovasc. Transl. Res.* **2017**, *10*, 116–127. [CrossRef]

21. Wang, E.Y.; Rafatian, N.; Zhao, Y.; Lee, A.; Lai, B.F.L.; Lu, R.X.; Jekic, D.; Davenport Huyer, L.; Knee-Walden, E.J.; Bhattacharya, S.; et al. Biowire model of interstitial and focal cardiac fibrosis. *ACS Cent. Sci.* **2019**, *5*, 1146–1158. [CrossRef] [PubMed]

22. Kofron, C.M.; Kim, T.Y.; King, M.E.; Xie, A.; Feng, F.; Park, E.; Qu, Z.; Choi, B.R.; Mende, U. Gq-activated fibroblasts induce cardiomyocyte action potential prolongation and automaticity in a three-dimensional microtissue environment. *Am. J. Physiol. Heart Circ. Physiol.* **2017**, *313*, H810–H827. [CrossRef] [PubMed]

23. Nurzynska, D.; Di Meglio, F.; Romano, V.; Miraglia, R.; Sacco, A.M.; Latino, F.; Bancone, C.; Della Corte, A.; Maiello, C.; Amarelli, C.; et al. Cardiac primitive cells become committed to a cardiac fate in adult human heart with chronic ischemic disease but fail to acquire mature phenotype: Genetic and phenotypic study. *Basic Res. Cardiol.* **2013**, *108*, 320. [CrossRef] [PubMed]

24. Trieschmann, J.; Bettin, D.; Haustein, M.; Koster, A.; Molcanyi, M.; Halbach, M.; Hanna, M.; Fouad, M.; Brockmeier, K.; Hescheler, J.; et al. The interaction between adult cardiac fibroblasts and embryonic stem cell-derived cardiomyocytes leads to proarrhythmic changes in in vitro cocultures. *Stem Cells Int.* **2016**, *2016*, 2936126. [CrossRef]

25. Pellman, J.; Zhang, J.; Sheikh, F. Myocyte-fibroblast communication in cardiac fibrosis and arrhythmias: Mechanisms and model systems. *J. Mol. Cell. Cardiol.* **2016**, *94*, 22–31. [CrossRef]

26. Wu, H.; Lee, J.; Vincent, L.G.; Wang, Q.; Gu, M.; Lan, F.; Churko, J.M.; Sallam, K.I.; Matsa, E.; Sharma, A.; et al. Epigenetic regulation of phosphodiesterases 2A and 3A underlies compromised beta-adrenergic signaling in an iPSC model of dilated cardiomyopathy. *Cell Stem Cell* **2015**, *17*, 89–100. [CrossRef]

27. Preininger, M.K.; Jha, R.; Maxwell, J.T.; Wu, Q.; Singh, M.; Wang, B.; Dalal, A.; McEachin, Z.T.; Rossoll, W.; Hales, C.M.; et al. A human pluripotent stem cell model of catecholaminergic polymorphic ventricular tachycardia recapitulates patient-specific drug responses. *Dis. Models Mech.* **2016**, *9*, 927–939. [CrossRef]

28. Ferrara, N.; Komici, K.; Corbi, G.; Pagano, G.; Furgi, G.; Rengo, C.; Femminella, G.D.; Leosco, D.; Bonaduce, D. Beta-adrenergic receptor responsiveness in aging heart and clinical implications. *Front. Physiol.* **2014**, *4*, 396. [CrossRef]

29. Krishnamurthy, P.; Subramanian, V.; Singh, M.; Singh, K. Beta1 integrins modulate beta-adrenergic receptor-stimulated cardiac myocyte apoptosis and myocardial remodeling. *Hypertension* **2007**, *49*, 865–872. [CrossRef]

30. Gravning, J.; Ahmed, M.S.; Qvigstad, E.; Krobert, K.; Edvardsen, T.; Moe, I.T.; Hagelin, E.M.; Sagave, J.; Valen, G.; Levy, F.O.; et al. Connective tissue growth factor/CCN2 attenuates beta-adrenergic receptor responsiveness and cardiotoxicity by induction of G protein-coupled receptor kinase-5 in cardiomyocytes. *Mol. Pharmacol.* **2013**, *84*, 372–383. [CrossRef]

31. Kim, T.J.; Sun, J.; Lu, S.; Zhang, J.; Wang, Y. The regulation of beta-adrenergic receptor-mediated PKA activation by substrate stiffness via microtubule dynamics in human MSCs. *Biomaterials* **2014**, *35*, 8348–8356. [CrossRef] [PubMed]

32. Zuppinger, C.; Gibbons, G.; Dutta-Passecker, P.; Segiser, A.; Most, H.; Suter, T.M. Characterization of cytoskeleton features and maturation status of cultured human iPSC-derived cardiomyocytes. *Eur. J. Histochem.* **2017**, *61*, 2763. [CrossRef] [PubMed]

33. Nguyen, N.; Nguyen, W.; Nguyenton, B.; Ratchada, P.; Page, G.; Miller, P.E.; Ghetti, A.; Abi-Gerges, N. Adult human primary cardiomyocyte-based model for the simultaneous prediction of drug-induced inotropic and pro-arrhythmia risk. *Front. Physiol.* **2017**, *8*, 1073. [CrossRef] [PubMed]

34. Nunes, S.S.; Miklas, J.W.; Liu, J.; Aschar-Sobbi, R.; Xiao, Y.; Zhang, B.; Jiang, J.; Masse, S.; Gagliardi, M.; Hsieh, A.; et al. Biowire: A platform for maturation of human pluripotent stem cell-derived cardiomyocytes. *Nat. Methods* **2013**, *10*, 781–787. [CrossRef]

35. Giacomelli, E.; Bellin, M.; Sala, L.; Van Meer, B.J.; Tertoolen, L.G.; Orlova, V.V.; Mummery, C.L. Three-dimensional cardiac microtissues composed of cardiomyocytes and endothelial cells co-differentiated from human pluripotent stem cells. *Development* **2017**, *144*, 1008–1017. [CrossRef]

36. Ronaldson-Bouchard, K.; Ma, S.P.; Yeager, K.; Chen, T.; Song, L.; Sirabella, D.; Morikawa, K.; Teles, D.; Yazawa, M.; Vunjak-Novakovic, G. Advanced maturation of human cardiac tissue grown from pluripotent stem cells. *Nature* **2018**, *556*, 239–243. [CrossRef]

 *cells* 

Article

Human Induced Pluripotent Stem Cell-Derived 3D-Neurospheres Are Suitable for Neurotoxicity Screening

Julianna Kobolak [1], Annamaria Teglasi [1], Tamas Bellak [1], Zofia Janstova [1], Kinga Molnar [2], Melinda Zana [1], Istvan Bock [1], Lajos Laszlo [2] and Andras Dinnyes [1,3,4,*]

[1] BioTalentum Ltd., H-2100 Gödöllő, Hungary; julianna.kobolak@biotalentum.hu (J.K.); annamaria.teglasi@biotalentum.hu (A.T.); tamas.bellak@biotalentum.hu (T.B.); Zofia.Janstova@biotalentum.hu (Z.J.); melinda.zana@biotalentum.hu (M.Z.); biotalentum.bock@yahoo.com (I.B.)

[2] Department of Anatomy, Cell and Developmental Biology, Eötvös Loránd University, H-1117 Budapest, Hungary; kinga.molnar@ttk.elte.hu (K.M.); laszlo@elte.hu (L.L.)

[3] Molecular Animal Biotechnology Laboratory, Szent István University, H-2101 Gödöllő, Hungary

[4] Translational Biomedicine Institute, University of Szeged, H-6720 Szeged, Hungary

* Correspondence: manuscript.dinnyes@biotalentum.hu; Tel.: +36-28-526-151

Received: 26 March 2020; Accepted: 29 April 2020; Published: 1 May 2020

Abstract: We present a hiPSC-based 3D in vitro system suitable to test neurotoxicity (NT). Human iPSCs-derived 3D neurospheres grown in 96-well plate format were characterized timewise for 6-weeks. Changes in complexity and homogeneity were followed by immunocytochemistry and transmission electron microscopy. Transcriptional activity of major developmental, structural, and cell-type-specific markers was investigated at weekly intervals to present the differentiation of neurons, astrocytes, and oligodendrocytes. Neurospheres were exposed to different well-known toxicants with or without neurotoxic effect (e.g., paraquat, acrylamide, or ibuprofen) and examined at various stages of the differentiation with an ATP-based cell viability assay optimized for 3D-tissues. Concentration responses were investigated after acute (72 h) exposure. Moreover, the compound-specific effect of rotenone was investigated by a panel of ER-stress assay, TUNEL assay, immunocytochemistry, electron microscopy, and in 3D-spheroid based neurite outgrowth assay. The acute exposure to different classes of toxicants revealed distinct susceptibility profiles in a differentiation stage-dependent manner, indicating that hiPSC-based 3D in vitro neurosphere models could be used effectively to evaluate NT, and can be developed further to detect developmental neurotoxicity (DNT) and thus replace or complement the use of animal models in various basic research and pharmaceutical applications.

Keywords: induced pluripotent stem cells; neurospheres; 3D culture; neurite outgrowth; neurotoxicity

---

1. Introduction

Environmental stressors, such as chemicals or the medical drugs could have a toxic effect on humans which may occur at any stage of their life, during fetal development, childhood, or adult life. The toxic effects of environmental agents coupled with inherited susceptibility of individuals make the toxicology prediction difficult. Conventional animal-based toxicity and safety tests have high costs, use large numbers of animals (mainly rats) and in many cases, they do not provide clearly translatable results for humans [1–3]. Consequently, in line with legislation, there is an increasing need to develop alternative testing methods which could handle thousands of drugs or chemicals with affordable time and cost and with human-relevant neurotoxicology (NT) outcome [4–7]. Development of new approach methods (NAM), would be important both for NT and developmental neurotoxicology (DNT) tests, providing data on the effect of chemicals and the potential adverse outcomes (AOs) [6,8–13].

An increasing number of studies use primary cell cultures and recently, pluripotent stem cells (PSCs) to create in vitro systems for NT and DNT screenings. It is widely accepted concept that 3D cell cultures can mimic better the original tissue environment including tissue-specific architecture, mechanical and biochemical features, cell-to-cell communication and signaling, and differentiation capability, while 2D cell culture systems are less complex and more artificial in this sense [14–17], although the existing assays provide meaningful readouts for specific neurodevelopmental processes [6,8,11]. Only one in vitro assay will not be able to cover the complexity of the in vivo development, therefore a battery of assays and fit-for-purpose applications should be used to cover the relevant processes [10,18]. Human induced pluripotent stem cell-derived (hiPSC) neurospheres, grown on 3D scaffolds or self-forming, can establish cell–cell interactions and model certain neurodevelopmental processes, therefore, might be used as an in vitro screening platform not only for NT but in DNT studies or drug development [19–21].

Neural stem cells (NSCs) are considered as a multipotent and self-renewing pool of cells in the mammalian central nervous system (CNS), occurring in vivo in the developing embryonic neural tissue as early as the neural tube formation happens. These cells have the capacity to differentiate into all neuronal cell types, as well as into glial cells, therefore they are able to emulate some fetal neurodevelopmental processes [22,23]. Moreover, NSCs can be differentiated in vitro from pluripotent stem cells (PSCs) thus providing an attractive and almost unlimited in vitro tool for toxicology studies including drug development [24,25].

Despite the developments mentioned above, still, a limited number of human iPSC derived 3D neuronal culture-based studies were published focusing on the development of NT or DNT models [21,26–31]. A limiting factor for the further development of such test methods is the lack of high-throughput screening (HTS) read-outs on 3D cell cultures [16], despite the developments using high-content image analysis (HCA) [27].

Here, we present a reliable model system where hiPSC-derived NSCs are differentiated towards subtypes of neurons, astrocytes, and oligodendrocytes, forming free-floating 3D neurospheres in 96-well plate format. Over 6 weeks the various differentiation stages are characterized, and three selected stages are exposed with different compounds to investigate their cytotoxic effect. The generated neuronal spheroids used for NT measurements at different time-points resembling various differentiation stages, therefore, provide an excellent platform for further DNT test system developments.

2. Methods

2.1. Chemicals and Plasticware

The chemicals were purchased from Sigma-Aldrich (St Louis, MO, USA) and all cell culture reagents and plasticware from Thermo Fisher Scientific Inc. (Waltham, MA, USA), unless otherwise specified.

2.2. Human iPSC Culture

Human iPSC line (Ctrl-2), derived from a healthy mid-age Caucasian female donor peripheral blood mononuclear cells (PBMCs), established and characterized earlier [32,33], was used in this study. Cells were cultured on BD Matrigel™ matrix (BD Biosciences, Franklin Lakes, NJ, USA) with mTeSR™1 medium (Stem Cell Technologies, Vancouver, Canada), using Gentle Cell Dissociation Reagent for passages, according to the manufacturer's instruction. Representative hiPSC colony morphology, pluripotency staining for POU5F1, NANOG, SSEA4, and TRA1-60 and karyotype analysis are presented in Figure S1. For mycoplasma screening, the Venor®GeM-Advance (Minerva Biolabs) Mycoplasma Detection Kit was used according to the manufacturer's protocol in every fifth passage during maintenance and before freezing. Cells were cultured at 37 °C in a humidified atmosphere containing 5% CO_2. In the current study cultures from passage 16 and 17 were used for differentiation.

2.3. Neuronal Differentiation and Maintenance

The Ctrl-2 hiPSC line was differentiated to neural stem cells (NSCs) by dual SMAD inhibition procedure [34], following the detailed protocol of Shi et al. [35]. Briefly, when hiPSC cultures reached 90% confluence, the culture was passaged onto poly-L-ornithine and laminin (POL/L; 0.002%/1 µg/cm^2) coated plates, using conventional iPSC passage (as detailed above). On the next day media was changed to neural induction medium (NIM; DMEM/F12: Neurobasal medium, supplemented with 1× N2, 2× B27, 2 mM glutamine, 1× non-essential amino acid (NEAA), 100 µM β-mercaptoethanol, 5 µg/mL insulin) supplemented with 10 µM SB431542, 500 ng/mL Noggin (R&D Systems, Inc., Minneapolis, MN, USA) and 5 ng/mL basic fibroblast growth factor (bFGF) to induce the neuroectodermal lineage. Neuronal induction media was applied until day 10 of the differentiation. The forming neural rosette-like structures were manually picked under a stereomicroscope (Olympus SZX2; Olympus Ltd. Tokyo, Japan) and re-plated onto POL/L (POL/L; 0.002%/1 µg/cm^2) plates. NSCs were expanded in neural maintenance medium (NMM; DMEM-F12: Neurobasal medium, supplemented with 1× N2, and 2× B27, 2 mM glutamine, 1× NEAA), supplemented with 10 ng/mL bFGF and 10 ng/mL Epidermal growth factor (EGF) and maintained in a monolayer on dishes coated with POL/L (0.002%/1 µg/cm^2). When reached 100% confluence, cells were passaged using Accutase (Sigma) and seeded as single cells (50,000 cells/cm^2) for further expansion on POL/L (0.002%/1 µg/cm^2) coated dishes. After 4 passages NSCs were frozen in 1 mL freezing medium (90% Fetal Bovine Serum, FBS; heat-inactivated, Thermo Fisher Sci.; Cat N.: 10500-064, LOT: 08F1180K; 10% DMSO; 2 million cells/vial), using Accutase passage. Before freezing the NSCs, a mycoplasma test was performed (see above), repeated every fifth passage during maintenance.

2.4. 3D Neurosphere Culture

After thawing, NSCs (generated as detailed above) were cultured on POL/L (0.002%/1 µg/cm^2) coated plates (50,000 cells/cm^2). When reached 100% confluence NSCs were propagated with Accutase treatment and single-cell suspension was plated onto a low-adherent 96-well plate with 10,000 cells/well in NMM medium. Neurospheres were formed within 48 h after plating and NMM half of the media was changed in every 3 days until analyzing the samples. Samples were collected at Day 0 (D0), Day 2 (D2), and weekly intervals from the end of the 1st week (D7) until the end of the 6th week (D14, D21, D28, D35, and D42).

2.5. Cryosectioning and Immunocytochemistry (ICC) Staining

3D neurospheres (untreated, vehicle or compound-treated) were fixed with 4% paraformaldehyde (PFA) in 0.1 mol/L phosphate buffer for 1 h at RT and washed 3 times with PBS. The fixed samples were cryoprotected in 30% sucrose in PBS containing 0.01% sodium azide at 4 °C until embedding in Shandon Cryomatrix gel (Thermo Fischer Scientific). The 16 µm parallel sections were made using cryostat (Leica CM 1850 Cryostat, Leica GmbH), mounted to Superfrost™ Ultra Plus Adhesion Slides (Thermo Fisher Scientific) and stored at −20 °C until use. After 10 min air-drying, the sections were blocked for 1 h at RT with blocking solution (3% BSA in PBS), supplemented with 0.2% TritonX-100. The sections were then incubated with primary antibodies (Table S1) overnight at 4 °C. Next day, sections were washed in PBS 3 times, and isotype-specific secondary antibodies (Table S1) were diluted in blocking buffer and applied for 1 h at RT. The sections were washed 3 times with PBS and covered using Vectashield® mounting medium containing DAPI (1.5 µg/mL; Vector Laboratories), that labelled the nuclei of the cells (at least 1 h at RT). Negative controls for the secondary antibodies were performed by omitting the primary antibodies. Immunoreactive sections were analyzed using a BX-41 epifluorescent microscope (objectives: 20× 0.50 NA; 40× 0.75 NA; Olympus) equipped with a DP-74 digital camera and its CellSens software (V1.18; Olympus). For confocal imaging, Olympus Fv10i-W compact confocal microscope system (objective: 60× 1.35 NA; Olympus) with Fv10i software

(V2.1; Olympus) was applied. All images were further processed using the GNU Image Manipulation Program (GIMP 2.10.0) and NIH ImageJ analysis software (imagej.nih.gov/ij).

Quantification of the immunocytochemistry data was performed using ImageJ software according to Tieng et al. [36]. Briefly, images were taken by confocal microscopy at 120× magnification. The numbers of Ki-67-, NESTIN-, GFAP-, AQP4-, TUBB3-, NF200kD-, MBP-, VAMP2-, and MAP2-immunoreactive pixels were measured in 5 neurosphere (highest diameter middle sections, 5 randomly selected fields/slide) at every time points. Data was normalized with DAPI positive nuclei number. Data were expressed as a percentage of marker/DAPI ratio ± SEM ($p < 0.05$).

2.6. Apoptosis Assay

Embedding and cryosectioning of 3D samples were performed as above. To detect apoptotic activity, the DeadEnd™ Colorimetric TUNEL System (Promega) was used on the middle cryosections (highest diameter) of the spheroids, following the instructions of the manufacturer. In brief, apoptosis was detected by immersing the slides in PBS for 5 min (at RT), adding 20 µg/mL Proteinase K solution and incubating for 10–30 min (at RT). After 5–10 min treatment in Equilibration buffer, recombinant terminal deoxynucleotidyl transferase (rTdT) was added to the reaction mixture. Next, the sections were incubated for 60 min at 37 °C inside of a humidified chamber to allow the end-labelling reaction to occur. The reaction was terminated by immersing the slides in saline-sodium citrate for 15 min (RT). Endogenous peroxidases were blocked by immersing the slides in 0.3% hydrogen peroxide in PBS for 3–5 min (RT). Streptavidin-HRP was added to slides, incubated for 30 min (RT), stained with diaminobenzidine (DAB) solution for 5 min until a light brown background appeared. For hematoxylin–eosin (HE) staining Mayer's Hematoxylin solution was used for 3 min. Sections were rinsed with tap water and placed into distilled water for 30 s, then into 96% alcohol for 30 s. One percent Eosin solution in distilled water was used for 3 min. Stained sections were dehydrated through alcohols, clear in xylene and mount in DPX. Microphotographs were made with DP-74 digital camera (Olympus) using a light microscope (BX-41, objectives: 20× 0.50 NA; 40× 0.75 NA; Olympus) and CellSens software (V1.18; Olympus). For counting the apoptotic and total (Hematoxylin-stained) number of cells, NIH ImageJ analysis software was used. Five Ctrl and five ROT-treated spheroids were randomly selected, and middle sections were analyzed from each differentiation stage (D21, D28, and D42) samples in three experiments ($n = 3$).

2.7. Transmission Electron Microscopy (TEM)

Neurospheres (untreated, vehicle or compound-treated) were fixed at different differentiation stages in a fixative solution containing 3.2% PFA, 0.2% glutaraldehyde, 1% sucrose, 40 mM $CaCl_2$ in 0.1 M cacodylate buffer (pH 7.4) for 12 h at 4 °C. Samples for ultrastructural analysis were embedded in 1.5% agar (dissolved in dH_2O), post-fixed in 1% ferrocyanide-reduced osmium tetroxide [37], then dehydrated using graded series of ethanol, finally embedded in Spurr low viscosity epoxy resin medium. Ultrathin sections were collected from the middle region of the spheroids (highest diameter) on copper slot grids coated with formwar (Agar Sci., Essex, UK) and counterstained with uranyl acetate and Reynolds's lead citrate. Sections were examined with a JEOL JEM 1011 transmission electron microscope (JEOL Ltd., Tokyo, Japan) equipped with a Morada 11-megapixel camera using iTEM software (Olympus).

2.8. RT-qPCR Analysis

For each sample, 12 spheroids were pooled, and 3 biological replicates were performed ($n = 3$). Total RNA was isolated using the RNeasy Plus Mini Kit (Qiagen, Hilden, Germany). For the reverse transcription, 600 ng of the isolated RNA was used applying the Maxima First Strand cDNA Synthesis Kit for RT-qPCR with dsDNase (Thermo Fisher Scientific) according to the manufacturer's instructions.

Gene-specific primers were designed using the Primer3 software [38], specified with mFOLD software [39] and Primer-BLAST software [40]. Primers were optimized using two-fold serial dilution standard curves (Table S2). As a reference, gene GAPDH was used (Table S2). Each real-time PCR reaction contained 5 ng RNA-equivalent cDNA template, 400 nM of each primer and 50% SYBR Green JumpStart Taq ReadyMix (Sigma Aldrich) in a total volume of 15 μL. PCR reactions were set up using QIAgility liquid handling robot and performed on a Rotor-Gene Q cycler (Qiagen). The cycling parameters were as follows: 94 °C for 3 min initial denaturation followed by 40 cycles of 95 °C for 5 s, 60 °C for 15 s, and 72 °C for 30 s. Melting curve analysis and agarose gel electrophoresis confirmed the specificity of the primers and the absence of gDNA contamination. Data of three replicates were analyzed for each gene, using the ddCT method [41].

2.9. XBP1-Assay of Endoplasmic Reticulum Stress

Upon accumulation of unfolded proteins in the endoplasmic reticulum (ER), a 26-nucleotide fragment from the X-box binding protein 1 mRNA (XBP1(U)) is removed with a special splicing mechanism [42]. This shorter mRNA (XBP1(S)) is a frequently used marker of ER-stress. To study the expression of XBP1(S), previously described primers [43] and our own primers were used [44] (Table S2). As a positive control, cells were treated with 5 μM and 10 μM Tunicamycin to induce ER-stress. RT-PCR reactions were performed and analyzed as above, using Phusion Hot Start II High-Fidelity DNA Polymerase and 20 ng of the cDNA samples.

2.10. Toxicity Treatments and ATP Viability Assay

Eleven well-known compounds were tested in 7 different concentrations to generate concentration-response curves. The tested compounds with the used concentrations are detailed in Table S3. The highest dosage was determined based on the solubility of the compounds, and care was taken not to reach above 0.1% DMSO levels after diluting the compound in the culture media. In each assay plate, 4 technical replicates were applied from each sample and 3 biological replicate assays were run ($n = 3$). The 3D cell cultures were exposed to the toxicants for 72 h (exposure schemes are detailed in Figure 5A or Figure 6A or Figure 7A). Vehicle control was used as a regular medium supplemented with 0.1% DMSO.

ATP viability assay was performed with CellTiter-Glo® 3D Cell Viability Assay, according to the manufacturer's protocol (Promega). Neurospheres were lysed with 100 μL CellTiter-Glo® 3D Reagent for 60 min at RT. Luminescence signal was recorded with a Thermo VarioScan Flash (Thermo Fisher Scientific) plate reader.

2.11. Diametric and Total Protein Determination of the Spheroids

3D spheroids grown in a 96-well plate were captured using the 4x objective (0.1 NA) of Olympus IX71 microscope and DP21 camera (Olympus). The images were analyzed by measuring the diameters of the spheroids using the Olympus CellSens Dimension software (V1.11). Each value represents the average of 3 experiments ($n = 3$) in each 96 spheroids were measured. The 3D spheroids were lysed individually with RIPA Lysis and Extraction Buffer supplemented with Halt™ Protease and Phosphatase Inhibitor Cocktail and Pierce™ Universal Nuclease for Cell Lysis, sonicated and the total protein concentration determined using a Pierce BCA Protein Assay Kit according to the manufacturer's instructions. Due to the small spheroid size of the D2, D7, and D14 samples, three spheroids were pooled, then the individual values were calculated accordingly. In total, three experiments were performed ($n = 3$), in which 24 spheroids were analyzed for each timepoint.

2.12. Neurite Outgrowth Assay

For neurite outgrowth measurement the method of Harris et al. was modified [45]. 3D spheroids were treated with different concentrations (vehicle, 0.1 µM, 0.5 µM, and 0.75 µM) of Rotenone (ROT) at D21 differentiation stage using the 72 h exposure scheme, similarly to the previous concentration-response experiments. After the treatment, the spheroids were plated on Matrigel-coated plates in NMM medium without compound, and 24 h later fixed with 4% PFA and immunostained against TUBB3, as described in detail under the immunostaining paragraph. First, the number of neurites was counted for each sample. Then, the "edge" of the spheroids was determined by drawing an "edge line" and every neurite length was measured from this "edge line" to the tip of each neurite. Total neurite length was calculated for each spheroid using the ImageJ software's Neurite Tracer plugin. In each experiment eight spheroids were treated in each experimental group (5 groups) and three independent experiments were performed ($n = 3$).

2.13. Statistical Analysis

All results were analyzed using Prism 5 (GraphPad Software, La Jolla, CA, USA) and handled in Microsoft Office 2010 (Microsoft, Redmond, WA, USA) software. For normalization of the concentration response, the average values of positive control were used as 0% and the average of vehicle control served as 100%. Four parameter curve fitting methods were used to determine EC10 and EC50 values, where the data on the graphs represent the average of three biological replicates. The "n" value corresponds to the number of biological replicates for each tested concentration. Analysis of data is presented as the mean ± SEM. Significance of data was determined with paired T test for RT-qPCR (* $p < 0.01$) and with One-way ANOVA for concentration response (* $p < 0.05$; ** $p < 0.01$).

3. Results

3.1. Three-Dimensional Spheroid Differentiation of iPSCs-Derived NSCs Revealed Complex Neuronal Cultures

Our first aim was to characterize the timewise differentiation of 3D neuronal spheroids free-floating in suspension culture, originating from hiPSCs. The neuronal differentiation capacity of the starting neural stem cell (NSCs) population, derived by the dual-SMAD inhibition protocol, was comprehensively characterized by us previously [32,33]. These cells expressed the major NPC markers (NESTIN, SOX1 (SRY-Box 1), PAX6 (Paired Box 6) investigated on protein level with ICC (see details in Figure S2). When the NPCs were terminally differentiated in a 2D culture system over 5 weeks, cortical neurons and glial cells were differentiated and formed a neuronal network, as we published recently [32]. In the present study differentiation in a 3D culture system was investigated in the course of 6 weeks (42 days).

In suspension culture upon withdrawing the two mitogens EGF and bFGF, NSCs formed compact 3D spheroids, the so-called neurospheres within 48 h (Figure 1A). Growth properties of the structures were investigated by measuring the diameter of the individual spheres and their total protein content. Results showed a continuous growth of the spheres during the 6 weeks culture period, in terms of diametric growth and protein content (Figure 1). While the diameter increased linearly, the protein content increased more dynamically after week 4, resembling the cellular and structural changes during differentiation (Figure 1B,C).

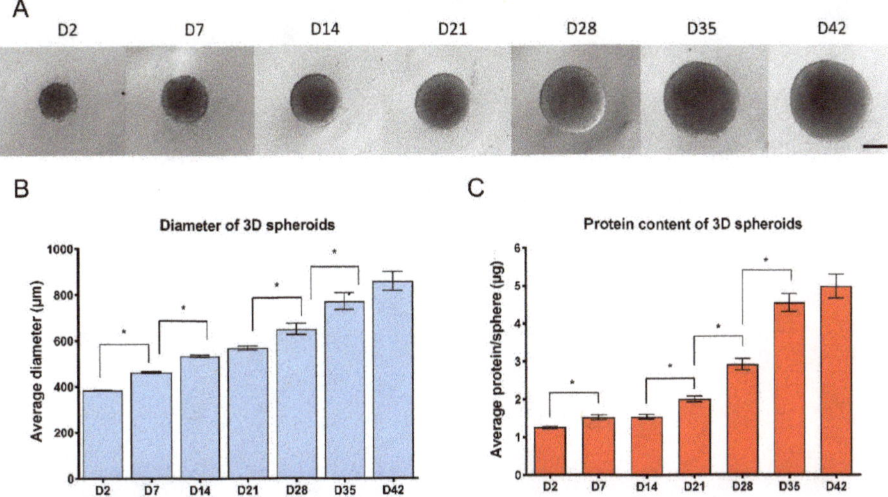

Figure 1. Growth properties of 3D neurospheres during the 6 weeks of differentiation (**A**) Representative microscopic view of 3D spheroids, investigated at weekly intervals (4×, scale bar: 200 μm). (**B**) The average diameter of 3D spheroids in μm measured by CellSens Dimension software (Olympus) ($n = 3$, in each experiment 96 spheroids were compared weekly trough 6-weeks). (**C**) Average total protein content of spheroids determined by Pierce BCA Protein Assay. Note that D2, D7 and D14 samples were measured by pooling three spheroids and individual values were calculated, while in all other timepoints spheroids were measured individually ($n = 3$, in each stage 24 spheroids were measured). ±SEM values are presented on graphs. (* $p < 0.05$).

This observation was in accordance with the expression of the cell proliferation marker Ki-67 (KI67), representing the proportion of dividing cells, monitored by ICC on protein level (Figure 2A,B) or by RT-qPCR on transcript level (Figure 3). The expression of Ki-67 and the number of Ki-67-positive cells were the highest in D2 and D7 samples, which showed a continuous decline over the differentiation period (Figure 2A,B or Figure 3). The forming 3D cell aggregates mainly expressed the early neuronal markers such as *PAX6*, *NESTIN*, and *SOX1*, which expressions decreased with maturation on a time-wise manner (RT-qPCR, Figure 3). Tubulin beta-3 chain (TUBB3) expression appeared very early in the samples, already at D2 as neuronal processes started to grow inside the spheres (ICC, Figure 2A,B). These changes were obvious when investigated at transcript level as well (RT-qPCR, *TUBB3*; Figure 3). It is important to note that no necrotic regions were detected in the center of the spheroids or in other regions, not even the highest diameter of spheroids reached 800 μm in average in D42 samples (Figure 1B or Figure 2 first line). This fact was corroborated in semithin sections where cell density proved to be high and consistent during the whole examined period. Layer organized fibers had become distinguishable on the surface from the 21st day (Figure S3a–g).

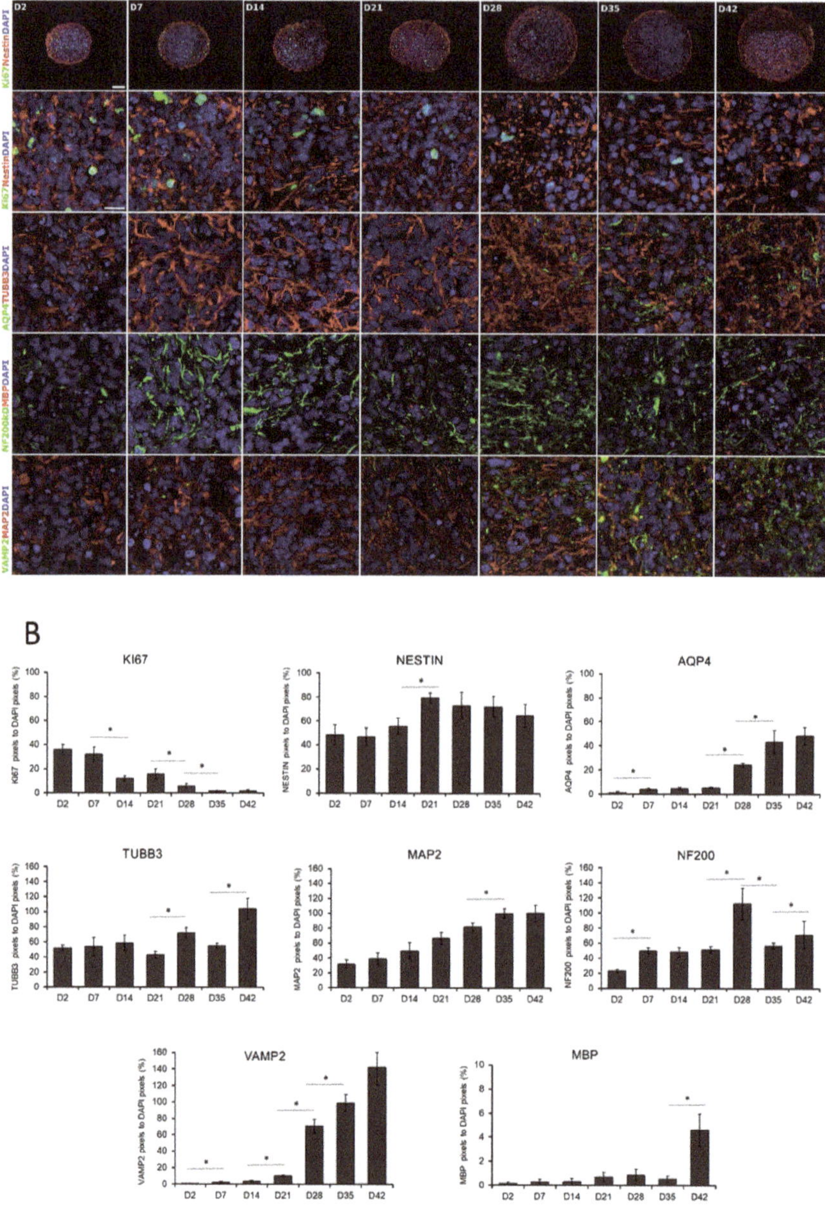

Figure 2. Immunocytochemical analysis of 3D spheroids. (**A**) Spheroids were fixed and cryosectioned then immunostained at weekly intervals from D2 until D42 stage. The first line represents the overview of the cryosectioned spheroids, while the rest of the panel shows higher magnifications. Relevant markers of proliferation (KI67), neural stem cells (NESTIN), neuronal differentiation (TUBB3 and MAP2), an intermediate filament of dendrites and axons (NF200), synaptic vesicles of neurons (VAMP2), astrocyte (AQP4), and oligodendrocyte (MBP) specific proteins were stained. Protein name IDs are indicated with colors, representing the color of the fluorophore used (e.g., green as Alexa 488; red as Alexa 594) Nuclei were counterstained with DAPI (in blue). Scale bar: 100 μm (first line only) and 25 μm.

(**B**) Quantitative analysis of the immunostainings on confocal images. The numbers of Ki-67, NESTIN, AQP4, TUBB3, NF200kD, MBP, VAMP2, and MAP2 immunoreactive pixels were measured in 5 neurospheres (middle sections, 5 randomly selected fields/slide) at every time points. Data was normalized with DAPI positive nuclei number. Data were expressed as percentage of marker/DAPI ratio ± SEM (* $p < 0.05$).

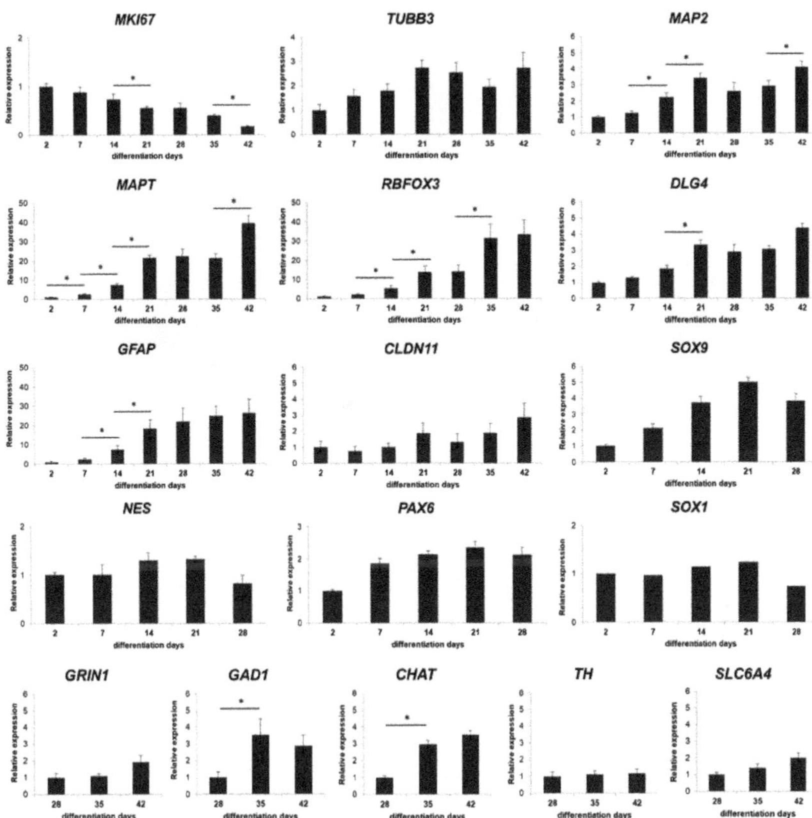

Figure 3. Real-time PCR measurements of relevant markers during the neuronal differentiation of the spheroids. Twelve 3D spheroids were pooled, lysed with RLT-buffer and used in RT-PCR analysis at each timepoints. Where the genes are already expressed in D2 samples, it is referred as 1. Day 2, 7, 14 and 21 values for *GRIN1, CHAT, TH, GAD1,* and *SLC6A4* are not indicated as these genes were not expressed at those time points. Graphs represent normalized relative expression values, analyzed for each gene using the ddCT method [41]. Mean values and ± SEM of three biological replicates ($n = 3$) are presented on graphs, significance are determined by paired T test (* $p < 0.01$). Note that y axis's scales alter according to the different relative expression values.

A weekly investigation of the gene expression revealed a continuous progress in neuronal differentiation and increase in the level of maturation markers, such as the dendritic marker Microtubule-associated protein 2 (*MAP2*) or Microtubule-associated protein tau (*MAPT*), and RNA Binding Fox-1 Homolog 3 (*RBFOX3*, also known as NeuN) (Figure 3), which were in accordance with the protein data of immunofluorescent staining of the 3D spheroids, MAP2 and Neurofilament 200 kDa (NF200); (ICC, Figure 2A,B).

Terminal differentiation resulted in the formation of neuronal networks and synapses, from the stage of D28, post-synaptic marker PSD95 (now DLG4, Discs Large MAGUK Scaffold Protein 4)

(Figures 3 and 4), and the major synaptic vesicle protein p38, Synaptophysin (SYNP) and Vesicle-associated membrane protein 2 (VAMP2) (Figures 2 and 4) expression appeared in maturing neurons. Synapse formation was assessed by transmission electron microscopy (TEM) revealing the presence of synaptic connections, presynaptic vesicles and postsynaptic density (Psd). Matured synapses were detected in D42 samples, where docked presynaptic vesicles were determined by identifiable contact point between the vesicle and the presynaptic membrane (Figure S3h,i).

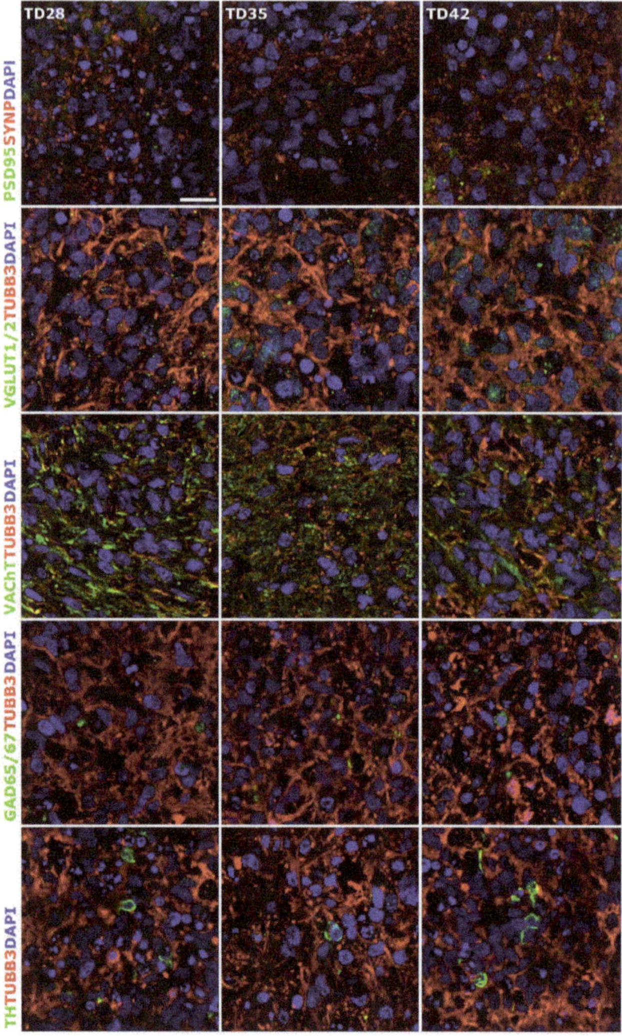

Figure 4. Immunocytochemical detection of neuronal subtypes in 3D neurospheres. Presence of synapses was determined with post-synaptic marker PSD95 and synaptic protein Synaptophysin (SYNP) double staining. Glutamatergic (VGLUT1/2), GABAergic (GAD65/67), cholinergic (VAChT) and dopaminergic (TH) neurons were detected in the developing 3D neurospheres from D28. All samples were stained with TUBB3 (in red) to label the neurites. Protein name IDs are indicated with colors, representing the color of the used fluorophore (e.g., green as Alexa 488; red as Alexa 594). Nuclei were counterstained with DAPI (in blue). Scale bar: 25 μm.

Neuronal subtypes were investigated by the expression of glutamatergic (VGLUT1/2), GABAergic (GAD65/67), cholinergic (VAChT) and dopaminergic (TH) neurons with ICC and qRT-PCR (*GRIN1*: Glutamate Ionotropic Receptor NMDA Type Subunit 1; *GAD1*: Glutamate Decarboxylase 1; *CHAT*: Choline *O*-Acetyltransferase; *TH*: Tyrosine Hydroxylase; *SLC6A4*: Solute Carrier Family 6 Member 4) (Figures 3 and 4). The expression of these markers was detectable mainly from D28 stage and gradually increased until the last investigated time point of the differentiation (D42), in line with the maturation process. It has to be remarked that the spontaneous differentiation of dopaminergic neurons was rare in the cultures, we identified only a few TH positive cells in the cultures (Figure 4).

During the neuronal differentiation, astrocytes started to emerge after week 2, based on the Aquaporin-4 (AQP4) or the Glial fibrillary acidic protein (GFAP) expression both at mRNA and protein level (Figure 2, Figure 3 and Figure S4). Oligodendrocyte differentiation followed a similar course, from D21 it was clearly detectable both at transcript (*Claudin 11*, *CLDN11*, formerly known as Oligodendrocyte marker 4; Figure 3) and protein level (Myelin basic protein, MBP; Figure 2A,B) which gradually increased from D35 with the maturation of the cells, however, still remained in a low expression level (Figure 2A,B or Figure 3). Our ICC results clearly show that the expression of mature neuronal and glial markers and neuron-neuron and neuron-glial interactions increased during the differentiation period, thus emulating the development of the human fetal neural tissue.

Overall, results at the transcript level were in accordance with those of the proteins detected by ICC, confirming the effective differentiation of hiPSC-derived NSCs towards the neural lineage. The 3D neurospheres represented a neuronal tissue-like differentiation, containing neurons, astrocytes and oligodendrocytes, what was presented by transcript-, protein-, and ultrastructural level as well. We can conclude that the 3D spheroid system provides a complex neuronal cell culture which can serve as a model of the early neuronal differentiation.

3.2. Early 3D Neurospheres as a Neurotoxicity Model

Our 3D spheroid-based model was tested in cytotoxicity assay, monitoring the viability of the cells within the neurospheres. Eleven compounds with well-known effects (drugs, pesticides, and chemicals) were selected and applied on the neurospheres in 7 different concentrations (compounds and concentrations are detailed in Table S3) at D21 by acute (72 h) exposure. Concentration-response curves were generated and evaluated. The exposure scheme is presented in Figure 5A.

According to the results, Paraquat (PQ; EC50: 1.89 log μM), Rotenone (ROT; EC50: −0.61 log μM), Mercury(II) chloride (HgCl$_2$; EC50: 1.87 log μM) and Doxorubicin (DOX; EC50: 0.67 log μM) caused maximum cell death, Hexachlorophene (HE; EC50: 1.26 log μM) and Colchicine (COL; EC50: −0.49 log μM) had a strong effect on the viability, but did not kill all the cells in the investigated concentration range. Acrylamide (ACR; EC50: 3.5 log μM) and Rifampicin (RIF; EC50: > 2 log μM) also reduced the viability, while Valproic acid (VPA; EC50: > 2.6 log μM) and Paracetamol (PAR; EC50: > 2 log μM) had minimal effect on the viability of 3D neurospheres at D21 stage. Ibuprofen (IBU; EC50: > 2 log μM) as a non-neurotoxic agent was applied as negative control which indeed, did not decrease the viability of D21 3D neurospheres (Figure 5B, Table S4). Overall, the different compounds induced different levels of cytotoxicity in a concentration-dependent manner on the 3D neurosphere cultures.

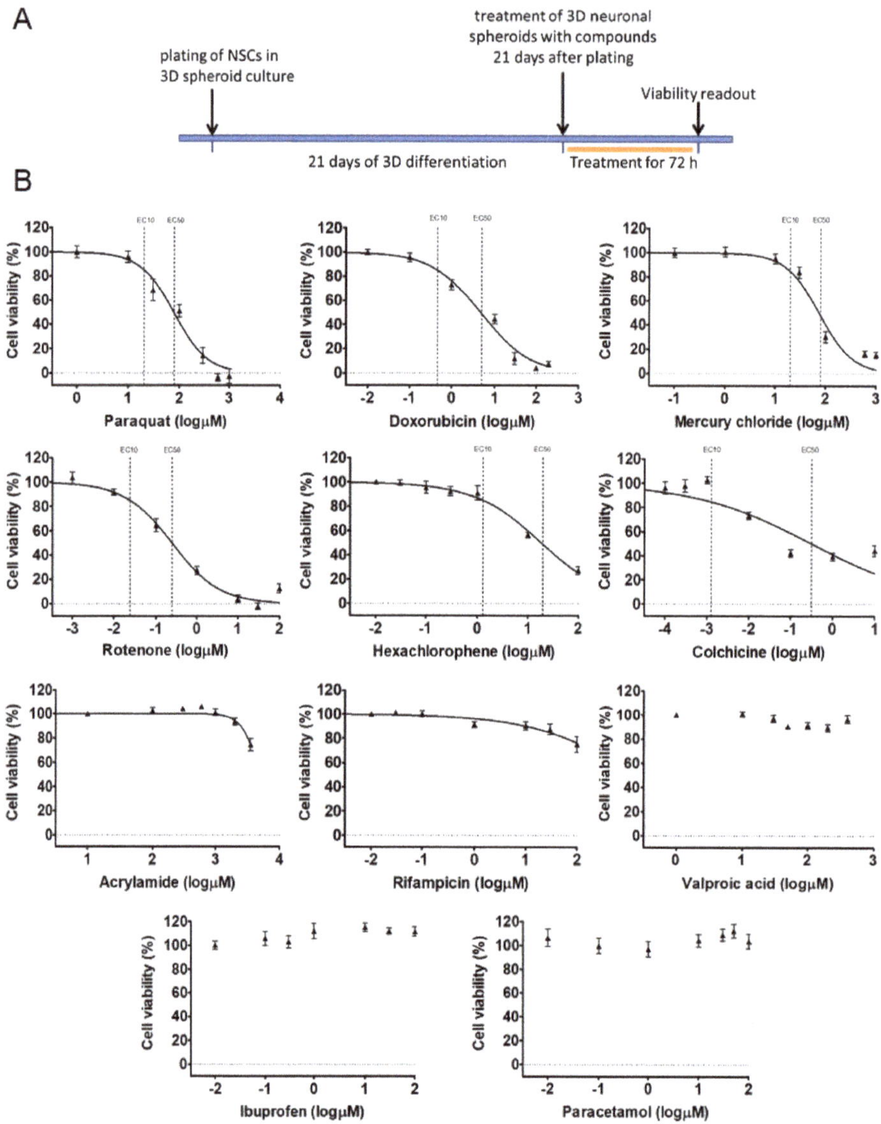

Figure 5. Cell viability measurement on D21 3D neurospheres after 72 h exposure. (**A**) Exposure scheme at D21 stage. (**B**) Concentration response curves of compounds, tested in 7 different concentrations (see concentrations listed in Table S3), representing the cell viability (%) of treated D21 3D neurospheres ($n = 3$). Concentration values are presented in log μM ± SEM. EC10 and EC50 values are presented on graphs where applicable.

3.3. *Different Age of the 3D Neurospheres Represent Distinct Differentiation Stages in the Cytotoxicity Model*

Next, we investigated the concentration-response of different differentiation stage-derived 3D neurospheres. D28 and D42 samples were analyzed in exposure schemes similar to that of the D21 samples (Figure 6A or Figure 7B) for all the previously tested compounds. As detailed above, in D28 samples differentiated astrocytes expressing GFAP and AQP4 are present (Figure 2 and Figure S4), axonal outgrowth is prevalent, and neuronal subtype-specific proteins start to appear (Figures 2–4). The D42 samples represent a more mature cell culture where synapsis of neurons was formed with established subtypes, mature astrocytes and a few oligodendrocytes are already present (Figure 2, Figure 4, Figures S3 and S4).

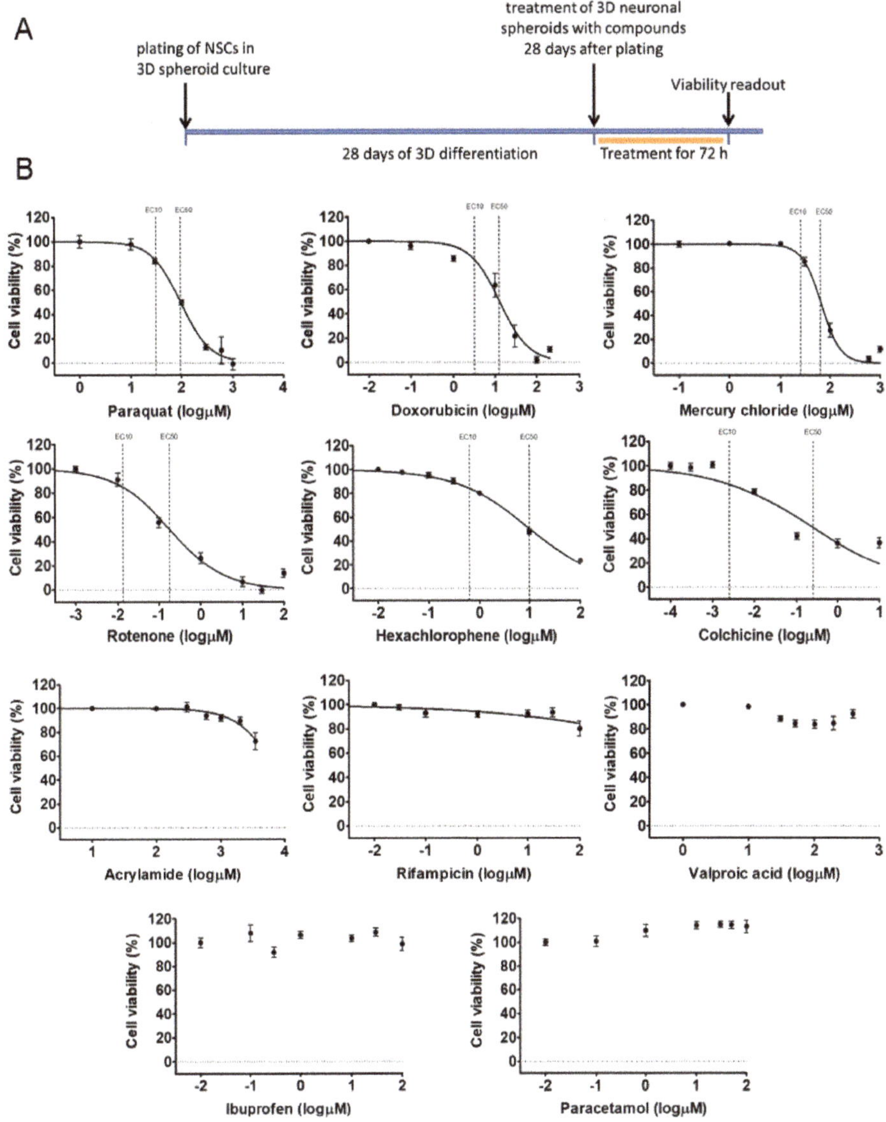

Figure 6. Cell viability measurement on D28 3D neurospheres after 72 h exposure with toxicants. (**A**) exposure scheme at D28 stage. (**B**) Concentration-response curves of compounds, tested in 7 different concentrations (see concentrations listed in Table S3), representing the cell viability (%) of treated D28 3D neurospheres (n = 3). Concentration values are presented in log μM ± SEM. EC10 and EC50 values are presented on graphs where applicable.

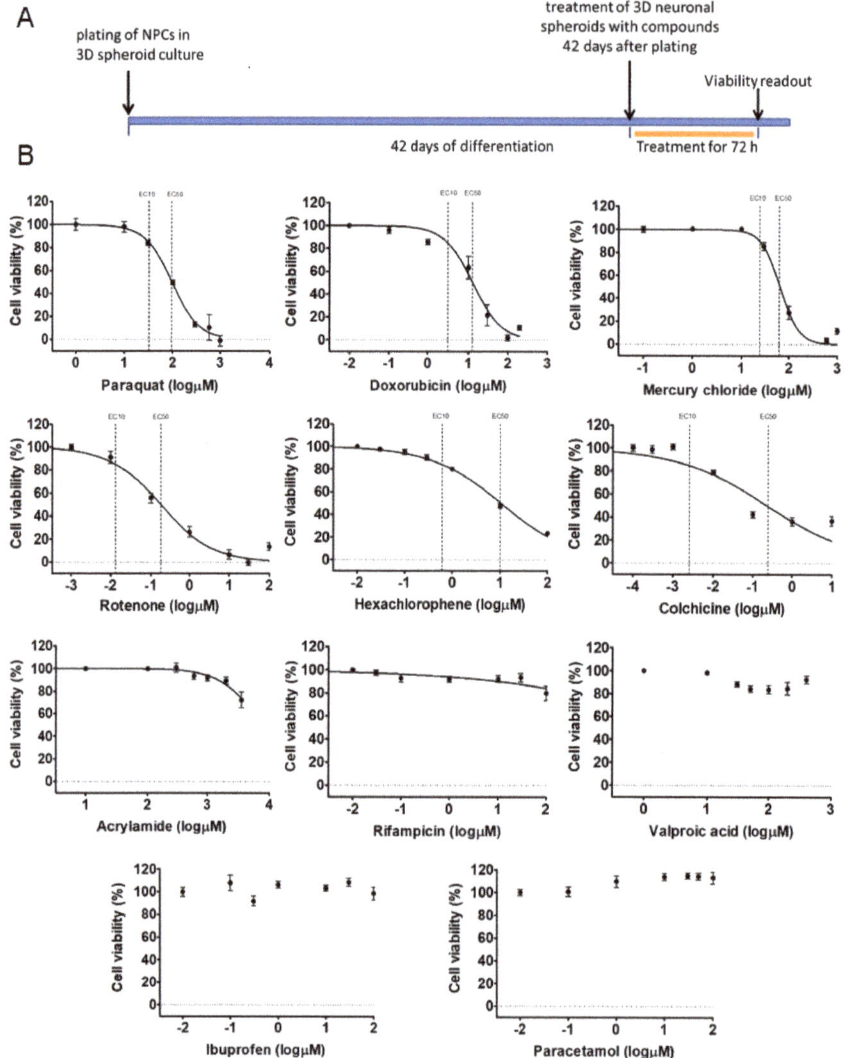

Figure 7. Cell viability measurement on D42 3D neurospheres after 72 h exposure with toxicants. (**A**) exposure scheme at D42 stage. (**B**) Concentration-response curves of compounds, tested in 7 different concentrations (see concentrations listed in Table S3), representing the cell viability (%) of treated D42 3D neurospheres ($n = 3$). Concentration values are presented in log µM ± SEM. EC10 and EC50 values are presented on graphs where applicable.

The experiments showed that both D28 and D42 stages are suitable for the viability assays without modifications in the culture system, using a 96-well plate format-based analysis. The EC50 and EC10 values for D28 and D42 samples were determined for each compound, as described above for D21 samples (Figure 6B or Figure 7B, Table S4). Some compounds (Figures 5–7, Table S4) showed less toxic effect on more matured samples (e.g., ACR: EC50^{D21} = 4.1 µM vs. EC50^{D42} = 3.6 µM) while it was the opposite action for other compounds (e.g., HgCl$_2$: EC50^{D21} = 1.7 µM vs. EC50^{D42} = 1.9 µM).

Based on the data obtained, a concentration-related ranking of the toxicant could be generated for each differentiation stage (Figure 8, Table S4).

A

	Ranking of compounds based on EC50 values		
Rank	**D21**	**D28**	**D42**
1	Colchicine	Colchicine	Colchicine
2	Rotenone	Rotenone	Rotenone
3	Doxorubicin	Doxorubicin	Hexachlorophene
4	Hexachlorophene	Hexachlorophene	Doxorubicin
5	Paraquat	Mercury chloride	Paraquat
6	Mercury chloride	Paraquat	Mercury chloride
7	Paracetamol	Paracetamol	Paracetamol
8	Ibuprofen	Ibuprofen	Ibuprofen
9	Rifampicin	Rifampicin	Rifampicin
10	Valproic acid	Valproic acid	Valproic acid
11	Acrylamide	Acrylamide	Acrylamide

B

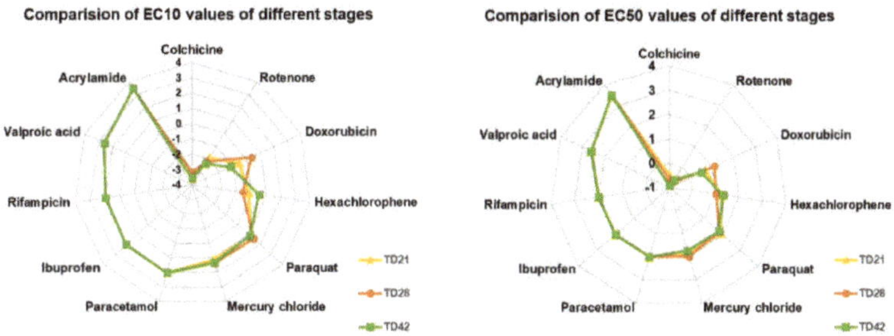

Figure 8. Comparison of the toxic effect of compounds. (**A**) Ranking of compounds at different differentiation stages based on EC50 values. Darker background color represents higher toxicity, while white color represents non-toxic compounds. (**B**) Radar chart showing the comparison of compounds based on the order of EC10 or EC50 values, at various differentiation stages (yellow line with rectangular: D21; orange line with circle: D28; green line with square: D42) (see also EC values in Table S4). Concentration values are presented in log µM.

In conclusion, we can state that a compound-specific concentration-response was detected in all stages. At the same time, a differentiation stage-relevant difference was observed, suggesting that the in vitro system can mimic the differential responses of the developing fetal human neuronal tissue, caused by distinct toxicants.

3.4. Compound-Specific Cellular Events Can be Detected in the 3D Neurospheres

ATP-based viability assay can detect cell death but is unsuitable for detecting other specific cellular events. Due to the lack of validated tests for 3D tissues, we decided to investigate the effect of a toxic compound at a subcellular level. We have chosen rotenone (ROT), a well-studied neurotoxic compound, known to interfere with the electron transport chain in mitochondria, to investigate the ultrastructure of the cells and especially mitochondria, followed by TEM. In parallel, a TUNEL assay was also applied to detect cell death. Based on the previously determined concentration-response curves and EC50 values, we treated the 3D neurospheres with 0.5 µM ROT for 72 h and analyzed the samples in the 3 differentiation stages. The data showed that ROT significantly increased the cell death in the spheroids (Figure 9A or Figure S6), compared to the vehicle-treated controls (Figure 9B), in accordance with the results of the ATP measurement (see Figure 5B or Figure 6B). Moreover, TEM revealed a change in the ultrastructure of the mitochondrial inner membrane, the cristae: Both shape and complexity were changed in the ROT treated samples, compared to the controls. Two types of mitochondria were detected, often in the same cell. Organelles with darker matrix were usually narrow and elongated, whereas lighter ones were more rounded. In control cells, cristae were straight, narrow, and long in both types. Matrix granules were observable more frequent in darker than in lighter mitochondria (Figure 9Ca–c). Effect of rotenone treatment was different on the two types of mitochondria. In darker organelles, cristae swelled and formed swirls. Several wide crista junctions became identifiable. In lighter mitochondria, cristae disintegrated and often became unrecognizable (Figure 9Cd–i). Matrix granules disappeared or showed decreased density (Figure 9Cc,i).

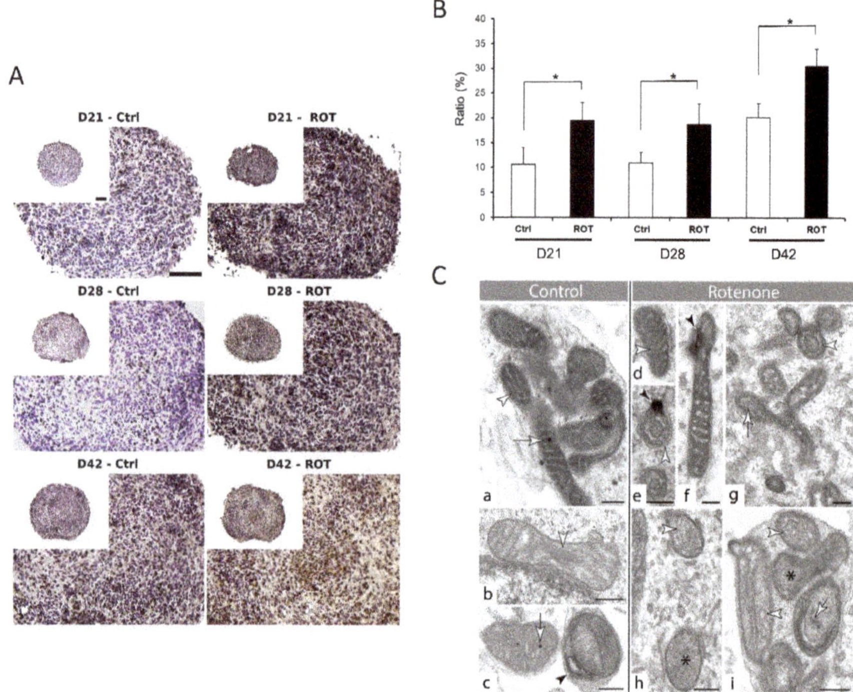

Figure 9. Effect of Rotenone (ROT) exposure on 3D neurospheres at three differentiation stages. (**A**) 3D neurospheres were treated with 0.5 µM ROT concentration for 72 h at three different maturation timepoints (D21, D28, and D42), fixed, sectioned and analyzed to detect the cellular effect of ROT by TUNEL assay (DeadEnd™ Colorimetric TUNEL System, Promega), compared to the vehicle (0.1% DMSO) treated control (scale bar: 100 µm). (**B**) ROT treatment revealed in average a 15% increase in the apoptotic cell number compared to the control in each stage (* $p < 0.05$). Average values are presented on graphs ($n = 3$). (**C**) Ultrastructure of mitochondria in control (panel a, b, c) and ROT treated (d–i) neurons in the 3D spheroids. See the alteration of the internal membrane (cristae) morphology (white arrowheads) in ROT treated cells (d, e, h, i). Black star: unidentifiable cristae morphology in lighter mitochondria (h, i); black arrowhead: membrane swirl in darker organelles (e, f); white arrows: matrix with and without matrix granules (control cells: a, c; treated cells: g, i) Note the density difference between these granules in control (c) and ROT treated mitochondria (panel i) (D21: a, d, f, h; D28: i; D42: b, c, g) (scale bar: 250 nm).

Finally, we used an ER-stress assay to detect if the observed change is compound-specific or an "overall" cell death event is detected with the ATP or TUNEL assay. The XBP1 assay was performed to identify ER-stress in the ROT treated samples, where ER stress was not expected. As a positive control, tunicamycin was used. The XBP1 assays demonstrated that ROT (using EC50 concentration) has no ER-stress inducing effect in the D21, D28 or D42 neurospheres (Figure S6B), while cell death was observed in the cryosectioned ROT-treated samples (Figure 9A), providing a strong evidence that compound-specific effects can be determined in 3D spheroids upon treatment. Immunocytochemical investigation reflected some disorganization of the treated group of spheroids compared to the controls, however other marked differences were not observed (Figure S6A).

3.5. Neurite Outgrowth Assay is Suitable to Determine the Effect of NT Compounds in the 3D Model

Functional readouts which connect the effect of a tested compound to a given cell type are essential to determine their tissue-specific effect. 3D spheroids represent a complex cell culture with a network of neurons, astrocytes, and oligodendrocytes, and this complexity makes more challenging to analyze minor differences in neurite length affected by chemical exposure due to the limitations of the current detection systems. Therefore, we investigated if neurite growth could be investigated using 3D neurospheres. We found that Matrigel-coated surface sufficiently supports the rapid and strong attachment of free-floating D21 3D neurospheres in order to generate a robust and reproducible procedure, where the effect of toxic compounds on neurite outgrowth can be investigated. The effect of different ROT concentrations were demonstrated comparing the total neurite outgrowth and the average number of neurites/spheroid ratio. ROT administration resulted significantly shorter and decreased number of neurites compared to the untreated and vehicle-treated control groups in all concentration levels in D21 spheroids (** $p < 0.01$) (Figure 10B).

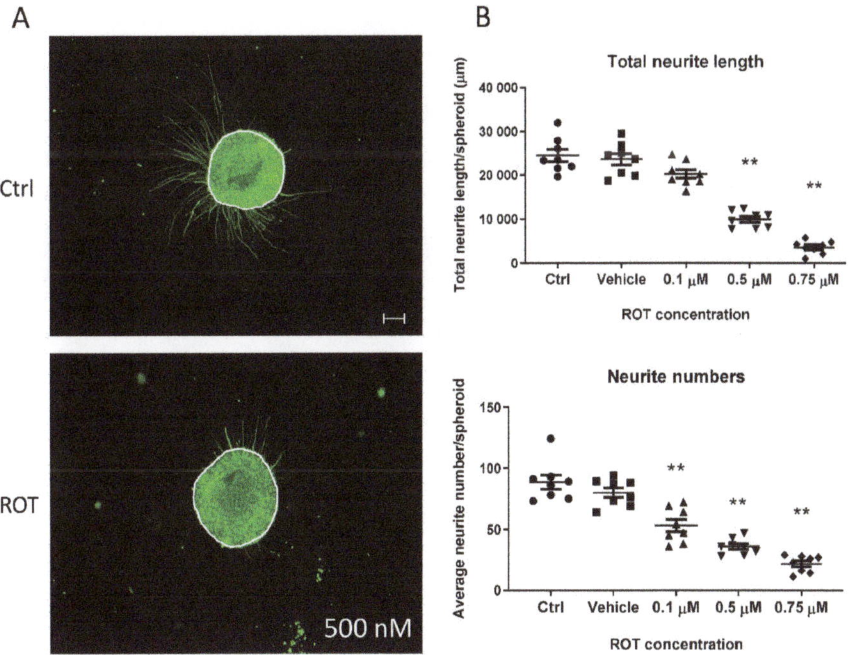

Figure 10. Neurite outgrowth measurement on D21 3D spheroids, exposed for 72 h with ROT. (**A**) Representative photograph of control (untreated) and ROT (0.5 µM) treated D21 spheroid immunolabeled with TUBB3 (in green). White lines represent the border of the spheroids where the neurite outgrowth was determined from, using ImageJ software (scale bar: 100 µm). (**B**) Total neurite length/spheroid (presented in µm ± SEM) and the average number of neurites/spheroids were determined 24 h after plating the treated spheroids. Different symbols denote treatment groups. (*n* = 3, in each experiment 8 spheroids were treated in each group) (** $p < 0.01$).

4. Discussion

The development of in vitro platforms for neurotoxicology screenings is driven by the urgent needs of the chemical, food, cosmetic, and pharma industries. Most NT studies are carried out in rodents or rodent derived primary cells, resulting in relatively high cost and lower translational value of the results due to the species differences [3,46]. Major international initiatives have started to convert the traditional animal-based neurotoxicity tests to in vitro assays using both mammalian brain cells and human cells to detect and predict chemical hazards [47,48]. However, there is only a limited number of human neuronal cell lines (e.g., carcinoma cell lines such as SH-SY5Y; BE2-M17 or immortalized cell lines like LUHMES) are available and hard to obtain primary human CNS tissue suitable for NT studies. Overall, the highly complex structure of the human brain makes in vitro modeling very difficult. Human iPSCs could fill this niche and offer the advantage that other cells and tissue types (e.g., kidney, liver, cardiac, neuronal, intestinal) sharing the same individual genetic background can be created using specific differentiation protocols in a replicable manner. This may provide a very effective in vitro tool for toxicologist for capturing the individual variability in the human population [49]. For example, in 2D neurotoxicity screening, the usage of human iPSC derived neuronal cultures, especially the commercially available QC controlled neuron and astrocyte cultures, where the differentiation of iPSCs is not required for the "users", is dynamically increasing [48].

In recent years, numerous in vitro models have been created in order to study the human CNS in a more physiological way, but the field of NT and DNT using 3D neuronal tissues did not develop as fast as disease or developmental modeling. It is due to the difficulty to find a compromise between biological complexity and technical reproducibility which are necessary for drug or toxicity screening [29].

An important approach was the application of human cell lines in NT assay development. For example, the LUHMES immortalized human fetal tissue-derived mesencephalic cell line can be efficiently differentiated towards dopamine-like neurons upon tetracycline administration [50,51] and provide a suitable system for neurotoxicology screenings [52,53] or Parkinson's disease-related drug testing [54] both in 2D or 3D cultures [45,55–57]. Despite the advantages in straightforward handling, they represent only one specific cell type of the CNS. New developments using stem cell-derived astrocytes or microglia in co-culture with LUHMES cells could reveal new potential both in NT and DNT tests [58], but also highlight that full-PSC derived systems could provide great advantages over the conventional cell lines.

To develop an efficient, highly reproducible neurotoxicology test system, an improved 3D culture of microtissues differentiated from human PSCs would be beneficial. For example, in neuronal disease modeling, a major step that opened new perspectives, was the development of cortical layer-organized 3D brain microtissues, providing a complex system forming under in vitro conditions from human PSCs [59,60]. However, for toxicological studies, the very low throughput potential of such complex systems is a significantly limiting factor at the moment.

Huang and his colleagues demonstrated that human brain organoids could be applied as an in vitro model for CNS drug screening to evaluate structural, cellular, and molecular changes. They used neurotoxic tranylcypromine in hiPSC-derived brain organoids leading to decreased proliferation activity and apoptosis induction [61]. HiPSC-derived cerebral organoids were treated with different concentrations of vincristine for 48 h, and the expansion of the treated organoids was measured, showing concentration-dependent neurotoxicity. Vincristine inhibited fibronectin, tubulin, and MMP10 expression in the cerebral organoids, which was specific for its well-known effect on microtubule dynamics [62]. Recently, iPSC-derived cortical neurons and astrocytes were co-cultured in 3D to detect calcium oscillations upon a chemical compound treatment, analyzing multiple parameters, highlighting the potential of such readouts in neurotoxicity assessment [31]. Another study has presented a novel 3D heterotypic glioblastoma-brain sphere (gBS) model applied for screening new anti-glioblastoma agents [63]. This new application highlighted the flexibility of iPSC-derived platforms to be used in disease modeling, drug testing, and toxicology.

In this study, we have analyzed the complexity of our hiPSC-derived 3D neurosphere system demonstrating intensive glial-neuron interaction with the astrocytes and oligodendrocytes present in the neurospheres. Complex characterization was performed with gene expression, protein level analysis in line with immunocytochemical investigations, including morphological evidence of neurocrine communication at the end of the examined period. Morphologically, synapses are composed of precisely opposed pre- and postsynaptic membranes decorated with electron-dense thickening, synaptic cleft filled with a fine meshwork of electron-dense material, and presynaptic vesicles [64]. One of the two populations of presynaptic vesicles belong to the active zone matrix, is the docked vesicles [65]. The membrane of them is in contact with the presynaptic membrane, therefore they form the readily releasable pool (RRP) of synaptic vesicles. [66].

Our system showed similarities in morphology and maturation properties with a previous study where 3D organoids were generated and characterized over 8 weeks [29]. In our system, there was no need to use BDNF, GDNF, or a special electrophysiology media; a basal medium was sufficient to promote the differentiation of complex spheroids within 6 weeks of culture. Although the spheroids showed a continuous growth and development/maturation, the homogeneity of the plates was not compromised, both the diameter and total protein content of the spheroids within the 96-well plate showed very low variation at a given time point, represented by the low SEM values (Figure 1). Low variability of samples and high homogeneity is crucial when the aim is the development of a reliable HTS for DNT studies.

Here, we performed a medium-throughput 96-well plate assay on 3D neurospheres to detect the cytotoxic effect of the selected compounds, based on ATP-release measurements. We tested drugs, pesticides and well-known chemicals with or without neurotoxic effect in different stages of the neuronal differentiation. Our results highlighted a compound-specific and differentiation stage-related effect of the tested chemicals, providing the possibility to determine EC50 and EC10 values for the compounds (Table S3). For example, colchicine has a strong toxic effect on neurospheres with very similar EC50 values in all differentiation stages. Its effect can be detrimental, it blocks basic cellular (protein assembly, endocytosis, exocytosis, cellular motility, etc.) and neuronal functions at the same time (assembly of tubulin filaments in the microtubules of neurites), which we supposed to happen in all differentiation stages in our neurospheres. A comparable neurotoxic effect was described on iPSC derived 2D cultures of NSCs, neurons, and astrocytes [67], and LUHMES cells [57]. By comparing EC10 or EC50 values, another example is doxorubicin, where D42 samples were the most sensitive for the treatment when intensive protein synthesis could happen both in neurons and glial cells (e.g., neurotransmitter synthesis, axonal growth, oligodendrocyte maturation), therefore the potential blocking of the transcription machinery might have a major effect on cell viability (see Table 1 for a summary).

Table 1. Known effects of the used compounds.

Compound Name (CAS Number)	Known Effects of the Compound
Acrylamide 79-06-1	- Water-soluble crystalline amide that polymerizes rapidly; - widely used in chemical industry (e.g., water treatment industry, paper industry, textile treatment industry) and cosmetics; - occurs in food upon heat treatment—both during home cooking or industrial processing of food— [68]; - neurotoxic: occupational exposures [69,70]; - could cross the placental barrier and appears in breast milk [71]; and - probable oral lethal human dose is between 50–500 mg/kg.

Table 1. *Cont.*

Compound Name (CAS Number)	Known Effects of the Compound
Colchicine 64-86-8	- Can bind to tubulin and inhibit tubulin polymerization leading to inhibition in mitosis; - interacts with the P-glycoprotein transporter (MDR1/ABCB1) and the CYP3A4 enzyme (both involved in toxin metabolism) [72,73]; - medication used to treat gout and Behçet's disease; - probable oral lethal dose in humans is less than 5 mg/kg; and - in vitro cytotoxicity limit: 0.02 µM.
Doxorubicin 25316-40-9	- Intercalating chemotherapy drug; - inhibiting the movement of topoisomerase II, which leads to the blocking of both replication and transcription; and - LD50: 21.8 mg/kg (rat, subcutaneous).
Hexachlorophene 70-30-4	- Disinfectant; - blocks the electron transport chain through acting on membrane-anchoring subunit of succinate dehydrogenase (SDHD) and Glutamate dehydrogenase 1, mitochondrial (GLUD1); - probable oral lethal dose in humans is not determined; LD50: 66 mg/kg (rat, oral); and - in vitro cytotoxicity limit: 1.86 µM.
Ibuprofen 15687-27-1	- Drug (pain killer); - effectively reduces fever, as a non-steroid anti-inflammatory drug (NSAID), acts on inhibiting cyclooxygenase (COX) enzymes (COX-1 and 2); and - overdose symptoms appear in individuals consumed more than 99 mg/kg; LD50: 636 mg/kg (rat, oral).
Mercury(II) chloride 7487-94-7	- Component of pesticides; - corrosive, toxic; - could accumulate in the kidney; - probable oral lethal dose is 5–50 mg/kg; and - in vitro cytotoxicity limit: 1.37 µM.
Paracetamol 103-90-2	- Drug (pain killer); - ioverdosing could cause liver toxicity; and - hepatic toxicity in humans occurred with acute overdoses more than 10 g; LD50: 2400 mg/kg (rat, oral).
Paraquat 7 5365-73-0	- Herbicide; - neurotoxic: occupational exposures leading to Parkinson's disease [74]; - widely investigated neurotoxic mechanism [75,76]; and - probable oral lethal dose in humans is 35 mg/kg.
Rifampicin 13292-46-1	- Antibiotic; - stops RNA synthesis in bacteria; - could cause liver toxicity; - robable lethal oral dose in humans is 14–60 gLD50: 1570 mg/kg (rat); and - in vitro cytotoxicity limit: 4.37 µM.

Table 1. *Cont.*

Compound Name (CAS Number)	Known Effects of the Compound
Rotenone 83-79-4	- Pesticide; - inhibits mitochondrial Complex I of the electron transport chain; - in rats Parkinson's disease- like symptoms were developed, →considered as an environmental risk factor for PD [77]; - probable lethal dose in human 0.3–0.5 g/kg; and - in vitro cytotoxicity limit: 0.22 µM.
Valproic acid 1069-66-5	- Drug used in epilepsy, bipolar disorders, and for the prevention of seizures; - known to block voltage-gated sodium channels and increase brain levels of gamma-aminobutyric acid (GABA); - well-known teratogen induces different congenital malformations and neural tube defects (NTDs); and - LD50: 670 mg/kg (rat, oral).

Note: probable lethal dose data is provided upon PubChem database (https://pubchem.ncbi.nlm.nih.gov/), in vitro multicellular cytotoxicity "comptox" data were collected from EPA's Chemical dashboard (https://comptox.epa.gov/dashboard).

Continuing the line with hexachlorophene that blocks the electron transport chain, by acting on GLUD1, it affects the turnover of an excitatory neurotransmitter, glutamate. This effect could explain the observed differentiation stage-related sensitivity differences of the neurospheres, still under an acute exposure scheme. Likewise, less cytotoxic effect was reported on iPSC derived 2D cultures of NSCs, early neurons (14 days old) and astrocytes when 10 or 100 µM HE was exposed for 24 h [67], while more pronounced cytotoxicity and neurite outgrowth inhibition of LUHMES cells were reported [57].

Valproate is known as a DNT positive compound. However, in our cell viability assay, it has no detectable impact on the investigated concentration range, which might be in correlation with the maturity of the treated cells, they are not in early neurodevelopment (neural tube stage), but maturing neurons and astrocytes. When LUHMES cells were exposed the EC50 value of VPA was determined as >100 µM, and significant inhibition of neurite outgrowth was detected. While the EC50 value was in a similar range with our 3D neurosphere-driven data, we did not investigate its effect on neurite outgrowth. Here, we also have to clarify that cytotoxicity read-out alone is not suitable to predict DNT, although it is suitable to predict the neurotoxic effect of compounds or detect necrosis in differentiating cultures, more relevant end-points (e.g., proliferation, migration, apoptosis, network formation, synaptogenesis, and growth of neurites) must be investigated for evaluating the DNT effect of a compound [18].

In the case of pesticide rotenone, which inhibits mitochondrial Complex I of the electron transport chain, strong cell death was detected. The effect of ROT exposure was investigated extensively. In iPSC-derived neuron-astroglia 2D cell cultures the activation of the Nrf2/ARE pathway was documented upon ROT-induced oxidative stress, which led to the activation of astrocytes and cell death of neurons [67,78,79]. Using LUHMES cells this robust cytotoxic effect was not obvious while a prominent neurite growth inhibition was reported [57]. Using "BrainSpheres" system [30], where ICC-based morphology analysis, ROS measurements and viability were compared in different developmental times and exposure schemes, dopaminergic-neuron selective toxicity and general cytotoxic concentration were determined [30]. Importantly, increased cell death and mitochondrial dysfunction were detected in our ROT exposed 3D neurospheres, similar to others' results [30]. In contrast, ibuprofen has no effect on neuronal cell viability in the investigated concentration range, as it was expected upon known in vitro and in vivo data.

Neurite outgrowth assays are well established in conventional 2D neuronal cultures, derived from cell lines, primary tissues, or iPSCs [57,80,81], however, the adaptation to 3D cultures is less tested. Determining the radial migration and neuronal density distribution within the migration area of NPCs grown as spheroids is rather described [27]. Here we adapted a simple system which provides a tissue-specific functional readout for neurospheres upon exposure, which can be automated using high content imaging (HCI) systems [45]. Based on our results, it is sensitive enough to distinguish different concentration of toxic compounds, does not require complicated read-out systems which was proved by testing the effect of ROT, previously confirmed to inhibit neurite growth of LUHMES cells [57]. Although these results agree with the reported neurite outgrowth inhibition effect, we cannot exclude the influence of cytotoxicity on the reduction of the overall neurite length we observed. Combining with LDH or lactate sampling, metabolite measurements from the media or terminal transcriptomic or proteomic assays, a complex dataset can be collected, and the effect of a given compound can be investigated systematically on the molecular level, as well to identify adverse effects. Similar to the RT-qPCR and ICC investigation that we used to characterize our 3D neurospheres these methods can be used when the effect of a given compound is investigated. Although we performed immunocytochemistry on treated spheroids using rotenone only, others successfully used such investigations to gain DNT readouts, for example detecting cell migration, neurite outgrowth, Ca^{2+} reabsorption, synaptogenesis, and PPAR pathway disruption [29,30,57,82,83].

Evaluating our concentration–response results, we faced with the lack of available in vitro human data, which makes the comparison challenging. This demonstrates that due to the lack of EC values and very different readouts, the appropriate comparison between the various test systems is often not possible. As summarized in Table 1, it is not possible to provide human in vitro NT values for every compound, the databases are incomplete. Although there are several initiatives and efforts on the field to collect and harmonize the available test methods and evaluate the data produced so far, there is still a lot to do in the field of in vitro neurotoxicity test method development [1,16].

5. Conclusions

We have established and evaluated a hiPSC-based 3D in vitro cellular model for the study of the neurotoxic effect of different compounds. Concentration–response was determined in different differentiation stages, using a well-known read-out the cell viability. For a certain compound, apoptotic activity, ER-stress assay and TEM was performed, providing possibilities to generate novel human CNS-relevant data for other compounds using hiPSC derived cells. Tissue-specific readouts such as neurite outgrowth were also investigated, however further, read-outs and DNT-relevant training compound set should be applied to evaluate the suitability of our 3D spheroids in DNT screening Nevertheless, due to the pluripotent nature of the hiPSC, this model offers an excellent tool for drug testing, gene therapy studies and toxicology studies parallelly on the same genotype using other cell or tissue types at the same time. Moreover, new advancement in gene manipulation such as CRISPR/Cas9 mediated gene targeting, makes it possible to target specific pathways and generate reporter cell lines for toxicological or other applications. Combining these new approaches with 3D cell culture-based assays could revolutionize the field of toxicology including DNT studies in the near future.

Supplementary Materials: The following are available online at http://www.mdpi.com/2073-4409/9/5/1122/s1, Figure S1: Characterization of CTRL-2 hiPSC line, Figure S2: Characterization of NPCs, Figure S3: Summary of the morphological differentiation of neurospheres, Figure S4: Representative immunostaining of cryosectioned neurospheres, Figure S5: The initial delta Ct distributions of relevant neuronal markers in D2 stage samples by Real-time PCR measurements, Figure S6: Effect of ROT treatment on 3D spheroids, Table S1: Antibodies used for immunocytochemistry and flow cytometry, Table S2: Primers used in the study, Table S3: Tested compounds and their concentration, Table S4: Effective concentration (EC) values of compounds tested on D21, D28 and D42 3D neurospheres.

Author Contributions: Data curation, M.Z. and I.B.; Funding acquisition, A.D.; Investigation, J.K., A.T., T.B., Z.J., K.M., I.B. and L.L.; Supervision, L.L. and A.D.; Writing—original draft, J.K.; Writing—review and editing, M.Z., L.L. and A.D. All authors have read and agreed to the published version of the manuscript.

Funding: This project has received funding from the European Union's Horizon 2020 research and innovation programme under grant agreement No 681002 (EU-ToxRisk).

Acknowledgments: We are grateful for the help of Balázs Mihalik in study design and assay development. We thank Mónika Truszka for the technical help in the preparation of samples for TEM.

Conflicts of Interest: The authors declare no conflict of interest.

Ethics approval and consent to participate: Written informed consents have been obtained from all subjects providing samples for iPSC derivation. Ethical approvals were obtained from the Hungarian Medical Research Council (in Hungarian: *Egészségügyi Tudományos Tanács; ETT*) to establish and maintain hiPSC lines (ETT-TUKEB 834/PI/09, 8-333/2009-1018EKU).

Consent for publication: Not applicable.

Availability of Data and Material: Additional files are made available online along with the manuscript.

Abbreviations

3D	three dimensional;
AO	adverse outcome;
BSA	bovine serum albumin;
CAS	chemical abstracts service;
CNS	central nervous system;
Ctrl	control;
D	day;
DNT	developmental neurotoxicity;
EC	effective concentration;
ER	endoplasmic reticulum;
FBS	fetal bovine serum;
HE	Hematoxylin-Eosin;
HTS	high-throughput screening;
hiPSC	human induced pluripotent stem cell;
ICC	immunocytochemistry;
iPSC	induced pluripotent stem cell;
NA	numeric aperture;
NAM	new approach method;
NEAA	non-essential amino acids;
NIM	neural induction media;
NMM	neural maintenance media;
NPC	neural progenitor cell;
NSC	neural stem cell;
O/N	overnight;
PBMC	peripheral blood mononuclear cells;
PBS	phosphate buffer saline;
PCR	polymerase chain reaction;
PFA	Paraformaldehyde;
POL/L	Poly-L-ornithine and Laminin;
PSC	pluripotent stem cell;
RT	room temperature;
rTdT	recombinant terminal deoxynucleotidyl transferase;
RT-PCR	reverse transcription polymerase chain reaction;
TEM	transmission electron microscopy;
TUNEL	terminal deoxynucleotidyl transferase dUTP nick end labeling.

References

1. Krewski, D.; Acosta, D., Jr.; Andersen, M.; Anderson, H.; Bailar, J.C., 3rd; Boekelheide, K.; Brent, R.; Charnley, G.; Cheung, V.G.; Green, S., Jr.; et al. Toxicity testing in the 21st century: A vision and a strategy. *J. Toxicol. Environ. Health Part B Crit. Rev.* **2010**, *13*, 51–138. [CrossRef]
2. Tsuji, R.; Crofton, K.M. Developmental neurotoxicity guideline study: Issues with methodology, evaluation and regulation. *Congenit. Anom. (Kyoto)* **2012**, *52*, 122–128. [CrossRef]
3. Hartung, T. Thoughts on limitations of animal models. *Parkinsonism Relat. Disord.* **2008**, *14*, S81–S83. [CrossRef] [PubMed]
4. Crofton, K.M.; Mundy, W.R.; Shafer, T.J. Developmental neurotoxicity testing: A path forward. *Congenit. Anom. (Kyoto)* **2012**, *52*, 140–146. [CrossRef] [PubMed]
5. Bal-Price, A.; Crofton, K.M.; Leist, M.; Allen, S.; Arand, M.; Buetler, T.; Delrue, N.; FitzGerald, R.E.; Hartung, T.; Heinonen, T.; et al. International STakeholder NETwork (ISTNET): Creating a developmental neurotoxicity (DNT) testing road map for regulatory purposes. *Arch. Toxicol.* **2015**, *89*, 269–287. [CrossRef] [PubMed]
6. Bal-Price, A.; Hogberg, H.T.; Crofton, K.M.; Daneshian, M.; FitzGerald, R.E.; Fritsche, E.; Heinonen, T.; Hougaard Bennekou, S.; Klima, S.; Piersma, A.H.; et al. Recommendation on test readiness criteria for new approach methods in toxicology: Exemplified for developmental neurotoxicity. *ALTEX* **2018**, *35*, 306–352. [CrossRef]
7. Westerink, R.H. Do we really want to REACH out to in vitro? *Neurotoxicology* **2013**, *39*, 169–172. [CrossRef]
8. Bal-Price, A.K.; Hogberg, H.T.; Buzanska, L.; Lenas, P.; van Vliet, E.; Hartung, T. In vitro developmental neurotoxicity (DNT) testing: Relevant models and endpoints. *Neurotoxicology* **2010**, *31*, 545–554. [CrossRef]
9. Sachana, M.; Rolaki, A.; Bal-Price, A. Development of the Adverse Outcome Pathway (AOP): Chronic binding of antagonist to N-methyl-d-aspartate receptors (NMDARs) during brain development induces impairment of learning and memory abilities of children. *Toxicol. Appl. Pharmacol.* **2018**, *354*, 153–175. [CrossRef]
10. Fritsche, E.; Grandjean, P.; Crofton, K.M.; Aschner, M.; Goldberg, A.; Heinonen, T.; Hessel, E.V.S.; Hogberg, H.T.; Bennekou, S.H.; Lein, P.J.; et al. Consensus statement on the need for innovation, transition and implementation of developmental neurotoxicity (DNT) testing for regulatory purposes. *Toxicol. Appl. Pharmacol.* **2018**, *354*, 3–6. [CrossRef]
11. Harrill, J.A.; Freudenrich, T.; Wallace, K.; Ball, K.; Shafer, T.J.; Mundy, W.R. Testing for developmental neurotoxicity using a battery of in vitro assays for key cellular events in neurodevelopment. *Toxicol. Appl. Pharmacol.* **2018**, *354*, 24–39. [CrossRef] [PubMed]
12. Alepee, N.; Bahinski, A.; Daneshian, M.; De Wever, B.; Fritsche, E.; Goldberg, A.; Hansmann, J.; Hartung, T.; Haycock, J.; Hogberg, H.; et al. State-of-the-art of 3D cultures (organs-on-a-chip) in safety testing and pathophysiology. *ALTEX* **2014**, *31*, 441–477. [CrossRef] [PubMed]
13. Bal-Price, A.K.; Sunol, C.; Weiss, D.G.; van Vliet, E.; Westerink, R.H.; Costa, L.G. Application of in vitro neurotoxicity testing for regulatory purposes: Symposium III summary and research needs. *Neurotoxicology* **2008**, *29*, 520–531. [CrossRef] [PubMed]
14. Pampaloni, F.; Reynaud, E.G.; Stelzer, E.H. The third dimension bridges the gap between cell culture and live tissue. *Nat. Rev. Mol. Cell Biol.* **2007**, *8*, 839–845. [CrossRef]
15. F-W Greiner, J.; Kaltschmidt, B.; Kaltschmidt, C.; Widera, D. Going 3D-Cell Culture Approaches for Stem Cell Research and Therapy. *Curr. Tissue Eng.* **2013**, *2*, 8–19. [CrossRef]
16. Pamies, D.; Hartung, T. 21st Century Cell Culture for 21st Century Toxicology. *Chem. Res. Toxicol.* **2017**, *30*, 43–52. [CrossRef]
17. Edmondson, R.; Broglie, J.J.; Adcock, A.F.; Yang, L. Three-dimensional cell culture systems and their applications in drug discovery and cell-based biosensors. *Assay Drug Dev. Technol.* **2014**, *12*, 207–218. [CrossRef]
18. Bal-Price, A.; Pistollato, F.; Sachana, M.; Bopp, S.K.; Munn, S.; Worth, A. Strategies to improve the regulatory assessment of developmental neurotoxicity (DNT) using in vitro methods. *Toxicol. Appl. Pharmacol.* **2018**, *354*, 7–18. [CrossRef]
19. Smirnova, L.; Hogberg, H.T.; Leist, M.; Hartung, T. Developmental neurotoxicity—Challenges in the 21st century and in vitro opportunities. *ALTEX* **2014**, *31*, 129–156. [CrossRef]

20. Schmidt, B.Z.; Lehmann, M.; Gutbier, S.; Nembo, E.; Noel, S.; Smirnova, L.; Forsby, A.; Hescheler, J.; Avci, H.X.; Hartung, T.; et al. In vitro acute and developmental neurotoxicity screening: An overview of cellular platforms and high-throughput technical possibilities. *Arch. Toxicol.* **2017**, *91*, 1–33. [CrossRef]

21. Hofrichter, M.; Nimtz, L.; Tigges, J.; Kabiri, Y.; Schroter, F.; Royer-Pokora, B.; Hildebrandt, B.; Schmuck, M.; Epanchintsev, A.; Theiss, S.; et al. Comparative performance analysis of human iPSC-derived and primary neural progenitor cells (NPC) grown as neurospheres in vitro. *Stem Cell Res.* **2017**, *25*, 72–82. [CrossRef] [PubMed]

22. Gage, F.H. Mammalian neural stem cells. *Science* **2000**, *287*, 1433–1438. [CrossRef] [PubMed]

23. Breunig, J.J.; Haydar, T.F.; Rakic, P. Neural stem cells: Historical perspective and future prospects. *Neuron* **2011**, *70*, 614–625. [CrossRef] [PubMed]

24. Thomson, J.A.; Itskovitz-Eldor, J.; Shapiro, S.S.; Waknitz, M.A.; Swiergiel, J.J.; Marshall, V.S.; Jones, J.M. Embryonic stem cell lines derived from human blastocysts. *Science* **1998**, *282*, 1145–1147. [CrossRef]

25. Takahashi, K.; Tanabe, K.; Ohnuki, M.; Narita, M.; Ichisaka, T.; Tomoda, K.; Yamanaka, S. Induction of pluripotent stem cells from adult human fibroblasts by defined factors. *Cell* **2007**, *131*, 861–872. [CrossRef]

26. Qian, X.; Nguyen, H.N.; Song, M.M.; Hadiono, C.; Ogden, S.C.; Hammack, C.; Yao, B.; Hamersky, G.R.; Jacob, F.; Zhong, C.; et al. Brain-Region-Specific Organoids Using Mini-bioreactors for Modeling ZIKV Exposure. *Cell* **2016**, *165*, 1238–1254. [CrossRef]

27. Schmuck, M.R.; Temme, T.; Dach, K.; de Boer, D.; Barenys, M.; Bendt, F.; Mosig, A.; Fritsche, E. Omnisphero: A high-content image analysis (HCA) approach for phenotypic developmental neurotoxicity (DNT) screenings of organoid neurosphere cultures in vitro. *Arch. Toxicol.* **2017**, *91*, 2017–2028. [CrossRef]

28. Hogberg, H.T.; Bressler, J.; Christian, K.M.; Harris, G.; Makri, G.; O'Driscoll, C.; Pamies, D.; Smirnova, L.; Wen, Z.; Hartung, T. Toward a 3D model of human brain development for studying gene/environment interactions. *Stem Cell Res. Ther.* **2013**, *4*, S4. [CrossRef]

29. Pamies, D.; Barreras, P.; Block, K.; Makri, G.; Kumar, A.; Wiersma, D.; Smirnova, L.; Zang, C.; Bressler, J.; Christian, K.M.; et al. A human brain microphysiological system derived from induced pluripotent stem cells to study neurological diseases and toxicity. *ALTEX* **2017**, *34*, 362–376. [CrossRef]

30. Pamies, D.; Block, K.; Lau, P.; Gribaldo, L.; Pardo, C.A.; Barreras, P.; Smirnova, L.; Wiersma, D.; Zhao, L.; Harris, G.; et al. Rotenone exerts developmental neurotoxicity in a human brain spheroid model. *Toxicol. Appl. Pharmacol.* **2018**, *354*, 101–114. [CrossRef]

31. Sirenko, O.; Parham, F.; Dea, S.; Sodhi, N.; Biesmans, S.; Mora-Castilla, S.; Ryan, K.; Behl, M.; Chandy, G.; Crittenden, C.; et al. Functional and Mechanistic Neurotoxicity Profiling Using Human iPSC-Derived Neural 3D Cultures. *Toxicol. Sci. Off. J. Soc. Toxicol.* **2019**, *167*, 58–76. [CrossRef] [PubMed]

32. Chandrasekaran, A.; Avci, H.X.; Ochalek, A.; Rosingh, L.N.; Molnar, K.; Laszlo, L.; Bellak, T.; Teglasi, A.; Pesti, K.; Mike, A.; et al. Comparison of 2D and 3D neural induction methods for the generation of neural progenitor cells from human induced pluripotent stem cells. *Stem Cell Res.* **2017**, *25*, 139–151. [CrossRef] [PubMed]

33. Zhou, S.; Szczesna, K.; Ochalek, A.; Kobolak, J.; Varga, E.; Nemes, C.; Chandrasekaran, A.; Rasmussen, M.; Cirera, S.; Hyttel, P.; et al. Neurosphere Based Differentiation of Human iPSC Improves Astrocyte Differentiation. *Stem Cells Int.* **2016**, *2016*, 4937689. [CrossRef] [PubMed]

34. Chambers, S.M.; Fasano, C.A.; Papapetrou, E.P.; Tomishima, M.; Sadelain, M.; Studer, L. Highly efficient neural conversion of human ES and iPS cells by dual inhibition of SMAD signaling. *Nat. Biotechnol.* **2009**, *27*, 275–280. [CrossRef] [PubMed]

35. Shi, Y.; Kirwan, P.; Livesey, F.J. Directed differentiation of human pluripotent stem cells to cerebral cortex neurons and neural networks. *Nat. Protoc.* **2012**, *7*, 1836–1846. [CrossRef] [PubMed]

36. Tieng, V.; Stoppini, L.; Villy, S.; Fathi, M.; Dubois-Dauphin, M.; Krause, K.H. Engineering of midbrain organoids containing long-lived dopaminergic neurons. *Stem Cells Dev.* **2014**, *23*, 1535–1547. [CrossRef]

37. White, D.L.; Mazurkiewicz, J.E.; Barrnett, R.J. A chemical mechanism for tissue staining by osmium tetroxide-ferrocyanide mixtures. *J. Histochem. Cytochem. Off. J. Histochem. Soc.* **1979**, *27*, 1084–1091. [CrossRef]

38. Rozen, S.; Skaletsky, H. Primer3 on the WWW for general users and for biologist programmers. *Bioinform. Methods Protoc.* **1999**, *132*, 365–386. [CrossRef]

39. Zuker, M. Mfold web server for nucleic acid folding and hybridization prediction. *Nucleic Acids Res.* **2003**, *31*, 3406–3415. [CrossRef]

40. Ye, J.; Coulouris, G.; Zaretskaya, I.; Cutcutache, I.; Rozen, S.; Madden, T.L. Primer-BLAST: A tool to design target-specific primers for polymerase chain reaction. *BMC Bioinform.* **2012**, *13*, 134. [CrossRef]

41. Livak, K.J.; Schmittgen, T.D. Analysis of relative gene expression data using real-time quantitative PCR and the 2(-Delta Delta C(T)) Method. *Methods* **2001**, *25*, 402–408. [CrossRef] [PubMed]

42. Yoshida, H.; Haze, K.; Yanagi, H.; Yura, T.; Mori, K. Identification of the cis-acting endoplasmic reticulum stress response element responsible for transcriptional induction of mammalian glucose-regulated proteins. Involvement of basic leucine zipper transcription factors. *J. Biol. Chem.* **1998**, *273*, 33741–33749. [CrossRef] [PubMed]

43. Van Schadewijk, A.; van't Wout, E.F.; Stolk, J.; Hiemstra, P.S. A quantitative method for detection of spliced X-box binding protein-1 (XBP1) mRNA as a measure of endoplasmic reticulum (ER) stress. *Cell Stress Chaperones* **2012**, *17*, 275–279. [CrossRef] [PubMed]

44. Kobolak, J.; Molnar, K.; Varga, E.; Bock, I.; Jezso, B.; Teglasi, A.; Zhou, S.; Lo Giudice, M.; Hoogeveen-Westerveld, M.; Pijnappel, W.P.; et al. Modelling the neuropathology of lysosomal storage disorders through disease-specific human induced pluripotent stem cells. *Exp. Cell Res.* **2019**, *380*, 216–233. [CrossRef] [PubMed]

45. Harris, G.; Eschment, M.; Orozco, S.P.; McCaffery, J.M.; Maclennan, R.; Severin, D.; Leist, M.; Kleensang, A.; Pamies, D.; Maertens, A.; et al. Toxicity, recovery, and resilience in a 3D dopaminergic neuronal in vitro model exposed to rotenone. *Arch. Toxicol.* **2018**, *92*, 2587–2606. [CrossRef] [PubMed]

46. Baumann, J.; Gassmann, K.; Masjosthusmann, S.; DeBoer, D.; Bendt, F.; Giersiefer, S.; Fritsche, E. Comparative human and rat neurospheres reveal species differences in chemical effects on neurodevelopmental key events. *Arch. Toxicol.* **2016**, *90*, 1415–1427. [CrossRef] [PubMed]

47. Tukker, A.M.; de Groot, M.W.; Wijnolts, F.M.; Kasteel, E.E.; Hondebrink, L.; Westerink, R.H. Is the time right for in vitro neurotoxicity testing using human iPSC-derived neurons? *ALTEX* **2016**, *33*, 261–271. [CrossRef]

48. Tukker, A.M.; Wijnolts, F.M.J.; de Groot, A.; Westerink, R.H.S. Human iPSC-derived neuronal models for in vitro neurotoxicity assessment. *Neurotoxicology* **2018**, *67*, 215–225. [CrossRef]

49. Fritsche, E.; Barenys, M.; Klose, J.; Masjosthusmann, S.; Nimtz, L.; Schmuck, M.; Wuttke, S.; Tigges, J. Current Availability of Stem Cell-Based In Vitro Methods for Developmental Neurotoxicity (DNT) Testing. *Toxicol. Sci. Off. J. Soc. Toxicol.* **2018**, *165*, 21–30. [CrossRef]

50. Lotharius, J.; Falsig, J.; van Beek, J.; Payne, S.; Dringen, R.; Brundin, P.; Leist, M. Progressive degeneration of human mesencephalic neuron-derived cells triggered by dopamine-dependent oxidative stress is dependent on the mixed-lineage kinase pathway. *J. Neurosci. Off. J. Soc. Neurosci.* **2005**, *25*, 6329–6342. [CrossRef]

51. Scholz, D.; Poltl, D.; Genewsky, A.; Weng, M.; Waldmann, T.; Schildknecht, S.; Leist, M. Rapid, complete and large-scale generation of post-mitotic neurons from the human LUHMES cell line. *J. Neurochem.* **2011**, *119*, 957–971. [CrossRef] [PubMed]

52. Zuberek, M.; Stepkowski, T.M.; Kruszewski, M.; Grzelak, A. Exposure of human neurons to silver nanoparticles induces similar pattern of ABC transporters gene expression as differentiation: Study on proliferating and post-mitotic LUHMES cells. *Mech. Ageing Dev.* **2018**, *171*, 7–14. [CrossRef] [PubMed]

53. Hollerhage, M.; Moebius, C.; Melms, J.; Chiu, W.H.; Goebel, J.N.; Chakroun, T.; Koeglsperger, T.; Oertel, W.H.; Rosler, T.W.; Bickle, M.; et al. Protective efficacy of phosphodiesterase-1 inhibition against alpha-synuclein toxicity revealed by compound screening in LUHMES cells. *Sci. Rep.* **2017**, *7*, 11469. [CrossRef] [PubMed]

54. Zhang, X.M.; Yin, M.; Zhang, M.H. Cell-based assays for Parkinson's disease using differentiated human LUHMES cells. *Acta Pharmacol. Sin.* **2014**, *35*, 945–956. [CrossRef]

55. Smirnova, L.; Harris, G.; Delp, J.; Valadares, M.; Pamies, D.; Hogberg, H.T.; Waldmann, T.; Leist, M.; Hartung, T. A LUHMES 3D dopaminergic neuronal model for neurotoxicity testing allowing long-term exposure and cellular resilience analysis. *Arch. Toxicol.* **2016**, *90*, 2725–2743. [CrossRef]

56. Tong, Z.B.; Hogberg, H.; Kuo, D.; Sakamuru, S.; Xia, M.; Smirnova, L.; Hartung, T.; Gerhold, D. Characterization of three human cell line models for high-throughput neuronal cytotoxicity screening. *J. Appl. Toxicol. JAT* **2017**, *37*, 167–180. [CrossRef]

57. Delp, J.; Gutbier, S.; Klima, S.; Hoelting, L.; Pinto-Gil, K.; Hsieh, J.H.; Aichem, M.; Klein, K.; Schreiber, F.; Tice, R.R.; et al. A high-throughput approach to identify specific neurotoxicants/developmental toxicants in human neuronal cell function assays. *ALTEX* **2018**, *35*, 235–253. [CrossRef]

58. Brull, M.; Spreng, A.S.; Gutbier, S.; Loser, D.; Krebs, A.; Reich, M.; Kraushaar, U.; Britschgi, M.; Patsch, C.; Leist, M. Incorporation of stem cell-derived astrocytes into neuronal organoids to allow neuro-glial interactions in toxicological studies. *ALTEX* **2020**. [CrossRef]

59. Lancaster, M.A.; Renner, M.; Martin, C.A.; Wenzel, D.; Bicknell, L.S.; Hurles, M.E.; Homfray, T.; Penninger, J.M.; Jackson, A.P.; Knoblich, J.A. Cerebral organoids model human brain development and microcephaly. *Nature* **2013**, *501*, 373–379. [CrossRef]

60. Kadoshima, T.; Sakaguchi, H.; Nakano, T.; Soen, M.; Ando, S.; Eiraku, M.; Sasai, Y. Self-organization of axial polarity, inside-out layer pattern, and species-specific progenitor dynamics in human ES cell-derived neocortex. *Proc. Natl. Acad. Sci. USA* **2013**, *110*, 20284–20289. [CrossRef]

61. Huang, J.; Liu, F.; Tang, H.; Wu, H.; Li, L.; Wu, R.; Zhao, J.; Wu, Y.; Liu, Z.; Chen, J. Tranylcypromine Causes Neurotoxicity and Represses BHC110/LSD1 in Human-Induced Pluripotent Stem Cell-Derived Cerebral Organoids Model. *Front. Neurol.* **2017**, *8*, 626. [CrossRef] [PubMed]

62. Liu, F.; Huang, J.; Liu, Z. Vincristine Impairs Microtubules and Causes Neurotoxicity in Cerebral Organoids. *Neuroscience* **2019**, *404*, 530–540. [CrossRef] [PubMed]

63. Plummer, S.; Wallace, S.; Ball, G.; Lloyd, R.; Schiapparelli, P.; Quinones-Hinojosa, A.; Hartung, T.; Pamies, D. A Human iPSC-derived 3D platform using primary brain cancer cells to study drug development and personalized medicine. *Sci. Rep.* **2019**, *9*, 1407. [CrossRef]

64. Sudhof, T.C. The presynaptic active zone. *Neuron* **2012**, *75*, 11–25. [CrossRef]

65. Bell, M.E.; Bourne, J.N.; Chirillo, M.A.; Mendenhall, J.M.; Kuwajima, M.; Harris, K.M. Dynamics of nascent and active zone ultrastructure as synapses enlarge during long-term potentiation in mature hippocampus. *J. Comp. Neurol.* **2014**, *522*, 3861–3884. [CrossRef] [PubMed]

66. Wang, S.S.H.; Held, R.G.; Wong, M.Y.; Liu, C.; Karakhanyan, A.; Kaeser, P.S. Fusion Competent Synaptic Vesicles Persist upon Active Zone Disruption and Loss of Vesicle Docking. *Neuron* **2016**, *91*, 777–791. [CrossRef] [PubMed]

67. Pei, Y.; Peng, J.; Behl, M.; Sipes, N.S.; Shockley, K.R.; Rao, M.S.; Tice, R.R.; Zeng, X. Comparative neurotoxicity screening in human iPSC-derived neural stem cells, neurons and astrocytes. *Brain Res.* **2016**, *1638*, 57–73. [CrossRef]

68. Mottram, D.S.; Wedzicha, B.L.; Dodson, A.T. Acrylamide is formed in the Maillard reaction. *Nature* **2002**, *419*, 448–449. [CrossRef]

69. Dearfield, K.L.; Abernathy, C.O.; Ottley, M.S.; Brantner, J.H.; Hayes, P.F. Acrylamide: Its metabolism, developmental and reproductive effects, genotoxicity, and carcinogenicity. *Mutat. Res.* **1988**, *195*, 45–77. [CrossRef]

70. Kjuus, H.; Goffeng, L.O.; Heier, M.S.; Sjöholm, H.; Øvrebø, S.; Skaug, V.; Paulsson, B.; Törnqvist, M.; Brudal, S. Effects on the peripheral nervous system of tunnel workers exposed to acrylamide and N-methylolacrylamide. *Scand. J. Work Environ. Health* **2004**, 21–29. [CrossRef]

71. Sorgel, F.; Weissenbacher, R.; Kinzig-Schippers, M.; Hofmann, A.; Illauer, M.; Skott, A.; Landersdorfer, C. Acrylamide: Increased concentrations in homemade food and first evidence of its variable absorption from food, variable metabolism and placental and breast milk transfer in humans. *Chemotherapy* **2002**, *48*, 267–274. [CrossRef] [PubMed]

72. Mundy, W.R.; Tilson, H.A. Neurotoxic effects of colchicine. *Neurotoxicology* **1990**, *11*, 539–547. [PubMed]

73. Goldschmidt, R.B.; Steward, O. Comparison of the neurotoxic effects of colchicine, the vinca alkaloids, and other microtubule poisons. *Brain Res.* **1989**, *486*, 133–140. [CrossRef]

74. Tanner, C.M.; Kamel, F.; Ross, G.W.; Hoppin, J.A.; Goldman, S.M.; Korell, M.; Marras, C.; Bhudhikanok, G.S.; Kasten, M.; Chade, A.R.; et al. Rotenone, paraquat, and Parkinson's disease. *Environ. Health Perspect.* **2011**, *119*, 866–872. [CrossRef] [PubMed]

75. Rappold, P.M.; Cui, M.; Chesser, A.S.; Tibbett, J.; Grima, J.C.; Duan, L.; Sen, N.; Javitch, J.A.; Tieu, K. Paraquat neurotoxicity is mediated by the dopamine transporter and organic cation transporter-3. *Proc. Natl. Acad. Sci. USA* **2011**, *108*, 20766–20771. [CrossRef] [PubMed]

76. Kanthasamy, A.; Jin, H.; Charli, A.; Vellareddy, A.; Kanthasamy, A. Environmental neurotoxicant-induced dopaminergic neurodegeneration: A potential link to impaired neuroinflammatory mechanisms. *Pharmacol. Ther.* **2019**, *197*, 61–82. [CrossRef]

77. Betarbet, R.; Sherer, T.B.; MacKenzie, G.; Garcia-Osuna, M.; Panov, A.V.; Greenamyre, J.T. Chronic systemic pesticide exposure reproduces features of Parkinson's disease. *Nat. Neurosci.* **2000**, *3*, 1301–1306. [CrossRef]

78. Pistollato, F.; Canovas-Jorda, D.; Zagoura, D.; Bal-Price, A. Nrf2 pathway activation upon rotenone treatment in human iPSC-derived neural stem cells undergoing differentiation towards neurons and astrocytes. *Neurochem. Int.* **2017**, *108*, 457–471. [CrossRef]

79. Zagoura, D.; Canovas-Jorda, D.; Pistollato, F.; Bremer-Hoffmann, S.; Bal-Price, A. Evaluation of the rotenone-induced activation of the Nrf2 pathway in a neuronal model derived from human induced pluripotent stem cells. *Neurochem. Int.* **2017**, *106*, 62–73. [CrossRef]
80. Radio, N.M.; Mundy, W.R. Developmental neurotoxicity testing in vitro: Models for assessing chemical effects on neurite outgrowth. *Neurotoxicology* **2008**, *29*, 361–376. [CrossRef]
81. Ryan, K.R.; Sirenko, O.; Parham, F.; Hsieh, J.H.; Cromwell, E.F.; Tice, R.R.; Behl, M. Neurite outgrowth in human induced pluripotent stem cell-derived neurons as a high-throughput screen for developmental neurotoxicity or neurotoxicity. *Neurotoxicology* **2016**, *53*, 271–281. [CrossRef] [PubMed]
82. Hellwig, C.; Barenys, M.; Baumann, J.; Gassmann, K.; Casanellas, L.; Kauer, G.; Fritsche, E. Culture of human neurospheres in 3D scaffolds for developmental neurotoxicity testing. *Toxicol. Int. J. Publ. Assoc. BIBRA* **2018**, *52*, 106–115. [CrossRef] [PubMed]
83. Pamies, D.; Bal-Price, A.; Chesne, C.; Coecke, S.; Dinnyes, A.; Eskes, C.; Grillari, R.; Gstraunthaler, G.; Hartung, T.; Jennings, P.; et al. Advanced Good Cell Culture Practice for human primary, stem cell-derived and organoid models as well as microphysiological systems. *ALTEX* **2018**, *35*, 353–378. [CrossRef] [PubMed]

 *cells*

Article

Vertically-Aligned Functionalized Silicon Micropillars for 3D Culture of Human Pluripotent Stem Cell-Derived Cortical Progenitors

Alessandro Cutarelli [1,†], Simone Ghio [2,†], Jacopo Zasso [1], Alessandra Speccher [3], Giorgina Scarduelli [4], Michela Roccuzzo [4], Michele Crivellari [2], Nicola Maria Pugno [5,6,7], Simona Casarosa [3], Maurizio Boscardin [2,*,‡] and Luciano Conti [1,*,‡]

[1] Laboratory of Stem Cell Biology, Department of Cellular, Computational and Integrative Biology-CIBIO, University of Trento, 38123 Trento, Italy; alessandro.cutarelli@unitn.it (A.C.); Jacopo.Zasso@unitn.it (J.Z.)

[2] Fondazione Bruno Kessler-Center for Material and Microsystem, 38123 Trento, Italy; simoneghio1@gmail.com (S.G.); crivella@fbk.eu (M.C.)

[3] Laboratory of Neural Development and Regeneration, Department of Cellular, Computational and Integrative Biology-CIBIO, University of Trento, 38123 Trento, Italy; a.speccher.1@unitn.it (A.S.); simona.casarosa@unitn.it (S.C.)

[4] Advanced Imaging Facility, Department of Cellular, Computational and Integrative Biology-CIBIO, University of Trento, 38123 Trento, Italy; Giorgina.Scarduelli@unitn.it (G.S.); michela.roccuzzo@unitn.it (M.R.)

[5] Laboratory of Bio-Inspired and Graphene Nanomechanics, Department of Civil, Environmental and Mechanical Engineering, University of Trento, 38123 Trento, Italy; nicola.pugno@unitn.it

[6] School of Engineering and Materials Science, Queen Mary University of London, London E1 4NS, UK

[7] Ket-Lab, Edoardo Amaldi Foundation, via del Politecnico snc, I-00133 Roma, Italy

[*] Correspondence: boscardi@fbk.eu (M.B.); Luciano.Conti@unitn.it (L.C.); Tel.: +39-0461-314458 (M.B.); +39-0461-285216 (L.C.)

[†] A.C. and S.G. contribute equally to this work and share first authorship.

[‡] M.B. and L.C. contribute equally to this work and share last authorship.

Received: 25 November 2019; Accepted: 23 December 2019; Published: 30 December 2019

Abstract: Silicon is a promising material for tissue engineering since it allows to produce micropatterned scaffolding structures resembling biological tissues. Using specific fabrication methods, it is possible to build aligned 3D network-like structures. In the present study, we exploited vertically-aligned silicon micropillar arrays as culture systems for human iPSC-derived cortical progenitors. In particular, our aim was to mimic the radially-oriented cortical radial glia fibres that during embryonic development play key roles in controlling the expansion, radial migration and differentiation of cortical progenitors, which are, in turn, pivotal to the establishment of the correct multilayered cerebral cortex structure. Here we show that silicon vertical micropillar arrays efficiently promote expansion and stemness preservation of human cortical progenitors when compared to standard monolayer growth conditions. Furthermore, the vertically-oriented micropillars allow the radial migration distinctive of cortical progenitors in vivo. These results indicate that vertical silicon micropillar arrays can offer an optimal system for human cortical progenitors' growth and migration. Furthermore, similar structures present an attractive platform for cortical tissue engineering.

Keywords: human cortical progenitors; 3D culture; silicon pillars; cell growth; hiPSC-derived neural progenitors; cerebral cortex

---

1. Introduction

In biology, developing tissues' microarchitecture is fundamental to allow correct cell differentiation and organization into appropriate structures that relate to specific physiological functions [1,2].

Standard in vitro cell culture models are mainly set in reductionist monolayer settings, conditions that intrinsically lack any structural architecture. This condition often represents a poor proxy to extrapolate cell growth in vivo, thus substantially affecting cell performance and biological assays outcomes [3,4]. Indeed, with respect to whole tissues, monolayer cultured cells are usually more responsive to toxic or therapeutic agents [5,6]. Additionally, cell culture on rigid surfaces can enhance cell proliferation, but might impact cell differentiation due to the partial cell interactions [7].

In this view, more appropriate three-dimensional (3D) cell culture environments can allow more physiological cell-to-cell contact, assisting cell growth and allowing better modelling of developmental processes [8–10]. A 3D environment also facilitates cells to organize into tissue-like structures that better mimic the in vivo function of cells, thus enhancing physiological relevance and predictive accuracy [11–14]. In the last few years, there has been a gradual development and adoption of technologies that enable tissue-like 3D cultures. Both scaffold-free organoid-based technologies and natural or synthetic scaffold-based culture systems have been developed [7]. In particular, since different tissue types show definite assemblies associated with their functional organization, scaffold-based methods allow assisted mimicking of complex tissue geometrical topographies, such as cerebral cortex, thus facilitating effective biofabrication of in vitro 3D tissue-like models [15–17]. With this purpose, silicon-based micro-fabricated culture substrates with well-defined continuous and discontinuous topographies, including the development of surfaces patterned with grooves, nanopillars or nanowires for the study of neural guidance and polarity, have been extensively developed in order to create scaffolds for a variety of applications [18–24].

In the present study, we exploited vertically-aligned silicon micropillar arrays to reproduce the developing cerebral cortex 3D architecture, where effective control over neural columns width, resembling the mammalian neocortex, is required for a large spectrum of applications. In the developing cerebral cortex, cortical progenitors are oriented from the ventricular to the pial surface with an apical-basal polarity. They divide to form radial glia and neuroblasts, the latter can migrate using the apico-basal oriented radial glia fascicles of the subventricular zone (SVZ) as a scaffold, thus forming the different cortical layers [25,26]. Here we focused on mimicking the radially-oriented cortical radial glia fibers, as they are the key players in controlling the expansion, radial migration and differentiation of cortical progenitors, thus allowing for the establishment of the correct multilayered cerebral cortex structure. We show that the scaffold material and structures are compatible with human cortical progenitors' maintenance. Immunofluorescence imaging analysis and RT-PCR results reveal that silicon vertical micropillar arrays efficiently promote the expansion and stemness preservation of human cortical progenitors, when compared to monolayer growth conditions. Furthermore, the precise orientation of the micropillars allows the radial migration, movement that is distinctive of cortical progenitors in vivo. These results indicate that vertically-aligned silicon micropillar arrays can offer an optimal system for human cortical progenitor growth and migration, and a potentially interesting platform for cortical tissue engineering.

2. Materials and Methods

2.1. Vertically-Aligned Silicon Micropillar Array Fabrication

Silicon slides containing vertically-aligned micropillar arrays have been realized on silicon surface through a CMOS like process at the Micro-Nano Facility of Fondazione Bruno Kessler (FBK). The silicon slides are composed of a central zone containing a matrix of vertically-aligned micropillars (height: 250–600 µm). Slides with different micropillar diameters have been obtained, with a diameter of 10 or 15 µm that leads to an aspect ratio that varies from 1:25 to 1:17. These significant proportions are particularly hard to achieve for pillar-like structures. The classical MEMS devices usually have aspect ratio of 1:12, for this reason the Bosch process used to realize the surfaces has been tuned in order to be able to reach higher aspect ratio.

The slide surface used for cellular seeding and culturing is 14 mm long and 5 mm wide these sizes were specifically designed to allow each slide to be accommodated in a well of a 12-well cell culture plate. The design foresees a range of distances between the micropillars, from 20 to 50 μm to evaluate the advantages of smaller and larger spaces between the micropillars for cellular deposition and growth. To realize the structures, the silicon wafer was first oxidized, patterned through soft lithography, then the silicon dioxide (SiO_2) was patterned through dry etching, and finally a deep reactive ion etching (DRIE) process was used to realize the micropillar-like structures. The DRIE process consists of a two-step process: first an isotropic plasma etch, second the deposition of a passivation layer that protects the lateral part of the structure. This two-step process is called Bosch process and it is repeated several times to realize nearly vertical wall. The time lapse of each of the two steps defines the roughness of the vertical wall as the process feed rate while the maximal depth that can be reached is defined by the power of the radio frequency bias that accelerates the ions toward the surface.

Since the fabrication process does not damage the mask, the same stamp can be employed for many subsequent cycles in a very reproducible manner.

During slides production, the micropillars on the open edge of the structures are systematically damaged during the DRIE process. This effect is connected with the presence of a blank space between slides, this void space increments the effective power of the ions impacting on the surfaces and then destroy or modify the shape of the first micropillars row. These defects have been eliminated in the cutting step. The cutting diamond disk was placed over the border, in order to cut away the defected micropillars and leaving a sharp edge.

2.2. Cell Cultures

Human cortical progenitors used in this study were differentiated from a commercial control hiPSC line (Gibco, Thermo Fisher Scientific, Monza, Italy) as previously described [27,28]. Briefly, hiPSCs commitment to a neural lineage and subsequently to the dorsal telencephalic lineage was performed by using N2B27 supplemented with human recombinant Noggin (500 ng/mL, Peprotech, London, UK), SB431542 (20 mM, Santa Cruz Biotechnologies, Heidelberg, Germany) and Fibroblast Growth Factor-2 (4 ng/mL, Peprotech). Cells were then detached and seeded on poly-ornithine and laminin-coated plastic dishes (Sigma-Aldrich, Milan, Italy) in medium supplemented with 10 μM Rock inhibitor Y27632 (Sigma-Aldrich). At day 10, neural rosettes containing cortical progenitor cells were manually collected and plated on poly-ornithine/laminin-treated culture dishes in N2B27 medium containing epidermal growth factor (10 ng/mL), fibroblast growth factor-2 (10 ng/mL) and brain-derived neurothrophic factor (20 ng/mL). Confluent cultures were passed as small multicellular clumps at a ratio of 1:3 using trypsin and amplified until passages 8–10. Cryopreserved stocks of human cortical progenitors were prepared from confluent cultures by trypsinization and resuspension in freezing medium (10% DMSO and 90% foetal calf serum).

Neuronal differentiation of hCPs was induced as previously described. Briefly, hCPs were plated at high density (10^5 cells per cm^2) on laminin-treated silicon vertical micropillar arrays. Cells were allowed to grow for four days and then switched in N2B27 medium without growth factors and cultured for 35 days. The medium was changed every four days.

To visualize the cells for fluorescent imaging analyses and time-lapse experiments, hiPSC-derived cortical progenitors were transduced with a lentiviral vector carrying eGFP cDNA under Cytomegalovirus (CMV) promoter and containing a puromycin selection cassette [29].

Procedure for generation and expansion of mouse NS cells (i.e., radial glia-like neural progenitors) were previously described [30–32]. In this study we used the LC1 NS cell line grown in standard conditions in expansion medium composed of Euromed-N medium (Euroclone) supplemented with 1% N2 (Thermo Fisher Scientific) and 20 ng/mL human recombinant epidermal growth factor (20 ng/mL) and fibroblast growth factor-2 (20 ng/mL).

Human cortical progenitors and mouse NS cells were seeded on silicon slides pre-coated with mouse laminin (Thermo Fisher Scientific).

2.3. Scanning Electron Microscopy (SEM)

Cultures were fixed for 30 min at 4 °C in 25% Glutaraldehyde and 0.1 M cacodylic acid in distilled water (pH 7.2). Then, samples were washed three times with 0.1 M cacodylic acid and dehydrated by the exposure to 50%, 70%, 90% and (2×) 100% *v/v* ethanol concentration, 10 min each. After being air-dried under an air flow, samples were gold coated by evaporation of a thin gold layer on top of the sample surface (thickness 6 nm, 1.5 nm Cr adhesion layer). Silicon micropillar-based devices deprived of cells did not require any treatment prior to SEM image acquisition. SEM micrographs were acquired by using a TESCAN VEGA III scanning electron microscope (Tescan Analytics, Fuveau, France) (operating voltage 4 kV, working distance 18 mm, stage tilting angle 45°).

2.4. Cell Growth/Viability Assay

Analysis of cell growth/viability of all cell types employed in this work (hCPs and mouse NS cells) was performed by MTT assay (Sigma-Aldrich). Briefly, MTT powder was dissolved into culture medium at a final concentration of 1.5 mg/mL. For culture incubation with MTT solution, cell medium was removed, cultures rinsed twice with PBS (Thermo Fisher Scientific) and incubated with MTT solution for 1h at 37 °C. Following incubation, MTT solution was removed, cells were air-dried and violet MTT precipitates dissolved with isopropanol. The absorbance was read at 570 nm wavelength with a microplate reader (Tecan Infinite M200PRO, Tecan Italia, Milan, Italy).

2.5. Immunocytochemistry

To process samples for immunofluorescence analyses, cultures were fixed in 4% paraformaldehyde for 30 min at room temperature (RT), permeabilized in PBS containing 0.5% Triton X-100 for 15 min at RT and then blocked in blocking solution (PBS containing 0.3% Triton X-100 and 5% FCS) for 1 h at RT. Samples were next incubated overnight at 4 °C with primary antibody diluted in antibody solution (PBS containing 0.2% Triton X-100 and 2% FCS), then washed three times with PBS and incubated for 2 h at RT with secondary antibodies. Samples were then counterstained with 1 μg/mL Hoechst 33,258 (Thermo Fisher Scientific) and further rinsed with PBS before proceeding with visualization. Fluorescent signals and Z-Stack of eGFP^{+ve} human cortical progenitors (12 slices of 7.7 μm each, shown at 7 fps) were detected using a Leica DMi8 microscope equipped with an Andor Zyla 4.2 PLUS, monochromatic, sCMOS sensor, 4.2 megapixel camera. Acquired images were processed with the open-source Fiji software (v2.0.0, open source under the GNU General Public License, Madison, WI, USA) [33].

Antibodies used in this study: primary mouse monoclonal anti-NESTIN antibody (R&D Systems, Minneapolis, MN, USA, 1:300), primary mouse monoclonal β3-TUBULIN antibody (Promega, Milan, Italy, 1:1000), primary rabbit polyclonal anti-SOX2 antibody (Millipore, Milan, Italy, 1:200), primary rabbit polyclonal anti-MAP2 antibody (Santa Cruz, Heidelberg, Germany, 1:200), primary mouse monoclonal anti-TBR2 antibody (ABCAM, Cambridge, UK, 1:500), primary mouse monoclonal anti CUX1 (ABCAM, 1:200), AlexaFluor-488 or -568 conjugated secondary antibodies (Thermo Fisher Scientific, 1:500).

2.6. Time Lapse Analysis

Time-lapse movies of live GFP-expressing cells migrating along micropillars were acquired with a Zeiss Axio Observer Z1 inverted microscope equipped with the Apotome 2 module for structured illumination and a 2.83 Megapixel AxioCam 503 mono D (all from Zeiss Italia, Castiglione Olona, Italy). Time-lapses were acquired as z-stacks (10 μm z-step) using a plan-apochromatic 10×/0.3 objective, with a frame interval of 30 min for 12.5 h. The movies shown are maximum intensity projections. Optimal focus selection was performed by manual extraction of each focus z-slices from original z-stack time-lapses to select the best focused z position for each time point, then adjusted for brightness and contrast and saved as 7 fps AVI files using Fiji software [33].

2.7. RNA Isolation and Quantitative RT-PCR (qRT-PCR)

Total RNA was isolated by using TRIzol Reagent (Thermo Fisher Scientific) following the manufacturer's protocol, then retro-transcribed with iScript cDNA Synthesis Kit (BioRad, Segrate, Italy). cDNA was used to verify the expression of specific target genes by qRT-PCR (quantitative RT PCR), using the SsoAdvanced Universal SYBR Green Supermix Kit. Specific primers sets were used (RT-qPCR data were analyzed according to the comparative ΔΔCt method and normalized by using β-Actin housekeeping gene. Sequence of primers used in this study:
Nestin forward 5′-GGAGAAGGACCAAGAACTG-3′, reverse 5′-ACCTCCTCTGTGGCATTC-3′; β3-tubulin forward 5′-TCAGCGTCTACTACAACGAGGC-3′, reverse 5′-GCCTGAAGAGATGTCCAAAGGC-3′; β-Actin forward 5′-GACAGGATGCAGAAGGAGATTACTG-3′, reverse 5′-CTCAGGAGGAGCAATGATCTTGAT-3′.

2.8. Statistical Analysis

Analyses were performed using either a two-sided unpaired Student's *t*-test or a one-way analysis of variance with a Dunnett's post hoc test. Experiments were repeated three times in triplicate and values were considered statistically significant for $p < 0.05$ (*), $p < 0.01$ (**), $p < 0.001$ (***), $p < 0.0001$ (****).

3. Results and Discussion

3.1. Generation of Vertically-Aligned Silicon Micropillar Array Structures

Biomimetic cortical-like 3D platforms have been created based on different approaches, such as cell spheroids, organoids and engineered constructs based on hydrogels [34–36]. These structures present advantages based on cell self-assembly or enabling spatial self-organization. Nonetheless fine control over the 3D micro-architecture, phenotype and reproducibility have been reported to be challenging [37]. To overcome these limitations, we sought to develop a novel scaffold-based approach mimicking the structural organization of the developing cerebral cortex by ideally controlling network topography.

Mammalian cerebral cortex is a complex layered structure organized in columns generated during developmental stages by means of cortical progenitors (CPs) and neuroblasts that move radially along radial glia fibers serving as scaffolds for directed columnar migration [25,26]. Our aim here is to fabricate a 3D culture platform that mimics in vitro these cerebral radial structures, being able to optimally support the outer growth (from the bottom to the top) of CPs such as that occurring in vivo during cortical neurogenesis. With this aim we designed micropatterned structures containing topographically ordered micropillar arrays. Micropatterned substrates with different geometries have been reported in the form of discontinuous micro-grooved configurations and discontinuous geometries like nanopillar arrays made from a several materials comprising polymers, such as PS, PLGA and PDMS together with hard materials, such as silicon and quartz [18,38–41]. In particular, in the neural field pillars and cone geometries at micron scale have been shown to control the outgrowth of neuronal processes, guiding neurite outgrowth alignment and cell growth [20,38,42,43].

Here, we produced a novel 3D neural cell culture platform based on silicon substrates displaying arrays of micropillars fabricated by lithographic patterning (see Methods) of crystalline silicon (Si) wafers. Such technique enables the fabrication of discontinuous micropillars exhibiting at the same time anisotropic geometry. The silicon structures consist of a rectangular seeding surface of 0.7 cm^2 containing vertically arranged micropillars (Figure 1A,B).

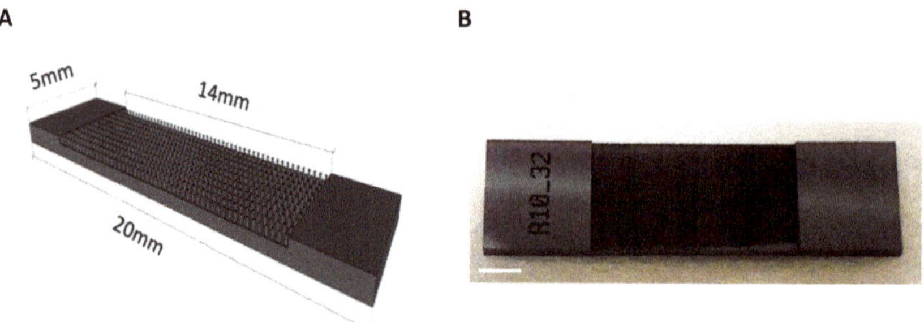

Figure 1. Design, size and morphology of 3D silicon micropillar array slides. (**A**) Schematic structure of a silicon slide containing vertically-aligned micropillars (micropillars are not in scale). (**B**) The image shows a 3D silicon slide. Scale bar: 1 mm.

Silicon-based nanostructures exhibit conceivable applications in several fields. As examples, plasmonic nanostructures based on array of silicon nanopillars could be used for surface enhanced Raman spectroscopy (SERS) or to control the wettability of a silicon surface [44–47]. Silicon micropillar arrays can be assembled by lithographic techniques allowing tight control over the size and density of the micropillars, differently from randomly generated rough surfaces as those presented in several other works [17,43,48,49]. Accordingly, they allow greatest control over the topographic structure of the system, thus warranting high reproducibility and robustness of the experiments. Additionally, differently from other materials, silicon surfaces hold particular technological impacts and potential.

The size of the slide was specifically designated to easily accommodate in a well of a tissue culture 12-multiwell plate (Figure S1A,B). By varying the reaction conditions of growth, we can produce pillars that have 10–15 μm diameter and height that can be regulated in the range of 200–600 μm. For this study we used silicon slides with 250 μm tall micropillars. Different micropillars topographies were successfully fabricated by using different masks. We produced 12 distinctive layouts of silicon micropillar arrays, coded as A1 to S6, with variable micropillar density and topographies (square grid-aligned and hexa grid-staggered micropillar arrays and with variable distances between the micropillars; see Table 1).

Table 1. List of different silicon slide arrays fabricated and tested in this work.

Slide Code	Pillar Diameter (μm)	Distance Among Pillars (μm)	Topography
A1	10	20	Alligned
A2	10	30	Alligned
A3	10	40	Alligned
A4	15	20	Alligned
A5	15	30	Alligned
A6	15	40	Alligned
S1	10	20	Staggered
S2	10	30	Staggered
S3	10	40	Staggered
S4	15	20	Staggered
S5	15	30	Staggered
S6	15	40	Staggered

Additionally, to set up the best conditions for cell growth, we fabricated both oxidized silicon (SiO_2) and silicon nitride (Si_3N_4) vertically-aligned silicon micropillar arrays.

The fabrication accuracy and morphology of the structures at the nanoscale were evaluated by measuring the distance between the micropillars and their diameter by scanning electron microscopy

(SEM). Cross-sectional SEM images were produced to derive the internal structure of the samples (Figure 2A,B) and top-view SEM images to derive the density of the silicon micropillar arrays.

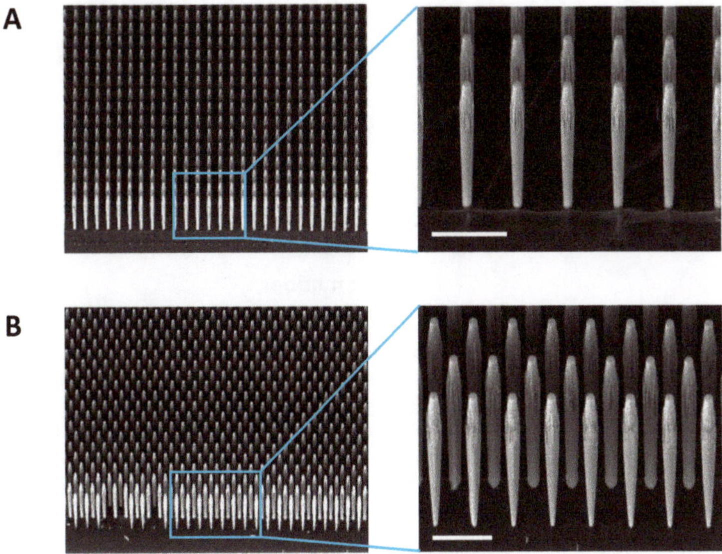

Figure 2. Silicon micropillars can be arranged with different topographies. Silicon micropillar arrays are perfectly vertically-aligned with a height that can be tuned in a 200–600 μm range. Specific spacing, density and morphology of the silicon pillars can be arranged by changing the mask of lithographic process. Low and high magnification of cross-sectional SEM images of aligned (**A**) and staggered (**B**) micropillar arrays. Scale bar: 50 μm.

We found that while the distance between micropillars was quite accurate, the diameter of the micropillars varied depending on the height of the structure. Indeed, the diameters at the bottom of the micropillars were reduced of about 40% (~5–7 μm) with respect to the top of the same micropillar.

This effect is particularly noticeable on the first row of micropillars, and we assumed that this is mainly due to the fact that the chemical-physical processes on which the Bosch process is based depending on the area exposed to the engraving process. Therefore, the etching rate is different in the areas between the micropillars than in the areas between the slides. Therefore, we assume that the first row of micropillars is etched much faster than the others and that this is the cause of the reduction of the diameter of the micropillar. Up to now, we were able to reach structures with a diameter of 15 μm and a height of 600 μm, nevertheless further tests to improve this aspect ratio are ongoing.

3.2. Culturing Human Cortical Progenitors on Vertically-Aligned Silicon Micropillar Arrays

To establish which human cortical progenitors (hCPs) density was the most comparable to the standard 2D monolayer method used as control and to detect potential toxic effects of silicon material, we first performed a cell viability assay. hCPs were seeded on laminin-coated samples A6–S6 with cell density ranging from 2×10^4 to 8×10^4 cells per 3D device (i.e., 2.8×10^4 and 1.1×10^5 cells per cm^2) and cultured for 48 h before being processed in an MTT assay (see Section 2). We found that cell viability was reduced for the lowest cell density (2×10^4 cells) and increased for the highest (8×10^4 cells), compared to the 2D culture. Indeed, 4×10^4 cells per 3D device showed no differences compared to the 2D monolayer culture (Figure 3).

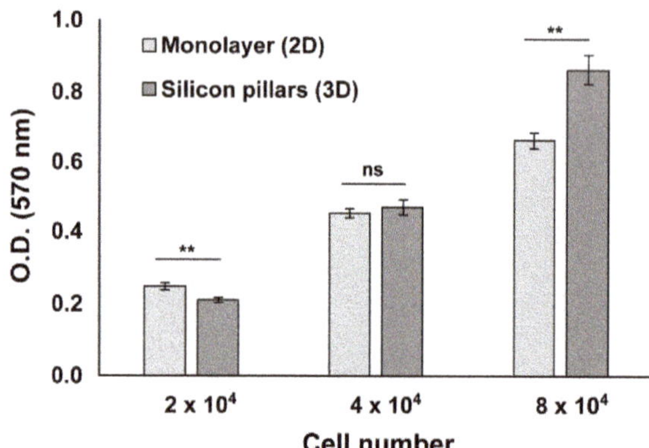

Figure 3. hCPs seeded on silicon micropillar arrays maintain their viability. MTT assay performed on hCPs plated on 3D silicon slide and in standard 2D monolayer shows that hCPs efficiently maintain their viability. Different cell densities were assessed 48 h after seeding. $p < 0.01$ (**), not significant (ns).

We found no significant difference between oxidized silicon or silicon nitride material (not shown), thus indicating that these 3D structures are fully compatible with living cells without affecting cell growth. Moreover, our silicon micropillar arrays might also allow for increased hCPs proliferation activity in long-term cultures by offering a 3D environment in which the cells can also exploit the third (vertical) dimension for growth. Additionally, we did not detect any significant difference between different topographic layouts and silicon types (oxidized silicon or silicon nitride) tested in terms of cell adhesion, viability and growth (not shown).

As further confirmation of these results, an analogous MTT analysis was performed on mouse radial glia-like NS cells plated at the same aforesaid cell densities (Figure S2A). NS cell analysis gave similar results to the hCPs, indicating that 4×10^4 cells per 3D device represents the preferred cell density. This was then selected as standard seeding density in the subsequent experiments in this work. The compatibility of oxidized silicon or silicon nitride for in vitro functional studies on neurons has been already reported, nevertheless, this is the first study to report compatibility with hCPs [50]. Furthermore, contrary to other studies employing silicon uncoated surfaces, we used surface topographies coated with laminin to optimize cell adhesion as also been reported by others [51–53]. In addition, these results demonstrate that our silicon micropillar arrays can be functionalized according to the cells' needs. We then performed SEM analysis in order to monitor the behaviour of the hCPs seeded on the arrays in terms of interaction with the vertical silicon micropillars. We found that cells at high density interact with each other and with the micropillars establishing a uniform network (Figure 4A).

Interestingly, hCPs plated at low density show an increased propensity to adhere to the micropillars rather than to the bottom flat surface of the device, with cells that can be found at different heights of the micropillars (Figure 4B,C). Cells formed 3D neural networks and suspended bridges throughout the micropillar height and along and between the micropillar walls. A similar formation of suspended neural process bridges has also been reported with microtowers, microfibres and 2PP-DLW fabricated microstructures [54–57]. In order to visualize the cells seeded on the silicon micropillar arrays with an inverted microscope, the device has to be set upside down, facing the microscope objectives. To facilitate this operation, we fabricated, by 3D-printing technique, a slide holder support to be placed on a glass bottom dish that allows to visualize the slide with a tilt of 45 degrees, so as to have a 3D visual prospect of the cells (Figure S3). Also, a glass bottom dish was used to place the holder (Figure S2B). To visualize live cultures, we generated eGFP⁺ᵛᵉ hCPs by infection with lentiviral particles allowing constitutive expression of eGFP cDNA (see Methods). eGFP⁺ᵛᵉ hCPs were used to live monitor the interactions of hCPs with the device and the single silicon micropillars by time-lapse analysis, and to capture the dynamics of cells movements among the micropillars. We found that hCPs move within the chip and interact with each other and with micropillars. In particular, hCPs can move along the whole micropillar from the bottom to the top (Movie S1) in a manner that resembles the in vivo migration along the radial glia during cortical neurogenesis, and extend processes embracing the micropillars (Movie S2). We also noticed the progressive neurite extension surrounding micropillars at the same height, which might represent the way for the cells to preferentially form a layered architecture.

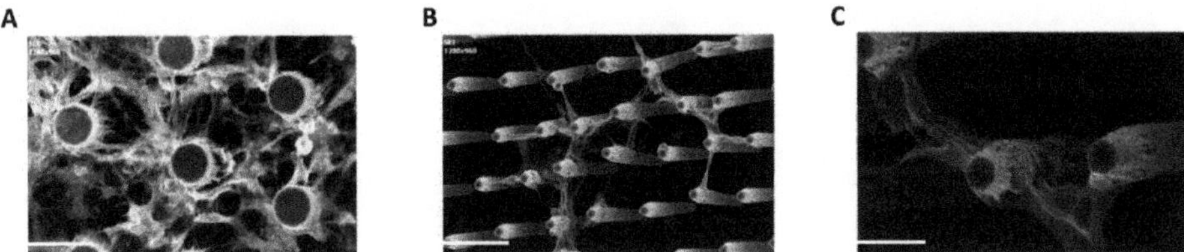

Figure 4. hCPs seeded on silicon slide establish close interactions with micropillars. (**A**) Top-view SEM image showing hCPs seeded at standard density on silicon micropillars. (**B,C**) Cross-sectional SEM images showing hCPs seeded at low density on silicon structures to visualize single cells-micropillars interactions. Scale bar: 20 μm (**A,B**) and 10 μm (**C**).

3.3. hCPs Seeded on Silicon Micropillar Arrays Proliferate and form 3D Layered Structures

We next assessed the proliferation of hCPs seeded on different silicon 3D devices in order to find the silicon type better supporting hCPs stable long-term growth. To this aim, hCPs were plated and maintained for several days on silicon oxide (SiO_2) and on silicon nitride (Si_3N_4) micropillar arrays, as well as on plastic dishes as standard monolayer growth conditions. Cultures were analysed by MTT-based growth assays at different time points. Resulting growth curves from the three experimental groups show a modest, yet statistically significant, reduction in cell growth within the first four days for both types of 3D silicon devices. After seven days cultures grown on Si_3N_4 devices still exhibit a significant reduction, whereas cultures on SiO_2 device show no difference with respect to the standard monolayer conditions. At the later time point considered, 14 days, a statistically significant difference was detected for the cells seeded on SiO_2 3D silicon device, which resulted in increased growth when compared to the other experimental groups (Figure 5).

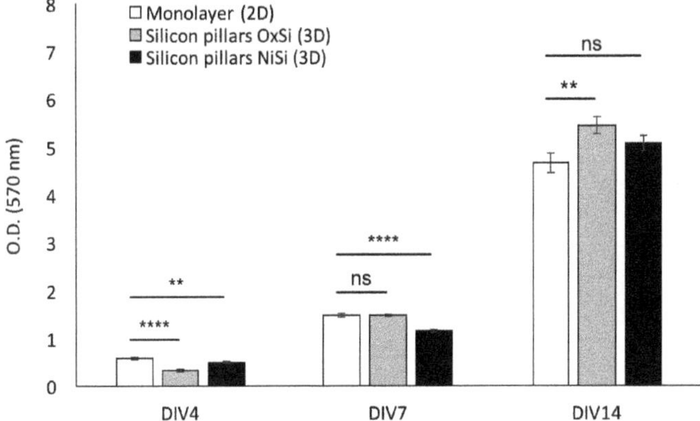

Figure 5. hCPs seeded on silicon slides maintain their proliferation capability. Cell growth analysis (MTT assays) performed on hCPs seeded on oxidized silicon (OxSi) or nitride silicon (NiSi) micropillar arrays. Standard 2D monolayer cultures were used as control. Cultures were assessed at different days in vitro (DIV) after seeding. $p < 0.01$ (**), $p < 0.0001$ (****), not significant (ns).

These results indicate that growth on silicon 3D silicon devices improves hCPs long-term maintenance. Additionally, since oxide silicon 3D arrays resulted to be more compatible and efficient in sustaining hCPs long-term survival and growth with respect with the nitride silicon devices, for the next experiments we employed the former type. Immunofluorescence imaging analysis of hCPs grown on silicon 3D micropillar arrays for two weeks show that cells are well distributed inside the device, filling the whole space among the micropillars (Figure 6A).

Additionally, SEM imaging shows that the cells form a complex three-dimensional multilayered-like structure by growing among the micropillars and interacting with each other, finally establishing regular horizontal layers (Figure 6B,C).

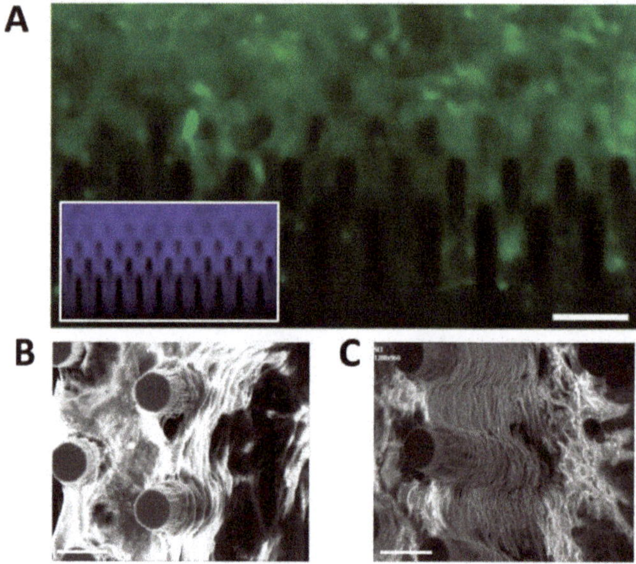

Figure 6. hCPs cultured on silicon micropillar arrays establish layered structures. (**A**) Picture of eGFP^{+ve} hCPs cultured for 14 days on silicon slides showing the establishment of a high-density 3D culture. Inset shows the same culture stained with Hoechst. (**B,C**) SEM images of hCPs maintained on silicon pillars device for 14 days. Cultures exhibit the generation of multiple cell layers. Scale bar: 50 μm (**A**) and 20 μm (**B,C**).

3.4. hCPs Grown on Silicon Micropillar Arrays Retain Their Multipotency and Regional Identity

We then assessed if growth on silicon 3D arrays might interfere with hCPs (i) multipotency and (ii) preservation of their cortical regional identity. To assess these issues, we first performed an immunostaining analysis for SOX2 and NESTIN, two key neural multipotent markers, on hCP cultures expanded for 14 days on silicon 3D arrays or in standard monolayer conditions. To this respect, we found that silicon 3D cultures show the great majority of cells to co-express SOX2 and NESTIN, comparably to the cultures maintained in standard monolayer conditions previously characterized to efficiently preserve hCPs multipotency (Figure 7A,B and Figure S4).

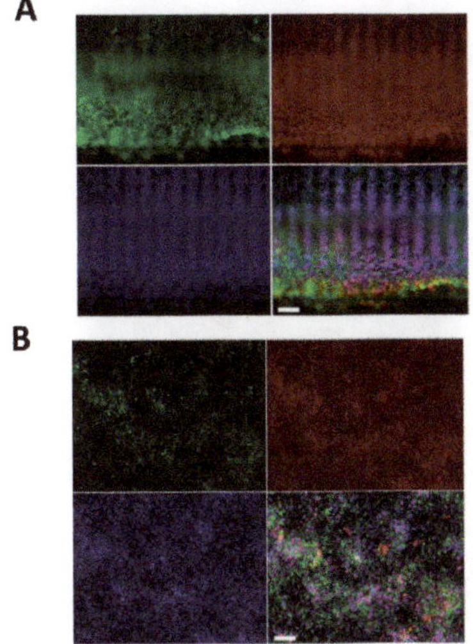

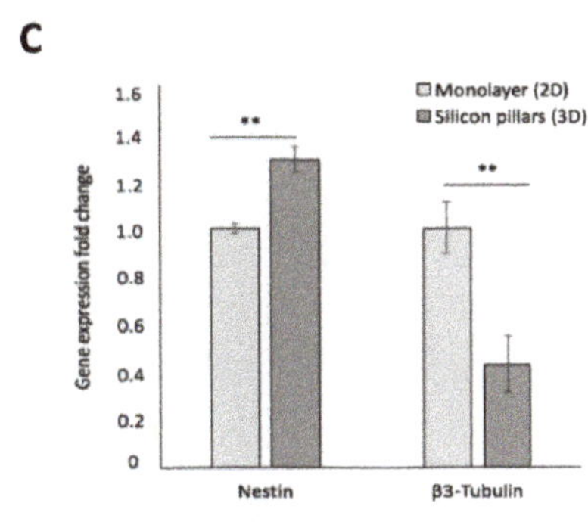

Figure 7. hCPs seeded on silicon slide maintain their neural immature identity. Immunostaining of hCPs for NESTIN (green), SOX2 (red) and nuclear staining with Hoechst (blue) in 3D culture (**A**) and 2D culture (**B**) cultured for 14 days. (**C**) Quantitative RT-PCR assay showing the expression levels of Nestin and β3-Tubulin transcripts in hCPs grown in 2D or 3D cultures. Scale bar: 100 μm (**A**) and 50 μm (**B**). $p < 0.01$ (**), not significant (ns).

To extend this result, we measured the transcript levels of Nestin and of the neuronal marker β3-Tubulin by quantitative RT-PCR (Figure 7C). This assay showed a 30% increase and a 60% decrease on the expression levels of Nestin and β3-Tubulin, respectively, in cultures maintained on silicon 3D arrays with respect to standard monolayer conditions (Figure 7C). These results further confirmed that growth on our silicon 3D devices does not alter hCP's multipotency. Furthermore, growth on silicon 3D arrays further lowers the occurrence of spontaneous neuronal differentiation in the cultures.

We then investigated the capability of hCPs cultured in silicon 3D arrays to retain their original cortical identity. To assess this issue, we first performed an immunostaining analysis for TBR2, a key maker of cortical progenitors, on hCP cultures expanded for 14 days on silicon 3D arrays or in standard monolayer conditions (Figure S5). We found that TBR2 immunoreactivity was maintained in nearly all of the cells in the silicon 3D cultures (Figure 8A; Movie S3). Finally, we tested the ability of hCPs plated in silicon 3D arrays to undergo neuronal maturation giving rise cortical glutamatergic neurons. To this end, hCPs were cultured for 5 days on 3D silicon devices and then exposed for 35 days to differentiative conditions and processed for immunofluorescent analysis for β3-TUBULIN, the mature neuronal marker MAP2 and the cortical neuronal marker CUX1. We found that differentiated cultures exhibited the presence of β3-TUBULIN[+ve] neuronal cells (Figure 8B, left panel) and the appearance of neurons co-expressing MAP2 and CUX1 (Figure 8B).

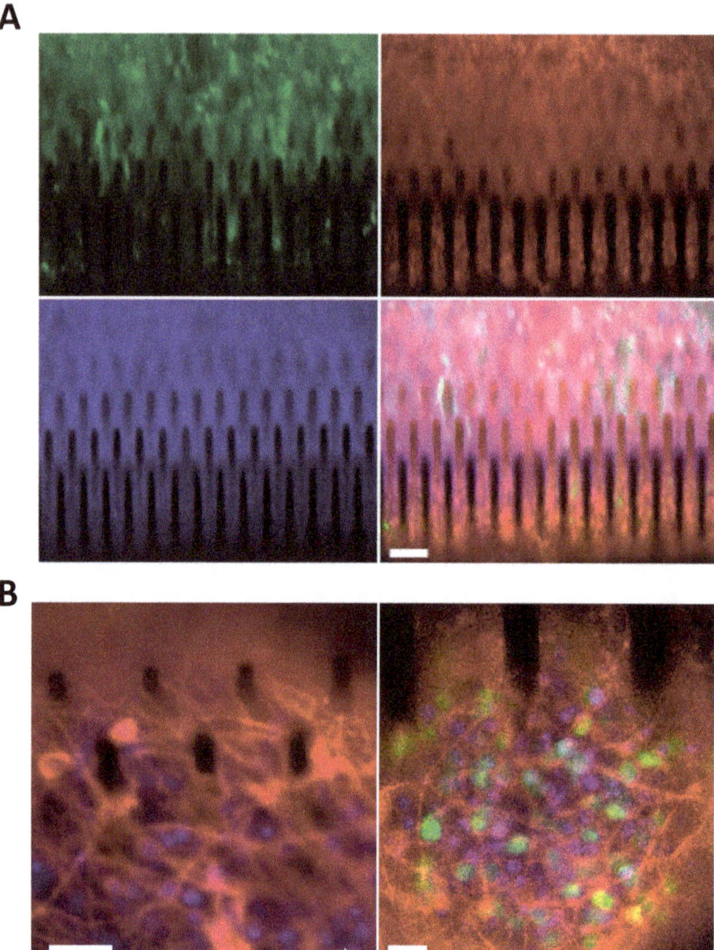

Figure 8. hCPs seeded on silicon micropillar arrays preserve their cortical regional identity and upon exposure to differentiative conditions, generate cortical glutamatergic neurons. (**A**) eGFP[+ve] hCPs cultured for 14 days on 3D silicon devices preserve the expression of the cortical progenitor marker TBR2 (red). Nuclei are stained with Hoechst (blue). Scale bar: 50 μm. (**B**) hCPs cultured for 5 days on 3D silicon devices and then exposed for 35 days to differentiative conditions mature into cortical glutamatergic neurons. Left: cultures stained for the pan-neuronal neuronal marker β3-TUBULIN (red). Nuclei are stained with Hoechst (blue). Scale bar: 10 μm. Right: cultures stained for the mature neuronal marker MAP2 (red) and for the cortical neuronal marker CUX1 (green). Nuclei are stained with Hoechst (blue). Scale bar: 10 μm.

These results demonstrate that hCPs grown on silicon micropillar arrays keep their multipotency and cortical identity and are able to generate glutamatergic cortical neurons. However, further experiments are required to further investigate whether the cortical neurons obtained in the 3D environment are able to organize themselves into defined upper and deeper cortical neuronal layers.

4. Conclusions

We have described a novel culture platform for hCPs growth based on 3D vertically-aligned silicon micropillar arrays. This structure mimics the radially-oriented cortical radial glia fibres that during embryonic development are essential to control the expansion, radial migration and differentiation of hCPs.

The silicon micropillar arrays can be arranged using different topographic organizations and report micropillar heights not tested with neural progenitors so far. The structures conceived practically combine the advantages of microscale topographies, without the need of complex fabrication techniques. Importantly, our fabrication process allows the production of biocompatible, 3D devices in a highly versatile and reproducible manner. In fact, the micrometre-sized base confers to micropillars good mechanical stability, while establishing a tight interface with the living hCPs.

Other scaffold-based methods have been reported for the biofabrication of in vitro 3D tissue-like models [15–17]. In particular, some of these systems, especially the ones based on 3D polymeric materials, i.e., microfibres and hydrogel scaffolds, offer a highly useful and strong method for creating large-scale 3D tissue cultures. Additionally, these systems have been shown to be extremely flexible in terms of production, biocompatibility, biodegradability, mechanical properties and functionalization with chemicals or oligopeptides in order to implement their adhesion and/or ECM mimicking properties [16]. The 3D vertically-aligned silicon micropillar arrays here described exhibited a lower flexibility with respect to these microfibers and hydrogel scaffolds, nevertheless they represent highly reproducible scaffolds in which both the height and the topography of the pillars can be finely regulated. Similar pillar-like structures have been previously reported by Limongi and colleagues and characterized for their ability to allow culturing of functional neuronal and glial cells in a 3D manner, allowing the formation of viable and functional neuronal networks [17]. In particular, the pillars there reported were different from ours since they were designed to reach limited height (cylindrical pillars of 10 µm in height and 10 µm in diameter) and to include a patterning in the nanoscale on their sidewall, leading to a spatial modulation in the z direction. Similarly, culturing platform containing arrays of microchannels arranged into ordered 2D arrays (with maximum height of ~3 µm and amplitudes ranging from ~10 to ~1.5 µm) have been reported [15]. These structures were shown to host neuronal cells strongly guiding their axons' growth direction and with additional advantage to enable coupling to devices for active sensing and stimulation at the local scale.

Our future efforts should reveal the actual potential of our 3D vertically-aligned silicon micropillar arrays to extensively support hCPs neuronal maturation in order to generate cortical-like tissue comparable to the self-assembled brain organoid technology but with the advantage of an increased reproducibility intrinsic to the scaffold-assisted process.

Additionally, to further exploit the potential of our 3D vertically-aligned silicon micropillar arrays, in the future we aim at coupling this system with a compartmentalized microfluidic device to reach a complete control of culture environment and reduce media volumes and related costs. The acquired knowledge will certainly pave the path towards the generation of valuable tools to study cortical development in humans and for cortical tissue engineering.

Supplementary Materials: The following are available online at http://www.mdpi.com/2073-4409/9/1/88/s1, Figure S1: Silicon devices were specifically designed to fit into 12 multiwell plates; Figure S2: Analysis of mouse NS cells viability seeded on 3D silicon micropillars arrays; Figure S3: Silicon slide holder device's picture. The slide lodges on a home-made holder inserted in a bottom glass dish; Figure S4: hCPs seeded on 3D silicon slide preserve their multipotency and create a cell network; Figure S5: hCPs grown in standard 2D conditions exhibit expression of cortical progenitor markers; Video S1: 12.5 hours time-lapse movie performed on eGFP^{+ve} hCPs seeded on silicon micropillars array. A single cell migrating from the bottom to the top of the micropillar is visible; Video S2: 12.5 hours time-lapse movie of eGFP+ve hCPs seeded on 3D silicon slide. Examples of cells extending neurites that interact with pillars are shown; Video S3: Z-stack of eGFP^{+ve} hCPs stained for TBR2 (red) and Hoechst (blue). Cells have been cultured for 14 days on silicon micropillars arrays before the time lapse analysis.

Author Contributions: S.G., M.B. and L.C. designed the pillar architecture. S.G. and M.B. developed the fabrication technology, realized the PDMS molds for pillar imprinting, and with help from M.C. produced the silicon micropillar array slides. A.C. prepared hCPs cultures and performed viability assay, growth curve analyses, morphological and immunohistochemical analyses on hCPs. A.C., J.Z. and A.S. prepared samples for the SEM analyses. J.Z. and A.S. prepared mNS cultures and performed viability assay on mouse NS cells. A.C. and J.Z. performed the time lapse experiments with the help of G.S. and M.R. G.S. designed and produced the slide holder devices. L.C., A.C., S.C., S.G., and M.B. planned the experiments, analysed the data and wrote the main manuscript, with contribution from all authors. N.M.P. and M.B. supervised S.G. activity. L.C. and M.B. conceived and supervised the work. All authors have read and agreed to the published version of the manuscript.

Funding: This work was supported by intramural funding from the University of Trento to L.C. and S.C. N.M.P. is supported by the European Commission under the Graphene Flagship Core 2 grant no. 785219 (WP14 'Composites') and FET Proactive 'Neurofibres' grant no. 732344, as well as by the Italian Ministry of Education, University and Research (MIUR) under the 'Departments of Excellence' grant L. 232/2016 and AR 901–01384—PROSCAN and PRIN-20177TTP3S.

Acknowledgments: The authors wish to thank Francesca Agostinacchio for helpful discussions and Mirco D'Incau for technical assistance for SEM imaging.

Conflicts of Interest: The authors declare no conflict of interest.

Data Availability: The raw data required to reproduce these findings are available upon request to the corresponding authors.

References

1. Davenport, R.J. What controls organ regeneration? *Science* **2005**, *309*, 84. [CrossRef] [PubMed]
2. Griffith, L.G.; Swartz, M.A. Capturing complex 3D tissue physiology in vitro. *Nat. Rev. Mol. Cell Biol.* **2006**, *7*, 211–224. [CrossRef] [PubMed]
3. Vergani, L.; Grattarola, M.; Nicolini, C. Modifications of chromatin structure and gene expression following induced alterations of cellular shape. *Int. J. Biochem. Cell Biol.* **2004**, *36*, 1447–1461. [CrossRef] [PubMed]
4. Thomas, C.H.; Collier, J.H.; Sfeir, C.S.; Healy, K.E. Engineering gene expression and protein synthesis by modulation of nuclear shape. *Proc. Natl. Acad. Sci. USA* **2002**, *99*, 1972–1977. [CrossRef] [PubMed]
5. Bhadriraju, K.; Chen, C.S. Engineering cellular microenvironments to cell-based drug testing improve cell-based drug testing. *Drug Discov. Today* **2002**, *7*, 612–620. [CrossRef]
6. Sun, T.; Jackson, S.; Haycock, J.W.; MacNeil, S. Culture of skin cells in 3D rather than 2D improves their ability to survive exposure to cytotoxic agents. *J. Biotechnol.* **2006**, *122*, 372–381. [CrossRef]
7. Knight, E.; Przyborski, S. Advances in 3D cell culture technologies enabling tissue-like structures to be created in vitro. *J. Anat.* **2015**, *227*, 746–756. [CrossRef]
8. Abbott, R.D.; Kaplan, D.L. Strategies for improving the physiological relevance of human engineered tissues. *Trends Biotechnol.* **2015**, *33*, 401–407. [CrossRef]
9. LaPlaca, M.C.; Vernekar, V.N.; Shoemaker, J.T.; Cullen, D.K. Three-dimensional neuronal cultures. In *3D Tissue Engineering*; Berthiaume, F., Morgan, J.R., Eds.; Artech House: Boston, MA, USA, 2010; Chapter 11; pp. 187–204.
10. Bercu, M.M.; Arien-Zakay, H.; Stoler, D.; Lecht, S.; Lelkes, P.I.; Samuel, S.; Or, R.; Nagler, A.; Lazarovici, P.; Elchalal, U. Enhanced survival and neurite network formation of human umbilical cord blood neuronal progenitors in three-dimensional collagen constructs. *J. Mol. Neurosci.* **2013**, *51*, 249–261. [CrossRef]
11. Cukierman, E.; Pankov, R.; Yamada, K.M. Cell interactions with three-dimensional matrices. *Curr. Opin. Cell Biol.* **2002**, *14*, 633–639. [CrossRef]
12. Mazzoleni, G.; Di Lorenzo, D.; Steimberg, N. Modelling tissues in 3D: The next future of pharmacotoxicology and food research. *Genes Nutr.* **2009**, *4*, 13–22. [CrossRef]
13. Vinci, M.; Gowan, S.; Boxall, B.; Patterson, L.; Zimmermann, M.; Court, W.; Lomas, C.; Mendiola, M.; Hardisson, D.; Eccles, S.A. Advances in establishment and analysis of three-dimensional tumor spheroid-based functional assays for target validation and drug evaluation. *BMC Biol.* **2012**, *10*, 29. [CrossRef]
14. Lai, Y.; Cheng, K.; Kisaalita, W. Three-dimensional neuronal cell cultures more accurately model Voltage Gated Calcium Channel functionality in freshly dissected nerve tissue. *PLoS ONE* **2012**, *7*, 45074. [CrossRef]
15. Cavallo, F.; Huang, Y.; Dent, E.W.; Williams, J.C.; Lagally, M.G. Neurite guidance and three-dimensional confinement via compliant semiconductor scaffolds. *ACS Nano* **2014**, *8*, 12219–12227. [CrossRef]

16. McNamara, M.C.; Sharifi, F.; Wrede, A.H.; Kimlinger, D.F.; Thomas, D.G.; Vander Wiel, J.B.; Chen, Y.; Montazami, R.; Hashemi, N.N. Microfibers as physiologically relevant platforms for creation of 3D cell cultures. *Macromol. Biosci.* **2017**, *17*, 1700279. [CrossRef]

17. Limongi, T.; Cesca, F.; Gentile, F.; Marotta, R.; Ruffilli, R.; Barberis, A.; Dal Maschio, M.; Petrini, E.M.; Santoriello, S.; Benfenati, F.; et al. Nanostructured superhydrophobic substrates trigger the development of 3D neuronal networks. *Small* **2013**, *9*, 402–412. [CrossRef]

18. Johansson, F.; Carlberg, P.; Danielsen, N.; Montelius, L.; Kanje, M. Axonal outgrowth on nano-imprinted patterns. *Biomaterials* **2006**, *27*, 1251–1258. [CrossRef]

19. Bucaro, M.A.; Vasquez, Y.; Hatton, B.D.; Aizenberg, J. Fine-tuning the degree of stem cell polarization and alignment on ordered arrays of high-aspect-ratio nanopillars. *ACS Nano* **2012**, *6*, 6222–6230. [CrossRef]

20. Dowell-Mesfin, N.M.; Abdul-Karim, M.A.; Turner, A.M.; Schanz, S.; Craighead, H.G.; Roysam, B.; Turner, J.N.; Shain, W. Topographically modified surfaces affect orientation and growth of hippocampal neurons. *J. Neural Eng.* **2004**, *1*, 78–90. [CrossRef]

21. Kwiat, M.; Elnathan, R.; Pevzner, A.; Peretz, A.; Barak, B.; Peretz, H.; Ducobni, T.; Stein, D.; Mittleman, M.; Ashery, U.; et al. Highly ordered large-scale neuronal networks of individual cells—Toward single cell to 3D nanowire intracellular interfaces. *ACS Appl. Mater. Interfaces* **2012**, *4*, 3542–3549. [CrossRef]

22. Yu, M.; Huang, Y.; Ballweg, J.; Shin, H.; Huang, M.; Savage, D.E.; Lagally, M.G.; Dent, E.W.; Blick, R.H.; Williams, J.C. Semiconductor Nanomembrane Tubes: Three-Dimensional Confinement for Controlled Neurite Outgrowth. *ACS Nano* **2011**, *26*, 2447–2457. [CrossRef] [PubMed]

23. Paul, F.; Huang, Y.; Cangellaris, O.V.; Huang, W.; Dent, E.W.; Gillette, M.U.; Williams, J.C. Toward intelligent synthetic neural circuits: Directing and accelerating neuron cell growth by self-rolled-up silicon nitride microtube array. *ACS Nano* **2014**, *25*, 11108–11117.

24. Mazzini, G.; Carpignano, F.; Surdo, S.; Aredia, F.; Panini, N.; Torchio, M.; Erba, E.; Danova, M.; Scovassi, A.I.; Barillaro, G. 3D Silicon Microstructures: A new tool for evaluating biological aggressiveness of tumor cells. *IEEE Trans. Nanobiosci.* **2015**, *14*, 797–805. [CrossRef] [PubMed]

25. McConnell, S.K. Constructing the cerebral cortex: Neurogenesis and fate determination. *Neuron* **1995**, *15*, 761–768. [CrossRef]

26. Solozobova, V.; Wyvekens, N.; Pruszak, J. Lessons from the embryonic neural stem cell niche for neural lineage differentiation of pluripotent stem cells. *Stem Cell. Rev. Rep.* **2012**, *8*, 813–829. [CrossRef]

27. Boissart, C.; Poulet, A.; Georges, P.; Darville, H.; Julita, E.; Delorme, R.; Bourgeron, T.; Peschanski, M.; Benchoua, A. Differentiation from human pluripotent stem cells of cortical neurons of the superficial layers amenable to psychiatric disease modeling and high-throughput drug screening. *Transl. Psychiatry* **2013**, *3*, 294. [CrossRef]

28. Corti, S.; Faravelli, I.; Cardano, M.; Conti, L. Human pluripotent stem cells as tools for neurodegenerative and neurodevelopmental disease modeling and drug discovery. *Expert Opin. Drug Discov.* **2015**, *10*, 615–629. [CrossRef]

29. Brilli, E.; Reitano, E.; Conti, L.; Conforti, P.; Gulino, R.; Consalez, G.G.; Cesana, E.; Smith, A.; Rossi, F.; Cattaneo, E. Neural stem cells engrafted in the adult brain fuse with endogenous neurons. *Stem Cells Dev.* **2013**, *22*, 538–547. [CrossRef]

30. Conti, L.; Pollard, S.M.; Gorba, T.; Reitano, E.; Toselli, M.; Biella, G.; Sun, Y.; Sanzone, S.; Ying, Q.L.; Cattaneo, E.; et al. Niche-independent symmetrical self-renewal of a mammalian tissue stem cell. *PLoS Biol.* **2005**, *3*, 283. [CrossRef]

31. Pollard, S.; Conti, L.; Smith, A. Exploitation of adherent neural stem cells in basic and applied neurobiology. *Regen. Med.* **2006**, *1*, 111–118. [CrossRef]

32. Pollard, S.M.; Conti, L. Investigating radial glia in vitro. *Prog. Neurobiol.* **2007**, *83*, 53–67. [CrossRef] [PubMed]

33. Schindelin, J.; Arganda-Carreras, I.; Frise, E.; Kaynig, V.; Longair, M.; Pietzsch, T.; Preibisch, S.; Rueden, C.; Saalfeld, S.; Schmid, B.; et al. Fiji: An open-source platform for biological-image analysis. *Nat. Methods* **2012**, *9*, 676–682. [CrossRef] [PubMed]

34. Gjorevski, N.; Ranga, N.; Lutolf, M.P. Bioengineering approaches to guide stem cell-based organogenesis. *Development* **2014**, *141*, 1794–1804. [CrossRef] [PubMed]

35. Kratochvil, M.J.; Seymour, A.J.; Li, T.L.; Sergiu, P.P.; Kuo, C.J.; Heilshorn, S.C. Engineered materials for organoid systems. *Nat. Rev. Mater.* **2019**, *4*, 606–622. [CrossRef]

36. Seliktar, D. Designing cell-compatible hydrogels for biomedical applications. *Science* **2012**, *336*, 1124–1128. [CrossRef]

37. Fang, Y.; Eglen, R.M. Three-dimensional cell cultures in drug discovery and development. *Slas Discov.* **2017**, *22*, 456–472.

38. Simitzi, C.; Efstathopoulos, P.; Kourgiantaki, A.; Ranella, A.; Charalampopoulos, I.; Fotakis, C.; Athanassakis, I.; Stratakis, E.; Gravanis, A. Laser fabricated discontinuous anisotropic microconical substrates as a new model scaffold to control the directionality of neuronal network outgrowth. *Biomaterials* **2015**, *67*, 115–128. [CrossRef]

39. Yao, L.; Wang, S.; Cui, W.; Sherlock, R.; O'Connell, C.; Damodaran, G.; Gorman, A.; Windebank, A.; Pandit, A. Effect of functionalized micropatterned PLGA on guided neurite growth. *Acta Biomater.* **2009**, *5*, 580–588. [CrossRef]

40. Rajnicek, A.; Britland, S.; McCaig, C. Contact guidance of CNS neurites on grooved quartz: Influence of groove dimensions, neuronal age and cell type. *J. Cell Sci.* **1997**, *110*, 2905–2913.

41. Cecchini, M.; Bumma, G.; Serresi, M.; Beltrami, F. PC12 differentiation on biopolymer nanostructures. *Nanotechnology* **2007**, *18*, 505103. [CrossRef]

42. Hanson, J.N.; Motala, M.J.; Heien, M.L.; Gillette, M.; Sweedler, J.; Nuzzo, R.G. Textural guidance cues for controlling process outgrowth of mammalian neurons. *Lab Chip* **2009**, *9*, 122–131. [CrossRef] [PubMed]

43. Mattotti, M.; Micholt, L.; Braeken, D.; Kovacic, D. Characterization of spiral ganglion neurons cultured on silicon micro-pillar substrates for new auditory neuro-electronic interfaces. *J. Neural Eng.* **2015**, *12*, 026001. [CrossRef] [PubMed]

44. Schmidt, M.S.; Hubner, J.; Boisen, A. Large Area Fabrication of Leaning Silicon Nanopillars for Surface Enhanced Raman Spectroscopy. *Adv. Mater.* **2012**, *24*, 11–18. [CrossRef] [PubMed]

45. Magno, G.; Belier, B.; Barbillon, G. Al/Si Nanopillars as Very Sensitive SERS Substrates. *Materials* **2018**, *11*, 1534. [CrossRef]

46. Vereecke, G.; Xu, X.M.; Tsai, W.K.; Yang, H.; Armini, S.; Delande, T.; Doumen, G.; Kentie, F.; Shi, X.; Simms, I.; et al. Partial Wetting of aqueous solutions on high aspect ratio nanopillars with hydrophilic surface finish. *ECS J. Solid State Sci. Technol.* **2014**, *3*, 3095–3100. [CrossRef]

47. Liu, C.; Sun, J.; Li, J.; Xiang, C.; Che, L.; Wang, Z.; Zhou, X. Long-range spontaneous droplet self-propulsion on wettability gradient surfaces. *Sci. Rep.* **2017**, *7*, 7552. [CrossRef]

48. Park, Y.S.; Yoon, S.Y.; Park, J.S.; Lee, J.S. Deflection induced cellular focal adhesion and anisotropic growth on vertically aligned silicon nanowires with differing elasticity. *NPG Asia Mater.* **2016**, *8*, 249. [CrossRef]

49. Tullii, G.; Giona, F.; Lodola, F.; Bonfadini, S.; Bossio, C.; Varo, S.; Desii, A.; Criante, L.; Sala, C.; Pasini, M.; et al. High-aspect-ratio semiconducting polymer pillars for 3D cell cultures. *ACS Appl. Mater. Interfaces* **2019**, *11*, 28125–28137. [CrossRef]

50. Medina Benavente, J.J.; Mogami, H.; Sakurai, T.; Sawada, K. Evaluation of silicon nitride as a substrate for culture of PC12 cells: An interfacial model for functional studies in neurons. *PLoS ONE* **2014**, *9*, 90189. [CrossRef]

51. Liliom, H.; Lajer, P.; Berces, Z.; Csernyus, B.; Szabò, A.; Pinke, D.; Low, P.; Fekete, Z.; Pongracz, A.; Schlett, K. Comparing the effects of uncoated nanostructured surfaces on primary neurons and astrocytes. *J. Biomed. Mater. Res. A* **2019**, *107*, 2350–2359. [CrossRef]

52. Fan, Y.W.; Cui, F.Z.; Hou, S.P.; Xu, Q.Y.; Chen, L.N.; Lee, I.S. Culture of neural cells on silicon wafers with nano-scale surface topograph. *J. Neurosci. Methods* **2002**, *120*, 17–23. [CrossRef]

53. Chen, W.S.; Guo, L.Y.; Tang, C.C.; Tsai, C.K.; Huang, H.H.; Chin, T.Y.; Yang, M.L.; Chen-Yang, Y.W. The effect of laminin surface modification of electrospun silica nanofiber substrate on neuronal tissue engineering. *Nanomaterials* **2018**, *8*, 165. [CrossRef] [PubMed]

54. Turunen, S.; Joki, T.; Hiltunen, M.L.; Ihalainen, T.O.; Narkilahti, S.; Kellomaki, M. Direct laser writing of tubular microtowers for 3D culture of human pluripotent stem cell-derived neuronal cells. *ACS Appl. Mater. Interfaces* **2017**, *9*, 25717–25730. [CrossRef] [PubMed]

55. Sharifi, F.; Patel, B.B.; Dzuilko, A.K.; Montazami, R.; Sakaguchi, D.S.; Hashemi, N. Polycaprolactone microfibrous scaffolds to navigate neural stem cells. *Biomacromolecules* **2016**, *17*, 3287–3297. [CrossRef] [PubMed]

56. Timashev, P.S.; Vedunova, M.V.; Guseva, D.; Ponimaskin, E.; Deiwick, A.; Mishchenko, T.A.; Mitroshina, E.V.; Koroleva, A.V.; Pimashkin, A.S.; Mukhina, I.V.; et al. 3D in vitro platform produced by two-photon polymerization for the analysis of neural network formation and function. *Biomed. Phys. Eng. Exp.* **2016**, *2*, 035001. [CrossRef]
57. Koroleva, A.; Gill, A.A.; Ortega, I.; Haycock, J.W.; Schilie, S.; Gittard, S.D.; Chichkov, B.N.; Claeyssens, F. Two-photon polymerization-generated and micromolding-replicated 3D scaffolds for peripheral neural tissue engineering applications. *Biofabrication* **2012**, *4*, 025005. [CrossRef]

Article

Tendon and Cytokine Marker Expression by Human Bone Marrow Mesenchymal Stem Cells in a Hyaluronate/Poly-Lactic-Co-Glycolic Acid (PLGA)/Fibrin Three-Dimensional (3D) Scaffold

Maria C. Ciardulli [1], Luigi Marino [1], Joseph Lovecchio [2], Emanuele Giordano [2], Nicholas R. Forsyth [3], Carmine Selleri [1], Nicola Maffulli [1,3,4] and Giovanna Della Porta [1,5,*]

[1] Department of Medicine, Surgery and Dentistry, University of Salerno, Via S. Allende, 84081 Baronissi (SA), Italy; mciardulli@unisa.it (M.C.C.); lmarino@unisa.it (L.M.); cselleri@unisa.it (C.S.); n.maffulli@qmul.ac.uk (N.M.)
[2] Department of Electrical, Electronic and Information Engineering "Guglielmo Marconi" (DEI), University of Bologna, Via dell'Università 50, 47522 Cesena (FC), Italy; joseph.lovecchio@unibo.it (J.L.); emanuele.giordano@unibo.it (E.G.)
[3] Guy Hilton Research Centre, School of Pharmacy and Bioengineering, Keele University, Stoke-on-Trent, Staffordshire ST4 7QB, UK; n.r.forsyth@keele.ac.uk
[4] Centre for Sport and Exercise Medicine, Queen Mary University of London, Barts and The London School of Medicine, London E1 4NL, UK
[5] Department of Industrial Engineering, University of Salerno, Via Giovanni Paolo II, 84084 Fisciano (SA), Italy
* Correspondence: gdellaporta@unisa.it; Tel.: +39-089965234

Received: 30 March 2020; Accepted: 18 May 2020; Published: 20 May 2020

Abstract: We developed a (three-dimensional) 3D scaffold, we named HY-FIB, incorporating a force-transmission band of braided hyaluronate embedded in a cell localizing fibrin hydrogel and poly-lactic-co-glycolic acid (PLGA) nanocarriers as transient components for growth factor controlled delivery. The tenogenic supporting capacity of HY-FIB on human-Bone Marrow Mesenchymal Stem Cells (hBM-MSCs) was explored under static conditions and under bioreactor-induced cyclic strain conditions. HY-FIB elasticity enabled to deliver a mean shear stress of 0.09 Pa for 4 h/day. Tendon and cytokine marker expression by hBM-MSCs were studied. Results: hBM-MSCs embedded in HY-FIB and subjected to mechanical stimulation, resulted in a typical tenogenic phenotype, as indicated by type 1 Collagen fiber immunofluorescence. RT-qPCR showed an increase of type 1 Collagen, scleraxis, and decorin gene expression (3-fold, 1600-fold, and 3-fold, respectively, at day 11) in dynamic conditions. Cells also showed pro-inflammatory (IL-6, TNF, IL-12A, IL-1β) and anti-inflammatory (IL-10, TGF-β1) cytokine gene expressions, with a significant increase of anti-inflammatory cytokines in dynamic conditions (IL-10 and TGF-β1 300-fold and 4-fold, respectively, at day 11). Mechanical signaling, conveyed by HY-FIB to hBM-MSCs, promoted tenogenic gene markers expression and a pro-repair cytokine balance. The results provide strong evidence in support of the HY-FIB system and its interaction with cells and its potential for use as a predictive in vitro model.

Keywords: hBM-MSCs; cytokines; tenogenic markers; cyclic strain; 3D microenvironment; PLGA carriers; bioreactor

---

1. Introduction

Tissue engineering strategies for tendon healing and regeneration are designed to improve existing therapies or provide new treatment possibilities. Three-dimensional (3D) bioengineered systems have the potential to promote our understanding of the physiopathology of tendinopathy and the role of stem cells in tendon regeneration. In this sense, 3D scaffold design and fabrication coupled to specific bioreactor arrangements could develop highly predictive 3D in vitro culture and differentiation systems to explore cell behaviors in response to defined external biochemical and mechanical stimuli [1]. The 3D scaffold provides a model of fidelity via its provision of a microenvironment with defined stiffness and elastic modulus as well as the necessary surfaces for cell attachment [2–4].

Detailed understanding of cell behavior when incorporated into specific biomaterials allows to develop designs with specific functionalization. These functionalization may, for instance, stimulate local stem cells, attract specific circulating nucleated blood cells, such as macrophages, and induce their polarization into M2 phenotype to accelerate tissue regeneration and healing following the biomaterials in vivo implantation [5]. For example, human Mesenchymal Stem Cells (hMSCs) are largely used in tissue engineering strategies, and their immune-modulatory activity in the development of tendon pathologies have been explored, but the precise mechanisms involved remain undetermined [6–10]. Neutrophils and macrophages infiltrate injured tendons, potentially interacting with MSCs and stimulating cytokine release at the site of repair and promoting degradation of the extracellular matrix (ECM), inflammation, apoptosis, and, in the later stages of acute tendon healing, they release anti-inflammatory cytokines to alleviate inflammation and promote tendon remodeling [11–13].

Among the biomaterials described for tendon tissue engineering [14], a promising emerging strategy is the use of a complex biomimetic matrix with a hydrogel component and extracellular matrix mimicking properties [15,16]. Tenocyte precursors can be harvested from different sources, including periosteum [17,18], bone marrow [19–21], tendon [21,22], and adipose tissue [21,23]. To overcome the intrinsic poor mechanical properties of the hydrogel, they can be merged with more force resistant biopolymers. Cells and biomaterials alone are not sufficient to achieve optimal levels of differentiation and matrix organization. Mechanical stimulation has a key role in tenogenic differentiation induction [19]. Scaffolds are therefore required to display an appropriate elastic behavior to deliver strain [24] or compression [25] inputs. Strain is a tenogenic differentiation signal [26–29], and several bioreactors have been used to impart tenogenic mechanical stimuli to cells in culture [19,30–35]. For example, Rinoldi et al. designed and fabricated 3D multilayered composite scaffolds, where an electrospun nanofibrous substrate was coated with a thin layer of GelMA-alginate composite hydrogel carrying MSCs. MSCs were subsequently differentiated by the addition of bone morphogenetic protein 12 (BMP-12) and, to mimic the natural function of tendons, the scaffolds were mechanically stimulated using a custom-built bioreactor [34]. Grier et al. described an aligned collagen-glycosaminoglycan scaffold able to enhance tenogenic differentiation of MSCs via cyclic tensile strain within a bioreactor, in the absence of growth factor supplementation [36]. Another protocol, proposed by Youngstrom et al., promoted tenogenic differentiation of MSCs cultivated on decellularized tendon scaffolds with the application of 3% cyclic strain for one hour per day for 11 days [31]. Additionally, several growth factors and other small molecules can stimulate transcriptional activation of genes involved in tenogenic differentiation [15,37,38]. Growth Differentiation Factor 5 (GDF-5), for instance, induces the expression of genes linked to the neo-tendon phenotype [39–41].

Tendinopathies associated with physical activity and age-related degeneration are a major medical issue [23], and recent healing and regeneration studies include the use of human Bone Marrow Mesenchymal Stem Cells (hBM-MSCs) [42–44]. hBM-MSCs are a multipotent population present in bone marrow that can be readily differentiated in vitro [45,46] into cells of three mesodermal lineages, namely adipocytes, chondrocytes and osteoblasts under appropriate conditions [47–50]. MSCs-based therapies include direct transplantation of MSCs populations, growth factor-loaded scaffolds for local MSCs recruitment or implantation of scaffolds containing in vitro culture-expanded MSCs populations [51,52].

We previously described an engineered multiphase three-dimensional (3D) scaffold as an in vitro model for tendon regeneration studies. The multiphase 3D construct was totally absorbable and consisted of a braided hyaluronate elastic band merged with a fibrin hydrogel containing hBM-MSCs and poly-lactic-co-glycolic acid nano-carriers (PLGA-NCs) themselves loaded with human

Growth Differentiation Factor 5 (hGDF-5) [53]. In that work, the PLGA nano-carriers were transient scaffold components to ensure sustained and controlled delivery of hGDF-5 with benefits beyond those associated with standard culture medium supplementation [53,54]. The study reported an early tenogenic commitment of hBM-MSCs after three days of cultivation under dynamic conditions.

In the present study, we describe the use of the same scaffold (named HY-FIB here) to investigate the effect of the 3D environment on hBM-MSCs for 11 days with or without mechanical stimulation and in the absence of any specific biochemical differentiation signal. HY-FIB was assembled with hBM-MSCs as previously described [53] including PLGA-NCs stratified within the 3D fibrin structure. Importantly, the PLGA-NCs carried an inactive form of human Growth Differentiation Factor 5 (ihGDF-5) enabling the overall 3D scaffold structure to be safely evaluated with or without mechanical input. Gene expression of type 1 Collagen, decorin, scleraxis, and tenomodulin were considered; type 3 collagen was also monitored, as negative control. Histology and quantitative immunofluorescence were used to monitor cell behavior and their interaction with the synthetic extracellular matrix. Moreover, to understand whether HY-FIB microenvironment configuration would stimulate any cell inflammation responses, the cells expression of cytokine markers was also monitored, including pro-inflammatory cytokines and anti-inflammatory ones. The results provide strong evidence that HY-FIB environment plus mechanical signaling, promoted tenogenic markers expression, collagen production and better pro-repair cytokine balance by hBM-MSCs.

2. Materials and Methods

2.1. hBM-MSCs Isolation and Harvesting

Human bone marrow mesenchymal stem cells (hBM-MSCs) were obtained from the bone marrow of three independent healthy donors (age 36, 38, 44 years). The donors gave written informed consent in accordance with the Declaration of Helsinki to the use of their filter residual bone marrow aspirate for research purposes, with approval from the University Hospital of San Giovanni di Dio e Ruggi d'Aragona (Salerno, IT). Review Board authorization number: (24988 achieved on April 9, 2015).

Briefly, total bone marrow aspirate was directly seeded at a concentration of 50,000 total nucleated cells/cm^2 in T75 plastic flask in Minimum Essential Medium Alpha (α-MEM) supplemented with 1% Glutagro^{TM}, 10% Fetal Bovine Serum (FBS), and 1% Pen/Strep and incubated at 37 °C in 5% CO_2 atmosphere and 95% relative humidity [55]. After 72h, non-adherent cells were removed by medium change, and the adherent cells were further fed twice a week with new medium. On day 14, colonies of adherent hBM-MSCs were detached and re-seeded at 4000 cells/cm^2 in the same culture conditions. Once the cell cultures reached 70–80% confluence, cells were detached using 0.05% trypsin-0.53mM EDTA and washed with PBS 1× (Corning Cellgro, Manassas, VA, USA), counted using Trypan Blue (Sigma-Aldrich, Milan, IT) and subcultured at a concentration of 4×10^3 cells/cm^2. Flow Cytometry analysis was performed on hBM-MSCs obtained at Passage 1 examining levels of CD90, CD105, CD73 CD14, CD34, CD45, and HLA-DR expression (Miltenyi Biotec, DE).

2.2. ihGDF-5 Effect on hBM-MSCs

These sets of experiments were performed to assure the absence of any effect of inactivated human GDF-5 (ihGDF-5, Cloud-Clone Corp., USA) on both tenogenic markers stimulation and cytokines expression by hBM-MSCs. Cells were seeded on coverslips in 12 well plates at a concentration of 4×10^3 cells/cm^2. Once the cultures reached 60% confluence, cells were treated with either 1.6 ng/mL or 100 ng/mL of ihGDF-5. Cells were fed twice a week with new medium and fresh ihGDF-5 supplementation for up to 16 days. Untreated cells for matched time-points studied were used for control purposes. Passage 3 cells were seeded in the 3D environment (~8×10^5 cells/mL) and were fed twice a week with new medium, without any growth factor added.

PLGA carriers were not tested because they cannot be supplemented to the cells planar monolayer culture, indeed, in static conditions their sedimentation on cells reduced oxygen exchange and prevented cells survival (data not shown).

2.3. Immunofluorescence and Immunohistochemical Assays

Cells were fixed with 3.7% formaldehyde for 30 min at room temperature (RT) followed by permeabilization with 0.1% Triton X-100 for 5 min and blocking with 1% Bovine Serum Albumine (BSA) for 1h. For type 1 and type 3 Collagen staining, cells were incubated overnight at 4 °C with a mouse monoclonal anti-type 1 Collagen antibody (1:100, Sigma-Aldrich) and a rabbit polyclonal anti-type 3 Collagen antibody (1:100, Santa Cruz Biotechnology). Following incubation with the primary antibody, cells were incubated for 1h at RT with the DyLight 649 anti-mouse IgG (1:500, BioLegend, CA) and the Alexa FluorTM 488 goat anti-rabbit IgG (1:500; Thermo Fisher Scientific, USA). Cell nuclei were stained with DAPI solution (1:1000) for 5 min. Images acquisition was at 20× magnification on a fluorescent microscope (Eclipse Ti-E Inverted Microscope; NIKON Instruments Inc., USA).

For 3D scaffold immunohistochemical analysis, slices were permeabilized with 0.1% Triton X-100 for 5 min, and non-specific staining blocked with 1% BSA for 1h at RT. For type 1 Collagen staining, slices were incubated overnight at 4 °C with a rabbit polyclonal anti-type 1 Collagen antibody (1:200, AbCam). Following incubation with the primary antibody, slices were incubated for 1h at RT with Alexa Fluor^{TM} 488 goat anti-rabbit IgG (1:400, Thermo Fisher Scientific, USA) antibody. Subsequently, cell nuclei were stained with DAPI solution (1:1000) and incubated for 5 min. Images were acquired as described above. Image quantification was performed using image analysis software (ImageJ, National Institutes of Health, USA) [56,57] by measuring the red and green areas where type 1 and type 3 collagen, respectively, are expressed. A minimum of 10 image fields was used for the image analysis at each time point. Signal intensity at each time point was normalized by the cell number (e.g., by amount of cell nuclei revealed by DAPI staining).

Sirius red staining was performed using the Picrosirius Red Stain Kit (Polysciences, Inc., USA). Sections of 15 μm of thickness were stained in hematoxylin for 8 min, then washed in water for 2 min. The sections were dipped into phosphomolybdic acid for 2 min, then washed in water for 2 m. Then they were dipped into Picrosirius Red F3BA Stain for 60 min and dipped into HCl 0.1M solution for 2 min. The sections were dehydrated in increasing ethanol gradient solutions (70–75–95–100%) and finally dipped into xylene for 5 min. Eukitt medium was used to mount the samples.

2.4. RNA Isolation and Gene Expression Profile by Quantitative Reverse Transcription PCR (RT-qPCR)

Total RNA was extracted from hBM-MSCs seeded into the 3D construct of each experimental group using QIAzol® Lysis Reagent (Qiagen, DE), chloroform (Sigma-Aldrich, Milan, IT) and the RNeasy Mini Kit (Qiagen, DE). For each sample, 300 ng of total RNA was reverse-transcribed using the iScriptTM cDNA synthesis kit (Bio-Rad, Milan, IT). Relative gene expression analysis was performed in a LightCycler® 480 Instrument (Roche, IT), using the SsoAdvancedTM Universal SYBR® Green Supermix (Bio-Rad) with the validated primers for COL1A1, COL3A1, DCN, IL-1β, IL-6, IL-10, IL-12A, SCX-A, TGF-β1, TNF, and TNMD (Bio-Rad), and following MIQE guidelines [58]. Amplification was performed in a 10 μL final volume, including 2 ng of complementary DNA (cDNA) as template. Specificity of the formed products was addressed via melting curve analysis. Triplicate experiments were performed for each condition explored, and data were normalized to glyceraldehyde-3-phosphate dehydrogenase (GAPDH) expression (reference gene), applying the geNorm method [59] to calculate reference gene stability between the different conditions (calculated with CFX Manager software; M <0.5). Fold changes in gene expression were determined by the $2^{-\Delta\Delta C_P}$ method, and are presented as relative levels versus hBM-MSCs just loaded within the HY-FIB system.

2.5. PLGA-NCs Fabrication, Size, Morphology, and ihGDF-5 Release Profile

PLGA nano-carriers (PLGA-NCs) were obtained using Supercritical Emulsion Extraction (SEE) technology enabling rapid polymer NCs production from multiple emulsions via dense gas extraction of the oily phase organic solvent utilizing a countercurrent packed tower operating in continuous mode [60]. In detail, ihGDF-5 (Cloud-Clone Corp., USA) was dissolved into 1% (*w/v*) human serum albumin (hSA; Sigma-Aldrich, Milan, IT) containing 0.06% polyvinyl alcohol (PVA). Human serum albumin (hSA) was included as an ihGDF-5 stabilizer. This solution was added to the oily phase formed in an Ethyl Acetate (EA, purity 99.9%) and PLGA (RG 504H, 38,000–54,000 kDa, Evonik, DE) at 5% (*w/w*) solution. All emulsions were processed immediately after their preparation.

SEE technology operative pressure and temperature conditions in the high-pressure column were set at 8 MPa and 38 °C, respectively, with a dense gas flow of Carbon Dioxide (CO_2) set at 1.4 kg/h with Liquid/Gas ratio of 0.1 (*w/w*) [61]. Carrier suspensions were collected at the bottom of the extraction column, washed, and lyophilized. Each run allowed the recovery of 98% of the loaded biopolymer. Empty and loaded NCs were produced using the same process conditions.

Carrier particle size distributions (PSDs) were measured using a laser granulometer (mod. Mastersizer S; Malvern Instruments Ltd., Worcestershire, UK), based on dynamic light scattering (DLS). Sizes are expressed as volume mean size (MS) with standard deviation (SD) in nanometers (nm). The shape and morphology of the PLGA-NCs were investigated by field emission-scanning electron microscopy (FE-SEM; mod. LEO 1525; Carl Zeiss, Oberkochen, D). Samples were placed on a double-sided adhesive carbon tape previously glued to an aluminum stub and coated with a gold film (250 A thickness) using a sputter coater (mod.108 A; Agar Scientific, Stansted, UK).

ihGDF-5 release profile was monitored in vitro from 20(±0.3) mg of carriers suspended in 1mL of α-MEM, placed in an incubator at 37 °C, and stirred continuously at 1× *g*. Every 24 h, samples were centrifuged at 160× *g* for 10 min and the supernatant completely removed and replaced with fresh media to maintain sink conditions. Released ihGDF-5 concentrations from collected samples were then measured with an Enzyme Linked Immunosorbent Assay (ELISA, Cloud-Clone Corp., USA). Release experiments were performed in triplicate (n = 3), and the curve describing the mean profile calculated as ng/g (protein released/PLGA-NCs) versus time.

2.6. HY-FIB Preparation and Characterization

For each sample, a mixture of 50 mg/mL fibrinogen from human plasma (Sigma-Aldrich, Milan, IT), 15,600 U/mL aprotinin (Sigma-Aldrich, Milan, IT), and α-MEM (Corning, NY, USA) supplemented with 10% FBS (referred to as growing media, GM) was added at a 1:1:1 ratio to 100 mg of PLGA-NCs (ihGDF-5 loading: 350 ng/g) and, then, to an average of 8×10^5 cells. A homogeneous cells/PLGA-NCs/fibrinogen suspension was then embedded into a mold (30 × 20 × 4.5 mm) where the braided band had been previously positioned. Free ends were left to enable HY-FIB fixing into the bioreactor. Upon addition of 100 U/mL thrombin (Sigma-Aldrich, Milan, IT), the mold was placed in a 37 °C humidified incubator for 30 min to allow fibrin polymerization. When the hydrogel was formed, the band was entrapped inside a uniformly distributed hydrogel. The construct was then transferred from the mold to either a standard polystyrene culture plate or to the bioreactor culture chamber, each containing 30 mL of the culture media, and placed in an incubator at 37 °C in a 5% CO_2 atmosphere and 95% relative humidity.

HY-FIB morphology was observed by field emission-scanning electron microscopy (FE-SEM; mod. LEO 1525; Carl Zeiss, Oberkochen, Germany). Samples were fixed in 4% PFA (4 °C, overnight) and then dehydrated by multiple passages across ethanol:water solutions (10 min each) with increasing concentrations of ethanol (10%, 20%, 30%, 50%, 70%, 90%), ending in a 100% dehydrating liquid (3 changes, 10 min each).

Samples were then lyophilized in a Critical Point Dryer (mod. K850 Emitech, Assing, Rome IT), placed on a double-sided adhesive carbon tape previously glued to an aluminum stub and coated with a gold film (250 A thickness) using a sputter coater (mod. 108 A; Agar Scientific, Stansted, United Kingdom) before observation.

HY-FIB mechanical characterization was performed according to the ASTM 1708 by a dynamometer (CMT 6000 SANS, Shenzen, China) equipped with a 1 kN load cell. The sample was conditioned in Dulbecco's Modified Essential Medium (DMEM) for 1 h, and then shaped to obtain a specimen with gauge length (Lo) of 22 mm and width (W) of 5 mm. Sample thickness (S) was measured with a thickness gauge brand at three different averaged points. Monoaxial deformation was applied to the sample at a speed of 10 mm/min, and force (F) and elongation (L) during traction were recorded. The elastic modulus and ultimate tensile strength (both expressed in MPa) were calculated from the stress/strain plot. For the immuno-histochemical analysis, at different time points, a portion of HY-FIB was fixed in 4% PFA (4 °C, overnight), cryo-protected in 30% sucrose overnight, mounted in OCT embedding compound, frozen at −20°C and then cut in slices of 10 μm of thickness using a cryostat. The remaining portion of HY-FIB was placed in QIAzol® Lysis Reagent for total RNA extraction.

2.7. Dynamic Culture

HY-FIB was clamped at both free ends, one motionless and one sliding (operated by a linear motor actuator) arm, into the bioreactor system culture chamber, described in detail elsewhere [23]. A maximal load, set by pre-tensioning, was relaxed to a minimum value cycling at a pre-determined frequency. In addition, continuous feedback signals provided by strain gauges located onto the fixed arm, allowed the maintenance of a defined load on the scaffold in response to physical system modifications, by automatic adjustment of the pre-tensioning position.

2.8. Finite Element Modeling

Finite Element Modeling (FEM) was implemented by using COMSOL Multiphysics Software (rel. 5.3a®) to assess the stress distribution into 3D constructs when a 10% mechanical cyclic strain stimulus is applied. All components were obtained using primitive geometries and Boolean operations. Linear elasticity equations were set as boundary conditions for the scaffold. A sensitivity study of the mesh obtained the most computationally efficient solution. The specific parameters used in the model are listed in Table 1. The simulation considered only the stress within the fibrin 3D environment neglecting any further contribution of the band, cells and PLGA-NCs.

Table 1. Finite Element Modeling (FEM) parameters used.

Parameters	Value	Unit
Young's modulus [1]	4.56	MPa
Poisson ratio [2]	0.25	-
Density	1050	Kg/m^3

[1] HY-FIB Young Modulus was measured experimentally [53] [2] Poisson ratio value was taken in the literature [62].

2.9. Statistical Analysis

Statistical analysis was performed using GraphPad Prism software (6.0 for Windows). Data obtained from multiple experiments are expressed as mean+/−SD and analyzed for statistical significance using ANOVA test, for independent groups. Differences were considered statistically significant when $p \leq 0.05$ [63].

3. Results

3.1. hBM-MSCs Cultivation in 2D Environment

hBM-MSCs were cultivated in a two-dimensional (2D) monolayer environment with medium supplemented with either 1.6 ng/mL or 100 ng/mL of inactive human Growth Differentiation Factor 5 (ihGDF-5) for up to 16 days. These two concentration conditions (two order of magnitude of difference) were chosen to ascertain absence of any effect of the inactive hGDF-5 form on cells expression of tenogenic markers (COL3A1, COL1A1, DCN, SCX-A and TNMD) and of cytokines (pro-inflammatory: IL-6, TNF, IL-12A, IL-1β; anti-inflammatory: IL-10, TGF-β1) by RT-qPCR (see Figure 1a–d).

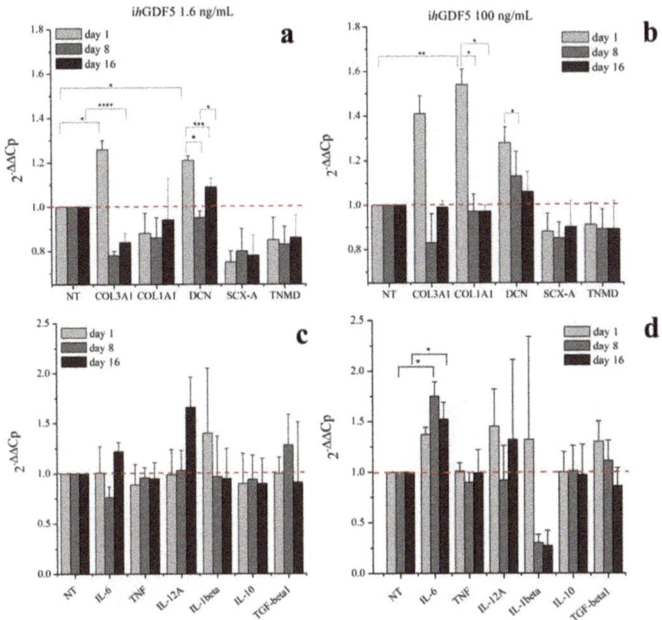

Figure 1. Gene expression profiles for tenogenic markers and pro-inflammatory and anti-inflammatory cytokines by hBM-MSCs treated with 1.6 ng/mL (**a,c**) and 100 ng/mL (**b,d**) of ihGDF-5 in monolayer 2D culture up to 16 days. mRNA levels of COL1A1, DCN, SCX-A, and TNMD were considered as tenogenic markers and COL3A1 selected as negative ones; pro-inflammatory (IL-6, TNF, IL-12A and IL-1β) and anti-inflammatory (IL-10 and TGF-β1) cytokines were monitored. ihGDF-5 not had significant impact on genes expression, especially at the lower concentration tested. Untreated cells for matched time-points were used as control. * ≤ 0.05; ** <0.01; *** <0.005; **** <0.001 N = 3 (biological replicates); n = 3 (technical replicates).

Transient and slight, though significant, upregulation of COL3A1 (0.4-fold), DCN (0.2-fold), and COL1A1 (0.5-fold) was observed at Day 1 in cultures supplemented with 100 ng/mL ihGDF-5 (Figure 1b). The reduced dose of 1.6 ng/mL induced low-level transient expression at Day 1 for only COL3A1 (0.3-fold) and DCN (0.2-fold) only (Figure 1a). No significant upregulation was noted for TNF, IL-12A, IL-1β, IL-10, or TGFβ at any time point or ihGDF-5 concentration tested (Figure 1c,d). Compared to controls, IL-6 displayed significant levels of elevation at Days 8 (0.8-fold) and 16 (0.5-fold) with 100 ng/mL ihGDF-5 (Figure 1d). Types 1 and 3 Collagen expression levels were monitored by immunofluorescence during the 16 day culture period, as illustrated in Figure 2a,b. Immunofluorescence quantitative data by image analysis were congruent with RT-qPCR outputs when 1.6 ng/mL ihGDF-5 was supplemented, in this case, both proteins signals were not significantly elevated, compared to untreated cells (see Figure 2c).

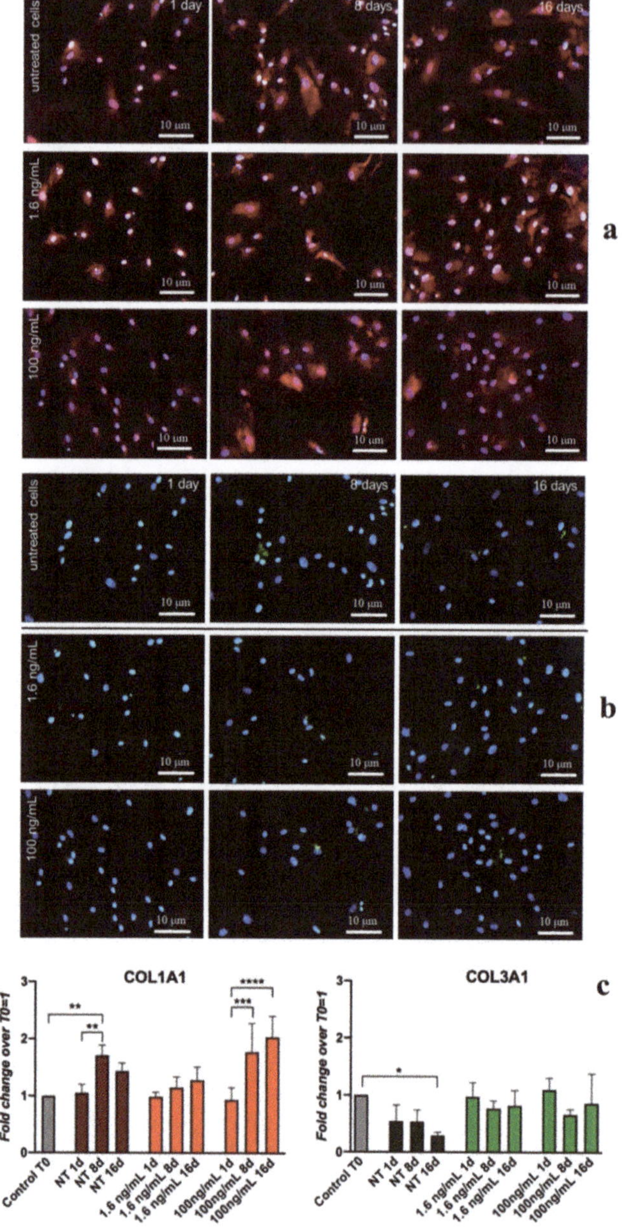

Figure 2. IF and quantitative-IF assays of type 1 Collagen (COL1A1) and type 3 Collagen (COL3A1) monitored along hBM-MSCs treatment with 1.6 ng/ and 100 ng/mL of ihGDF-5 for 16 days. Type 1 collagen was stained in red; type 3 collagen was stained in green; cell nuclei highlighted with DAPI in blue (**a,b**). Quantitative signal detection was performed via ImageJ software (**c**). A slight up-regulation of COL1A1 was observed when hBM-MSC were in routine culture for 16 days. Color intensity in each time point was normalized by the cell number. * ≤0.05; * *<0.01; *** <0.005; **** <0.001. n = 10 (image fields for each time point).

Quantitative image analysis displayed COL1A1 signal increase (1 fold) at Day 16 only with 100 ng/mL ihGDF-5 supplementation. This last data is in contrast with gene expression ones. COL1A1 and COL3A1 are the major components of the extracellular matrix in connective tissues, and their slight up-regulations was reported when hBM-MSC were in routine culture for 16 days [64]. However, in our case, ihGDF-5 seemed not to impact on their production, especially at the lower concentration tested. This preliminary information is important to confirm the inactivity of the biochemical input in regards to the gene expression and proteins that will be monitored in the 3D experiments.

3.2. PLGA-NCs Fabrication and ihGDF-5 Controlled Deliver

Poly-lactic-co-glycolic-acid nano-carriers (PLGA-NCs) displayed a spherical morphology with a mean size of 230 ± 80 nm (Figure 3a). PLGA-NCs had an ihGDF-5 loading of 350 ng/g and provided a daily released peptide mean concentration of 1.6 ng/mL/day (Figure 3b), when an amount of 100 mg were inserted within HY-FIB over 11 days of culture. As highlighted above, these ihGDF-5 concentration levels did not stimulate sustained impacts on hBM-MSCs gene expression in 2D monolayer culture (see Figure 1a,c).

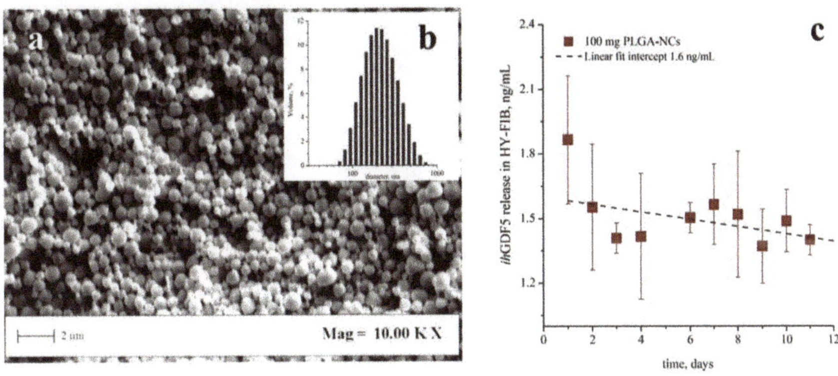

Figure 3. Poly-lactic-co-glycolic-acid (PLGA) transient carriers field emission-scanning electron microscopy (FE-SEM) image, particle size distribution, and ihGDF-5 release profile. FE-SEM images indicated spherical morphology of carriers (**a**); the size distribution set at 230 ± 80 nm their mean size (**b**). Release profiles performed in vitro at 37 °C indicated a ihGDF-5 mean concentration of 1.6 ng/mL/day released from the 100 mg of PLGA-NCs loaded within HY-FIB over 11 days of culture (**c**).

Therefore, by excluding any non-specific ihGDF-5 induction (released within the 3D scaffold by the NCs), we could now observe cell behaviors arising from the HY-FIB microenvironment in both static and dynamic conditions.

3.3. hBM-MSCs Cultivation in HY-FIB 3D Microenvironment

The HY-FIB assembly featured a braided band (3 × 10cm) joined to a fibrin hydrogel (on a band surface of 6 cm^2) containing 8 × 10^5 hBM-MSCs and 100 mg of ihGDF-5/PLGA-NCs. The picture and schematic representation of the 3D system is shown (Figure 4a,b). Field Emission Scanning Electron Microscopy (FE-SEM) images of the scaffold illustrate hyaluronate fibers, embedded within a fibrin hydrogel (Figure 4c), which provided an entrapment surface for both NCs and hBM-MSCs (Figure 4d).

HY-FIB was exposed to 10% deformation over a 1 Hz frequency for 4 h a day during the dynamic culture experiments via a cyclic strain bioreactor, illustrated in Figure 4e [24]. In greater detail, a HY-FIB braided band was held at one end by a motionless arm and at the other end by a sliding one. Motion was driven by a linear motor and transmitted through the braided band to cells embedded within the fibrin hydrogel. The motionless arm comprises a base, attached to the side wall of the culture chamber, housing the electronic components for load monitoring, and from which extended a cantilevered shelf whose deformation is measured by four strain gauges. The whole system was housed within an incubator to ensure the appropriate CO_2 gaseous environment to control the pH of the cell culture media and 37 °C operational temperature.

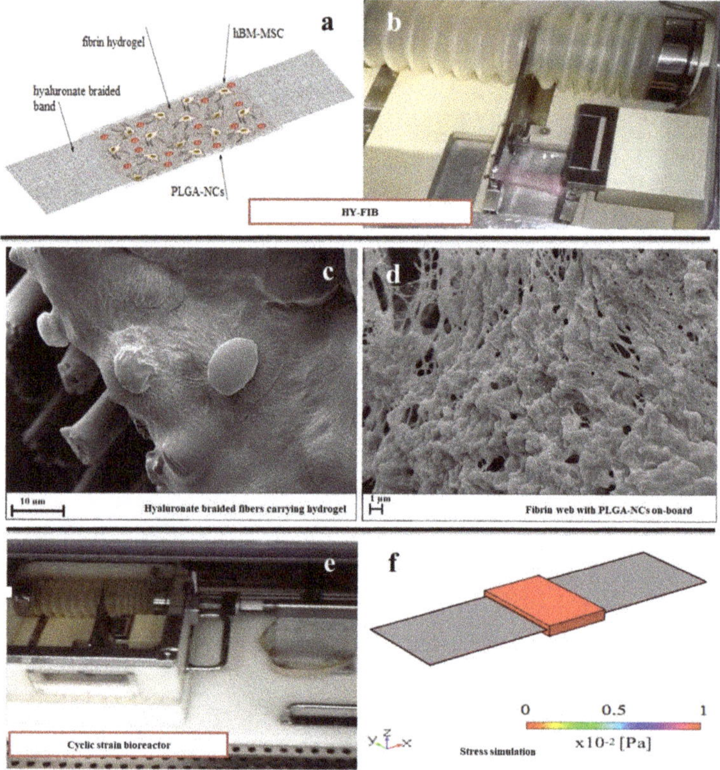

Figure 4. HY-FIB three-dimensional (3D) scaffold features and cyclic strain bioreactor. Schematic HY-FIB representation (**a**) and image of 3D scaffold (**b**). SEM images of hyaluronate braided fibers (10 µm mean diameter size) (**c**) joined to a fibrin web which entrapped both NCs and hBM-MSCs (**d**). Cyclic strain bioreactor (**e**) and in-silico study of stress distribution over HY-FIB upon mechanical strain of 10% (**f**). The simulation involved only the stress of the fibrin 3D environment, neglecting any further contribution.

The stress delivered to the cells immobilized within the system was explored via computational analysis that estimated a mean shear stress value estimated at 9×10^{-2} Pa within the fibrin 3D environment (Figure 4f). This order of magnitude of stress value was reported for tenogenic induction [30]; larger deformation for longer times were excluded to focus the study on 3D environment assembled.

HY-FIB samples were collected at Day 1, 2, 5, and 11 to monitor tenogenic and cytokine marker expression. Time points at Day 1 and Day 2 were added for 3D culture to monitor the effect of HY-FIB on cells behavior alone or in combination with cyclic strain culture. Indeed, in static conditions COL1A1 and DCN both displayed significant upregulation of 3.8 fold (COL1A1) and 2.6-fold (DCN) at Days 1 and 2 before dropping progressively to elevated but non-significant levels (Figure 5a and Figure S1 in Supplementary Materials), confirming an HY-FIB effect on this gene expression in the first days of culture. In dynamic conditions, COL1A1 levels displayed responses similar to the static culture in the first two days but progressively rising thereafter to significant levels (2.9 fold) at Day 11, probably due to strain input. DCN expression levels in response to dynamic culture were to be elevated throughout, achieving significance at Day 11 (3-fold) (Figure 5b and Figure S1 of Supplementary Materials).

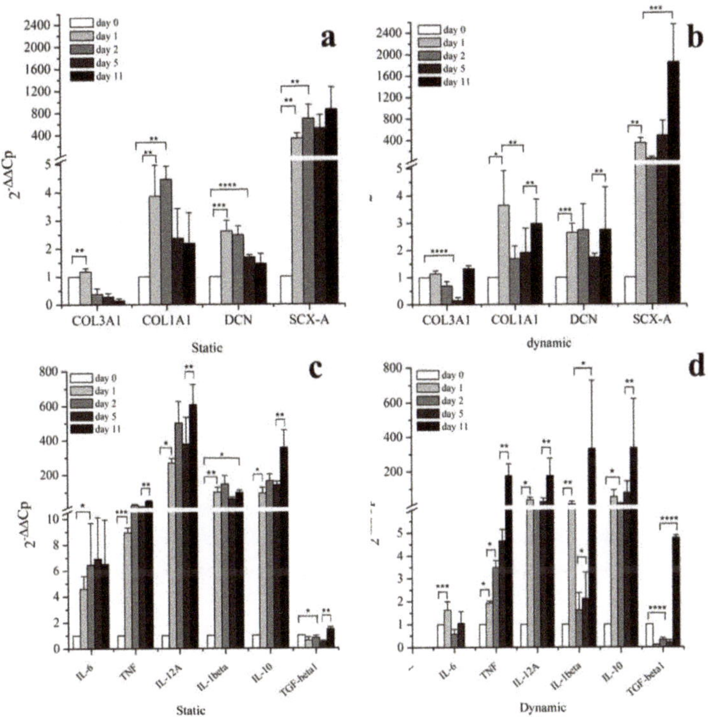

Figure 5. Gene expression profiles of tenogenic markers and pro-inflammatory and anti-inflammatory cytokines from hBM-MSCs within HY-FIB environment in static (**a,c**) and dynamic culture (**b,d**) up to 11 days. Days 1, 2, 5, and 11 were selected as time points to study the mRNA levels of positive tenogenic markers (COL1A1, DCN, SCX-A, and TNMD), negative ones (COL3A1) and pro-inflammatory (IL-6, TNF, IL-12A and IL-1β) and anti-inflammatory (IL-10 and TGF-β1) cytokines. Effect of HY-FIB environment on cells behavior was visible along the first two days of culture; a better over expression of tenogenic markers and anti-inflammatory cytokines was observed in dynamic culture at Day 11. hBM-MSCs within HY-FIB at time zero were used as control. * <0.05; ** <0.01; *** <0.005; **** <0.001. N = 3 (biological replicates); n = 3 (technical replicates).

SCX-A displayed significant upregulation (~340-fold) in both static and dynamic conditions at Day 1, suggesting an effect of HY-FIB system, on this gene expression. SCX-A levels were substantially elevated in both static and dynamic condition at all following time points studied, even if a larger and significant increase was observed in dynamic condition; an increase of 800-fold in static and of 1600-fold in dynamic culture conditions was monitored at Day 11 (Figure 5a,b and Figure S1 of Supplementary Materials). Tenomodulin gene expression was also tested by RT-qPCR, but no expression was detected, probably because it is an event occurring during late differentiation [65]. Sustained, significant, downregulation of COL3A1 was observed in either static or dynamic conditions which instead either decreased progressively (static) or decreased through to Day 5 before reestablishing Day 0 levels at Day 11 (dynamic).

The data suggested an overall effect of the 3D environment on cells behavior clearly visible along the first two days of culture; furthermore, a statistically significant COL1A1, DCN, and SCX-A over-expression was observed after 11 days when mechanical strain was provided (Figure S1 in Supplementary Materials).

Cytokine transcript expression data is illustrated in Figure 5c,d, for static and dynamic culture, respectively. HY-FIB system has an effect also on cytokines gene expression, as observed in all time points monitored with respect to Day 0, within static culture. Indeed, pro-inflammatory cytokines IL-6 (~6-fold), TNF (~10-fold), IL-12A (≤600-fold), and IL-1β (~200-fold) displayed rapid and significant upregulation that was maintained for the entirety of the experimental duration. Anti-inflammatory TGF-β1 on the other hand displayed either no change (Day 11) or significant down-regulation (other time points) while IL-10 exhibited an overall similar profile to Il-1β culminating in marked upregulation at day 11 (~300-fold) (Figure 5c).

Dynamic culture conditions had a distinct and significant effect on IL-6 with expression levels achieving a peak upregulation of 1.5-fold at Day 1 and decreasing to undetectable levels by Day 11 (Figure S2 in Supplementary Materials). TNF and IL-1β were both gradually upregulated before achieving ~200-fold and ~300-fold, respectively, upregulations at day 11 (compared to 10-fold and 100-fold in static conditions). IL-12 displayed a similar profile of upregulation in dynamic vs. static culture conditions while achieving maximal levels that were 3X less in dynamic. Anti-inflammatory IL-10 expression levels were consistent across both dynamic and static conditions. In contrast to static culture, TGF-β1 was significantly downregulated until day 5, and it underwent a 5-fold increase at Day 11 in dynamic culture conditions (Figure 5d and Figure S2 in Supplementary Materials).

Histological characterization of HY-FIB scaffold in both static and dynamic culture at Days 5 and 11 are reported in Figure 6; the overall scaffold structure was stained with Sirius Red for collagen highlighting. Despite fibrin hydrogel matrix, collected at Day 0, was only light pink stained, the same matrix was clearly stained in red at Day 5 and 11 in both samples taken from static and dynamic culture. However, a less homogeneous matrix organization and staining was observed in the samples taken from static culture. This data is in agreement with gene expression indications and confirmed that both HY-FIB alone and HY-FIB plus cyclic strain had an effect on cells phenotype commitment.

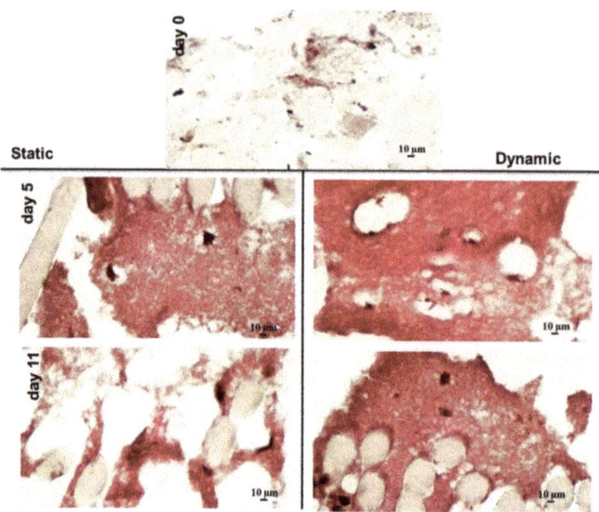

Figure 6. Histology characterization of the overall HY-FIB scaffold structure with Sirius Red staining. HY-FIB scaffolds in both static and dynamic culture at Days 5 and 11 are reported; the overall scaffold

structure was stained with Sirius Red for collagen highlighting. Fibrin hydrogel was light pink stained in the sample collected at Day 0. Fibrin matrix was clearly stained in red at Day 5 and 11 in both samples from static and dynamic culture. Less homogeneous scaffold matrix structure and staining was observed in samples taken from static culture.

The expression of type 1 Collagen, a tenogenic matrix-associated marker, was monitored by immunofluorescence over the culture time (see Figure 7). At day 1, empty areas surrounding the cells are present, probably due to absence of uniform fibrin hydrogel. These spaces were then progressively filled with the protein, presumably via secretion into the extracellular environment. The level of staining observed under static conditions decreased after Day 1 and was maintained at ~50% of original levels thereafter while levels were maintained consistent to Day 0 in dynamic culture. Moreover, in the dynamic condition a more uniform cells distribution was noted throughout the hydrogel matrix, especially at day 11.

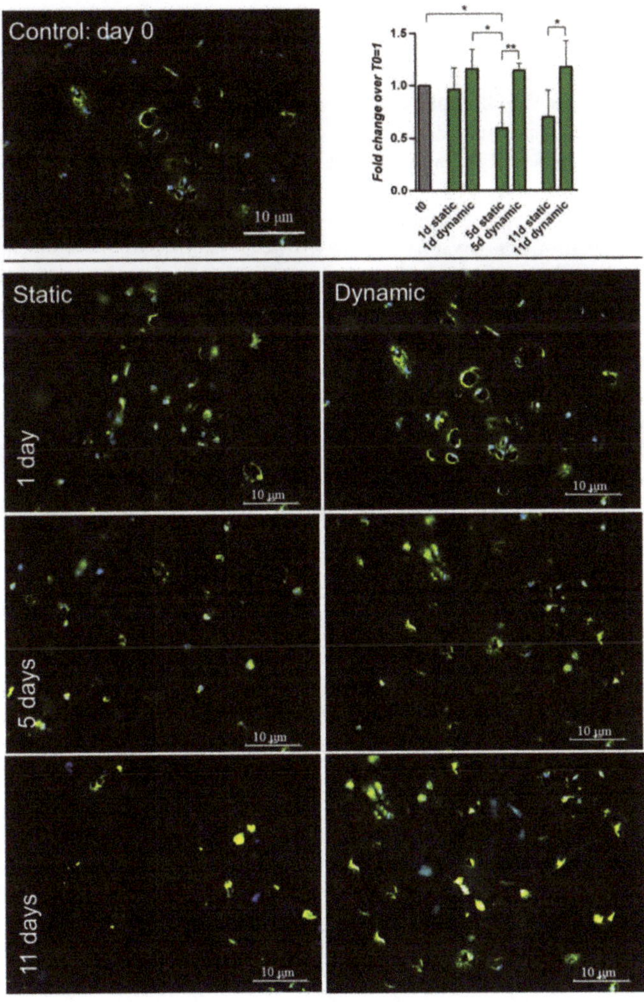

Figure 7. IF and quantitative-IF assays of type 1 Collagen (COL1A1) in 3D static and dynamic2 culture of hBM-MSCs for 11 days. At day 1, holes within the hydrogel structure are evident nearby cells

surroundings. These areas were progressively filled by COL1A1 protein (stained in green) and it happened more uniformly during the dynamic cultivation (see day 11). Fluorescence quantification by ImageJ software, reported in the plot, confirmed the presence of more abundant signal (0.2–0.3 fold changes) in dynamic condition; signal intensity in each time point was normalized by cell number (e.g., by amount of cell nuclei revealed by DAPI staining). * ≤0.05; ** <0.01. N = 3 (biological replicates); n = 3 (technical replicates).

4. Discussion

The HY-FIB system is engineered to support delivery of PLGA nanocarriers (PLGA-NCs) within the hydrogel matrix, enabling controlled delivery of specific molecules within 3D environment, e.g., drugs or other biological signals. The active form of hGDF-5 loaded into PLGA-NCs for controlled delivery within HY-FIB environment was investigated in a previous study [53]. Here, we investigated the effect of the HY-FIB 3D environment (hyaluronate band + PLGA carriers + fibrin gel) on hBM-MSCs tenogenic and cytokine marker gene expression in both static and dynamic, mechanical, input scenarios. We adopted the previous HY-FIB configuration including PLGA-NCs, but on this occasion, we delivered an inactive form of hGDF5. In this manner, the biochemical input provided by the growth factor was excluded, but the complete HY-FIB configuration was maintained, and we were thus able to investigate the impact of mechanical input alone.

HY-FIB braided fibers enabled a defined mechanical stimulation of 9×10^{-2} Pa provided to hBM-MSCs during the 4h/day dynamic culture regime. The mean shear stress was calculated by FEM modeling [62], assuming a system homogeneous behavior at a density of 1050 kg/m^3 and Young's modulus of 4.56 Mpa [53]. A Poisson ratio value of 0.25 was adopted as described elsewhere [62]. Further mechanical inputs with different intensities and durations were not investigated, not being the aim of the present work. Stress values resembling reduced physiological activity, similar that the ones used here, have been reported to direct stem cell commitment to a tenogenic phenotype [31,36,66].

COL1A1 is the major component of tendon tissue (75–85% of the dry mass of tendon), and is responsible for its mechanical strength [64]. In the static group, COL1A1 showed a ≥3-fold upregulation during the first and second day of cultivation. These data seem to suggest an overall effect of the 3D environment on cells behavior. COL1A1 expression was progressively reduced to a 2-fold upregulation at Day 11, in static environment. In dynamic conditions, its mRNA levels showed similar behavior during the first two days of cultivation (an increase up to 3-fold-changes, then reduced at Day 2). However, its expression was subsequently increased again to 2.8-fold at Day 11. Decorin (DCN), a small leucine-rich proteoglycan implicated in the regulation of fibrillogenesis, is a fundamental component of the tendon extracellular matrix (ECM) [67]. Compared to the static condition, a significant enhancement, up to 2.5-fold, of the mRNA level of DCN was shown when hBM-MSCs were cultured for 11 days with mechanical stimulation.

Scleraxis-A (SCX-A) is a neotendon marker, expressed in pro-tendon sites in the developing embryo. Specifically, SCX-A is a tendon-specific basic helix-loop-helix transcription factor responsible for the transition of MSCs into tendon progenitors [68]. We observed substantial increases in SCX-A expression, up to 800-fold in static and 1600-fold in dynamic conditions after 11 days, demonstrating a stimulatory effect via the 3D system organization and consistent with previous observations [69–71].

COL3A1 mRNA level seems to be downregulated after 2 days in the static group and after 5 days in the dynamic group. Its downregulation can be considered a positive indication of proper cell differentiation; indeed, it seems that COL3A1 is the main responsible of fibrotic and scarred tissue arrangement and has been consistently reported at the rupture site of human tendons. [64].

From both histology and immunofluorescence assays, we noted that the area surrounding the cells was progressively filled by type 1 Collagen and, at Day 11, the extracellular matrix seemed to undergo remodeling (Figures 6 and 7). Moreover, in dynamic conditions a more homogeneous cell distribution within the hydrogel matrix was observed in the IF images. These findings support the concept that 3D cultivation provides cues to the hBM-MSCs, and that dynamic signaling enables the adoption of a more uniform behavior including type 1 Collagen protein deposition in the externally available space of the fibrin hydrogel. A near total absence of type 3 Collagen was found, except for a very small fluorescence signal at day 1, in both static and dynamic conditions (data not shown). These data suggest that tenogenic commitment of hBM-MSCs cultured within HY-FIB environment may be enhanced when dynamic stretching is applied.

MSCs secrete a variety of cytokines and growth factors that promote cell recruitment, migration, proliferation, and differentiation. MSCs are also immunomodulatory, which may allow them to exert beneficial effects on the local immune cell population at the site of muscle injury [72]. To better understand the hBM-MSCs inflammatory response when cultured within HY-FIB, cytokine expression was monitored along the 11 days of culture. The balance between pro- and anti-inflammatory soluble factors in the tendon healing process exerts a major impact on successful resolution of inflammation [73]. Recent analysis of tendinopathy biopsies showed a distinct inflammatory infiltrate in the initial phase of tendinopathy with a high content of pro-inflammatory factors such as IL-6, TNF-α and IL-17 [74].

To exclude a role for ihGDF5 in cytokine expression induction we evaluated their expression in hBM-MSCs undergoing 2D planar cultivation as a negative control. Indeed, in disc degeneration models using in vitro three-dimensional cultures, human annulus cells display increased expression of pro-inflammatory cytokines, such as IL-1β and TNF-α, while exposure to TNF-α and IL-1β resulted in significant downregulation of GDF-5 [75]. Therefore, it is plausible that the GDF-5 may upregulate the expression of pro-inflammatory genes in hBM-MSCs leading to the maintenance of an autocrine feedback. However, when the ihGDF-5 was added, no statistically significant expression of pro-inflammatory cytokines was observed; therefore, ihGDF-5 did not exert any effect on cytokines expression.

The addition of PLGA-NCs enabled an informed analysis of the inductive role of the HY-FIB overall structure. Previous studies have noted that cytotoxicity of SEE-fabricated PLGA-NCs on blood mononucleate viability, monitored with MTT assay [76], was not affected after either 24 or 48 h. Here, the overall HY-FIB system (loaded with SEE fabricated NCs) does not evidence any toxic effect on hBM-MSC cultivated within it for 11 days, providing an indirect indication about SEE technology as suitable process for biomedical carriers production.

In general, we observed that pro-inflammatory gene expression was higher in static than in dynamic conditions at all-time points. On the contrary, the anti-inflammatory cytokines IL-10 was consistently upregulated in both static and dynamic conditions; while TGF-β1 was downregulated at all the time points tested except day 11, when, it showed a marked increase (4-fold) only in dynamic environment. The described behavior confirmed that MSCs respond to a variety of biophysical cues; indeed, as suggested by Qazi et al., 3D culture of MSCs on biomaterials can promote cell-cell interactions and enhance the paracrine effects of MSCs [77]. Moreover, as concluded by Ogle et al., historically, biomaterial-based therapies to promote tissue regeneration were designed to minimize the host inflammatory response. Recently, the roles that monocytes and macrophages can play in tissue repair have been highlighted. In this context, material properties and their possibility of specific biomolecule controlled delivery has been engineered to achieve a given biological response that can be tuned not only to a better integration with biological systems but also in regulating the inflammatory response [5].

The overall and statistically significant balance of pro- vs. anti-inflammatory cytokines expressed by cells provided indications regarding the importance of dynamic culture for 3D in vitro model systems. For instance, IL-6, a well-known pleiotropic cytokine delivered by tissues in response to physio-pathological changes such as physical exercise, infection, and injury, was reported to deeply alter skeletal muscle milieu, by affecting the activity and quality of cellular interactors during tissue regeneration and leading to the fibrotic response [78]. In our 3D model system, IL-6 gene expression was considerably reduced in hBM-MSCs that underwent dynamic 3D HY-FIB cultivation when compared to the same cells cultivated in static condition.

It is worth of note that there is no specific literature on cytokines response by hBM-MSCs cultivated within 3D scaffold. Almost all published studies described cell-specific differentiation toward a given phenotype, without considering how cytokines expression may be related to a 3D in vitro scaffold system. In this sense, the present investigation is the first study, which suggests cytokines expression as a further variable to monitor cell behavior and reaction when loaded into a 3D in-vitro model. Moreover, improved balance in anti-inflammatory cytokines observed for HY-FIB plus cyclic strain may be considered an indication of better cells response to the 3D in vitro system designed and proposed.

5. Conclusions

The 3D cell culture yielded evidence of type 1 Collagen expression observed by both immunohistology and gene assay. When the same 3D system was cultivated under cyclic strain stimulation, the mechanical input stimulated a statistically significant increase in tenogenic markers expression when compared to the same cells assembled into the 3D system, but cultivated in a static culture. Further studies may involve a deeper understanding of the relation between collagen type I production, cell commitment and mechanical input strain percentage or duration; in this sense, HY-FIB system can be considered a good instrument for this study. The 3D culture system activated also the expression of pro-inflammatory cytokines, and, when cyclic strain was applied, pro-inflammatory cytokine gene over-expression by hBM-MSCs was better balanced against over-expression of anti-inflammatory cytokines. It remains to be determined what the involvement and the immunomodulatory activity of hBM-MSCs are, and the role of implantable biomaterials in the stimulation of inflammatory reactions. For instance, the stimulation of local inflammation is reported as an important event in triggering repair in avascular tissues, such as cartilage and tendons [5].

On the other hand, the presence of PLGA-NCs within the fibrin hydrogel would allow the delivery of specific biomolecules that may be studied for the ability to further modulate inflammation reactions or promote regeneration/repair events. In this sense, HY-FIB provides a potential strategic approach to address a range of issues via the provision of a tightly controlled in vitro protocol. The 3D scaffold is a potential system to organize the sustained release of different biochemical signals and opens concrete perspectives for developing 3D bioengineered models to understand specific molecular and cellular composition of damaged systems.

Supplementary Materials: The following materials are available online at http://www.mdpi.com/2073-4409/9/5/1268/s1, Figure S1: Profiles of positive (SCX-A, COL1A1, DCN) and negative (COL3A1) tenogenic markers between HY-FIB static vs. dynamic culture. * < 0.05; ** < 0.01; *** < 0.005; **** < 0.001, Figure S2: Profiles of pro-inflammatory (IL-6, TNF, IL-12A, IL-1β) and anti-inflammatory (IL-10, TGFβ1) cytokines markers between HY-FIB static vs. dynamic culture. * < 0.05; ** < 0.01; *** < 0.005; **** < 0.001.

Author Contributions: M.C.C. developed the experimental activity and optimized the protocols and methodology; she was also responsible for the paper draft preparation; L.M. isolated the stem cells and characterized them with formal analysis and validated methodology; J.L. followed the cyclic strain bioreactor protocols and provided the FEM data; E.G. supervised the bioreactor protocols and data acquisitions; N.R.F. provided contribution in data supervision and paper writing; C.S. provided the bone marrow aspirate and the methodology for hBM-MSCs cultivation; N.M. helped in the interpretation of the data, reviewed the manuscript, and was responsible for funding acquisition and management; G.D.P. was responsible for all experimental data production, curation and supervision, paper writing and editing, and research project administration. All authors have read and agreed to the published version of the manuscript.

Funding: This research was funded by: (i) University of Salerno: FARB-Della Porta Year: 2019-22; (ii) American Orthopedic Foot & Ankle Society (AOFAS) Research Committee, Grant ID#: 2019-133-S, Grant Project Title: Nano-FT3C+: An Innovative Liposome-Based Formulation for Thyroid Hormone Controlled Delivery. An In Vitro Study on Tendinopathic Human Achilles Tendon Tenocytes.

Acknowledgments: The authors acknowledge MiUR within the framework of PON-RI 2014/2020. Action I.1–"Innovative PhDs with industrial characterization" Cicle XXXIII (D.D. n 0001377 June 5th, 2017) for the PhD fellowship entitled: "Scaffold innovation for the cure of tendon disorders: new generation of poly-hyaluronate functionalized biocomposites. The authors acknowledge Devis Galesso and R&D team at Fidia Farmaceutici S.p.A (Abano Terme, PD, Italy) for their participation into the PhD training program. We also thanks Annamaria Giordano for help in cells culture during her laboratory internship.

Conflicts of Interest: The authors declare no conflict of interest. The funders had no role in the design of the study; in the collection, analyses, or interpretation of data; in the writing of the manuscript, or in the decision to publish the results.

Patents: The SEE technology for nanocarriers fabrication was described in the US Patent US/8628802 B2 Jan 2014. Inventors: Reverchon E., Della Porta G., Continuous process for microspheres production by using expanded fluids. Applicant: University of Salerno.

References

1. Ahmed, S.; Chauhan, V.M.; Ghaemmaghami, A.M.; Aylott, J.W. New generation of bioreactors that advance extracellular matrix modelling and tissue engineering. *Biotechnol. Lett.* **2019**, *41*, 1–25. [CrossRef] [PubMed]

2. Okamoto, M. The role of Scaffolds in Tissue Engineering. In *Handbook of Tissue Engineering Scaffolds: Volume One*; Elsevier: Amsterdam, The Netherlands, 2019; pp. 23–49. ISBN 978-0-08-102563-5.

3. D'Angelo, M.; Benedetti, E.; Tupone, M.G.; Catanesi, M.; Castelli, V.; Antonosante, A.; Cimini, A. The Role of Stiffness in Cell Reprogramming: A Potential Role for Biomaterials in Inducing Tissue Regeneration. *Cells* **2019**, *8*, 1036. [CrossRef] [PubMed]

4. Cipollaro, L.; Ciardulli, M.C.; Della Porta, G.; Peretti, G.M.; Maffulli, N. Biomechanical issues of tissue-engineered constructs for articular cartilage regeneration: In vitro and in vivo approaches. *Br. Med. Bull.* **2019**, *132*, 53–80. [CrossRef] [PubMed]

5. Ogle, M.E.; Segar, C.E.; Sridhar, S.; Botchwey, E.A. Monocytes and macrophages in tissue repair: Implications for immunoregenerative biomaterial design. *Exp. Biol. Med.* **2016**, *241*, 1084–1097. [CrossRef] [PubMed]

6. Vinhas, A.; Rodrigues, M.T.; Gomes, M.E. Exploring Stem Cells and Inflammation in Tendon Repair and Regeneration. In *Cell Biology and Translational Medicine, Volume 2*; Turksen, K., Ed.; Springer International Publishing: Cham, Germany, 2018; Volume 1089, pp. 37–46. ISBN 978-3-030-04169-4.

7. Rees, J.D.; Stride, M.; Scott, A. Tendons—Time to revisit inflammation. *Br. J. Sports Med.* **2014**, *48*, 1553–1557. [CrossRef] [PubMed]

8. Dakin, S.G.; Martinez, F.O.; Yapp, C.; Wells, G.; Oppermann, U.; Dean, B.J.F.; Smith, R.D.J.; Wheway, K.; Watkins, B.; Roche, L.; et al. Inflammation activation and resolution in human tendon disease. *Sci. Transl. Med.* **2015**, *7*, 311ra173. [CrossRef] [PubMed]

9. John, T.; Lodka, D.; Kohl, B.; Ertel, W.; Jammrath, J.; Conrad, C.; Stoll, C.; Busch, C.; Schulze-Tanzil, G. Effect of pro-inflammatory and immunoregulatory cytokines on human tenocytes. *J. Orthop. Res.* **2010**, *28*, 1071–1077. [CrossRef]

10. Manning, C.N.; Havlioglu, N.; Knutsen, E.; Sakiyama-Elbert, S.E.; Silva, M.J.; Thomopoulos, S.; Gelberman, R.H. The early inflammatory response after flexor tendon healing: A gene expression and histological analysis: Early inflammatory response after flexor tendon healing. *J. Orthop. Res.* **2014**, *32*, 645–652. [CrossRef]

11. Marsolais, D.; Coté, C.H.; Frenette, J. Neutrophils and macrophages accumulate sequentially following Achilles tendon injury. *J. Orthop. Res.* **2001**, *19*, 1203–1209. [CrossRef]

12. Chisari, E.; Rehak, L.; Khan, W.S.; Maffulli, N. The role of the immune system in tendon healing: A systematic review. *Br. Med. Bull.* **2020**, *133*, 49–64. [CrossRef]

13. Giordano, L.; Della Porta, G.; Peretti, G.M.; Maffulli, N. Therapeutic potential of microRNA in tendon injuries. *Br. Med. Bull.* **2020**, *133*, 79–94. [CrossRef] [PubMed]

14. Oryan, A.; Moshiri, A.; Meimandi Parizi, A.; Maffulli, N. Implantation of a Novel Biologic and Hybridized Tissue Engineered Bioimplant in Large Tendon Defect: An In Vivo Investigation. *Tissue Eng. Part A* **2013**, *20*, 447–465. [CrossRef] [PubMed]

15. Gaspar, D.; Spanoudes, K.; Holladay, C.; Pandit, A.; Zeugolis, D. Progress in cell-based therapies for tendon repair. *Adv. Drug Deliv. Rev.* **2015**, *84*, 240–256. [CrossRef] [PubMed]

16. Lui, P.P.Y.; Wong, O.T. Tendon stem cells: Experimental and clinical perspectives in tendon and tendon-bone junction repair. *Muscles Ligaments Tendons J.* **2012**, *2*, 163–168.

17. Chang, C.-H.; Chen, C.-H.; Liu, H.-W.; Whu, S.-W.; Chen, S.-H.; Tsai, C.-L.; Hsiue, G.-H. Bioengineered Periosteal Progenitor Cell Sheets to Enhance Tendon-Bone Healing in A Bone Tunnel. *Biomedical* **2012**, *35*, 473. [CrossRef]

18. Ito, R.; Matsumiya, T.; Kon, T.; Narita, N.; Kubota, K.; Sakaki, H.; Ozaki, T.; Imaizumi, T.; Kobayashi, W.; Kimura, H. Periosteum-derived cells respond to mechanical stretch and activate Wnt andBMP signaling pathways. *Biomed. Res.* **2014**, *35*, 69–79. [CrossRef]

19. Doroski, D.M.; Levenston, M.E.; Temenoff, J.S. Cyclic Tensile Culture Promotes Fibroblastic Differentiation of Marrow Stromal Cells Encapsulated in Poly(Ethylene Glycol)-Based Hydrogels. *Tissue Eng. Part A* **2010**, *16*, 3457–3466. [CrossRef]

20. Chaudhury, S. Mesenchymal stem cell applications to tendon healing. *Muscles Ligaments Tendons J.* **2012**, *2*, 222–229.

21. Youngstrom, D.W.; LaDow, J.E.; Barrett, J.G. Tenogenesis of bone marrow-, adipose-, and tendon-derived stem cells in a dynamic bioreactor. *Connect. Tissue Res.* **2016**, *57*, 454–465. [CrossRef]

22. Bi, Y.; Ehirchiou, D.; Kilts, T.M.; Inkson, C.A.; Embree, M.C.; Sonoyama, W.; Li, L.; Leet, A.I.; Seo, B.-M.; Zhang, L.; et al. Identification of tendon stem/progenitor cells and the role of the extracellular matrix in their niche. *Nat. Med.* **2007**, *13*, 1219–1227. [CrossRef]

23. Yang, G.; Rothrauff, B.B.; Lin, H.; Gottardi, R.; Alexander, P.G.; Tuan, R.S. Enhancement of tenogenic differentiation of human adipose stem cells by tendon-derived extracellular matrix. *Biomaterials* **2013**, *34*, 9295–9306. [CrossRef] [PubMed]

24. Govoni, M.; Lotti, F.; Biagiotti, L.; Lannocca, M.; Pasquinelli, G.; Valente, S.; Muscari, C.; Bonafè, F.; Caldarera, C.M.; Guarnieri, C.; et al. An innovative stand-alone bioreactor for the highly reproducible transfer of cyclic mechanical stretch to stem cells cultured in a 3D scaffold: A novel stand-alone bioreactor to induce cell muscle phenotype. *J. Tissue Eng. Regen. Med.* **2014**, *8*, 787–793. [CrossRef] [PubMed]

25. Lovecchio, J.; Gargiulo, P.; Vargas Luna, J.L.; Giordano, E.; Sigurjonsson, O.E. A standalone bioreactor system to deliver compressive load under perfusion flow to hBMSC-seeded 3D chitosan-graphene templates. *Sci. Rep.* **2019**, *9*, 16854. [CrossRef] [PubMed]

26. Barber, J.G.; Handorf, A.M.; Allee, T.J.; Li, W.-J. Braided Nanofibrous Scaffold for Tendon and Ligament Tissue Engineering. *Tissue Eng. Part A* **2013**, *19*, 1265–1274. [CrossRef] [PubMed]

27. Wang, T.; Lin, Z.; Day, R.E.; Gardiner, B.; Landao-Bassonga, E.; Rubenson, J.; Kirk, T.B.; Smith, D.W.; Lloyd, D.G.; Hardisty, G.; et al. Programmable mechanical stimulation influences tendon homeostasis in a bioreactor system. *Biotechnol. Bioeng.* **2013**, *110*, 1495–1507. [CrossRef]

28. Delaine-Smith, R.M.; Reilly, G.C. Mesenchymal stem cell responses to mechanical stimuli. *Muscles Ligaments Tendons J.* **2016**, *2*, 169–180.

29. Chen, Y.-J.; Huang, C.-H.; Lee, I.-C.; Lee, Y.-T.; Chen, M.-H.; Young, T.-H. Effects of Cyclic Mechanical Stretching on the mRNA Expression of Tendon/Ligament-Related and Osteoblast-Specific Genes in Human Mesenchymal Stem Cells. *Connect. Tissue Res.* **2008**, *49*, 7–14. [CrossRef]

30. Govoni, M.; Muscari, C.; Lovecchio, J.; Guarnieri, C.; Giordano, E. Mechanical Actuation Systems for the Phenotype Commitment of Stem Cell-Based Tendon and Ligament Tissue Substitutes. *Stem Cell Rev. Rep.* **2016**, *12*, 189–201. [CrossRef]

31. Youngstrom, D.W.; Rajpar, I.; Kaplan, D.L.; Barrett, J.G. A bioreactor system for in vitro tendon differentiation and tendon tissue engineering. *J. Orthop. Res.* **2015**, *33*, 911–918. [CrossRef]

32. Testa, S.; Costantini, M.; Fornetti, E.; Bernardini, S.; Trombetta, M.; Seliktar, D.; Cannata, S.; Rainer, A.; Gargioli, C. Combination of biochemical and mechanical cues for tendon tissue engineering. *J. Cell. Mol. Med.* **2017**, *21*, 2711–2719. [CrossRef]

33. Woon, C.Y.L.; Kraus, A.; Raghavan, S.S.; Pridgen, B.C.; Megerle, K.; Pham, H.; Chang, J. Three-Dimensional-Construct Bioreactor Conditioning in Human Tendon Tissue Engineering. *Tissue Eng. Part A* **2011**, *17*, 2561–2572. [CrossRef] [PubMed]

34. Rinoldi, C.; Fallahi, A.; Yazdi, I.K.; Campos Paras, J.; Kijeńska-Gawrońska, E.; Trujillo-de Santiago, G.; Tuoheti, A.; Demarchi, D.; Annabi, N.; Khademhosseini, A.; et al. Mechanical and Biochemical Stimulation of 3D Multilayered Scaffolds for Tendon Tissue Engineering. *ACS Biomater. Sci. Eng.* **2019**, *5*, 2953–2964. [CrossRef]

35. Nam, H.Y.; Pingguan-Murphy, B.; Abbas, A.A.; Merican, A.M.; Kamarul, T. Uniaxial Cyclic Tensile Stretching at 8% Strain Exclusively Promotes Tenogenic Differentiation of Human Bone Marrow-Derived Mesenchymal Stromal Cells. *Stem Cells Int.* **2019**, *2019*, 1–16. [CrossRef] [PubMed]

36. Grier, W.G.; Moy, A.S.; Harley, B.A. Cyclic tensile strain enhances human mesenchymal stem cell Smad 2/3 activation and tenogenic differentiation in anisotropic collagen-glycosaminoglycan scaffolds. *Eur. Cell Mater.* **2017**, *33*, 227–239. [CrossRef] [PubMed]

37. Kay, A.G.; Dale, T.P.; Akram, K.M.; Mohan, P.; Hampson, K.; Maffulli, N.; Spiteri, M.A.; El Haj, A.J.; Forsyth, N.R. BMP2 repression and optimized culture conditions promote human bone marrow-derived mesenchymal stem cell isolation. *Regen. Med.* **2015**, *10*, 109–125. [CrossRef]

38. Correia, S.I.; Pereira, H.; Silva-Correia, J.; Van Dijk, C.N.; Espregueira-Mendes, J.; Oliveira, J.M.; Reis, R.L. Current concepts: Tissue engineering and regenerative medicine applications in the ankle joint. *J. R. Soc. Interface* **2013**, *11*, 20130784. [CrossRef]

39. Keller, T.C.; Hogan, M.V.; Kesturu, G.; James, R.; Balian, G.; Chhabra, A.B. Growth/differentiation factor-5 modulates the synthesis and expression of extracellular matrix and cell-adhesion-related molecules of rat Achilles tendon fibroblasts. *Connect. Tissue Res.* **2011**, *52*, 353–364. [CrossRef]

40. Hogan, M.; Girish, K.; James, R.; Balian, G.; Hurwitz, S.; Chhabra, A.B. Growth differentiation factor-5 regulation of extracellular matrix gene expression in murine tendon fibroblasts. *J. Tissue Eng. Regen. Med.* **2011**, *5*, 191–200. [CrossRef]

41. Tan, S.-L.; Ahmad, R.E.; Ahmad, T.S.; Merican, A.M.; Abbas, A.A.; Ng, W.M.; Kamarul, T. Effect of Growth Differentiation Factor 5 on the Proliferation and Tenogenic Differentiation Potential of Human Mesenchymal Stem Cells in vitro. *Cells Tissues Organs* **2012**, *196*, 325–338. [CrossRef]

42. Lui, P.P. Stem cell technology for tendon regeneration: Current status, challenges, and future research directions. *SCCAA* **2015**, *8*, 163. [CrossRef]

43. MacLean, S.; Khan, W.S.; Malik, A.A.; Snow, M.; Anand, S. Tendon Regeneration and Repair with Stem Cells. *Stem Cells Int.* **2012**, *2012*, 1–6. [CrossRef] [PubMed]

44. Liu, L.; Hindieh, J.; Leong, D.J.; Sun, H.B. Advances of stem cell based-therapeutic approaches for tendon repair. *J. Orthop. Transl.* **2017**, *9*, 69–75. [CrossRef] [PubMed]

45. Caplan, A.I. Adult mesenchymal stem cells for tissue engineering versus regenerative medicine. *J. Cell. Physiol.* **2007**, *213*, 341–347. [CrossRef] [PubMed]

46. Akram, K.M.; Samad, S.; Spiteri, M.A.; Forsyth, N.R. Mesenchymal stem cells promote alveolar epithelial cell wound repair in vitro through distinct migratory and paracrine mechanisms. *Respir. Res.* **2013**, *14*, 9. [CrossRef]

47. Pontikoglou, C.; Deschaseaux, F.; Sensebé, L.; Papadaki, H.A. Bone Marrow Mesenchymal Stem Cells: Biological Properties and Their Role in Hematopoiesis and Hematopoietic Stem Cell Transplantation. *Stem Cell Rev. Rep.* **2011**, *7*, 569–589. [CrossRef]

48. Dominici, M.; Le Blanc, K.; Mueller, I.; Slaper-Cortenbach, I.; Marini, F.C.; Krause, D.S.; Deans, R.J.; Keating, A.; Prockop, D.J.; Horwitz, E.M. Minimal criteria for defining multipotent mesenchymal stromal cells. *Cytotherapy* **2006**, *8*, 315–317. [CrossRef]

49. Chamberlain, G.; Fox, J.; Ashton, B.; Middleton, J. Concise Review: Mesenchymal Stem Cells: Their Phenotype, Differentiation Capacity, Immunological Features, and Potential for Homing. *Stem Cells* **2007**, *25*, 2739–2749. [CrossRef]

50. Mosna, F.; Sensebé, L.; Krampera, M. Human Bone Marrow and Adipose Tissue Mesenchymal Stem Cells: A User's Guide. *Stem Cells Dev.* **2010**, *19*, 1449–1470. [CrossRef]

51. Kumar, D.; Gerges, I.; Tamplenizza, M.; Lenardi, C.; Forsyth, N.R.; Liu, Y. Three-dimensional hypoxic culture of human mesenchymal stem cells encapsulated in a photocurable, biodegradable polymer hydrogel: A potential injectable cellular product for nucleus pulposus regeneration. *Acta Biomater.* **2014**, *10*, 3463–3474. [CrossRef]

52. Bullough, R.; Finnigan, T.; Kay, A.; Maffulli, N.; Forsyth, N.R. Tendon repair through stem cell intervention: Cellular and molecular approaches. *Disabil. Rehabil.* **2008**, *30*, 1746–1751. [CrossRef]

53. Govoni, M.; Berardi, A.C.; Muscari, C.; Campardelli, R.; Bonafè, F.; Guarnieri, C.; Reverchon, E.; Giordano, E.; Maffulli, N.; Della Porta, G. An Engineered Multiphase Three-Dimensional Microenvironment to Ensure the Controlled Delivery of Cyclic Strain and Human Growth Differentiation Factor 5 for the Tenogenic Commitment of Human Bone Marrow Mesenchymal Stem Cells. *Tissue Eng. Part A* **2017**, *23*, 811–822. [CrossRef] [PubMed]

54. Della Porta, G.; Ciardulli, M.C.; Maffulli, N. Microcapsule Technology for Controlled Growth Factor Release in Musculoskeletal Tissue Engineering. *Sports Med. Arthrosc. Rev.* **2018**, *26*, e2–e9. [CrossRef] [PubMed]

55. Giordano, R.; Canesi, M.; Isalberti, M.; Isaias, I.; Montemurro, T.; Viganò, M.; Montelatici, E.; Boldrin, V.; Benti, R.; Cortelezzi, A.; et al. Autologous mesenchymal stem cell therapy for progressive supranuclear palsy: Translation into a phase I controlled, randomized clinical study. *J. Transl. Med.* **2014**, *12*, 14. [CrossRef] [PubMed]

56. Jensen, E.C. Quantitative Analysis of Histological Staining and Fluorescence Using ImageJ: Histological Staining/Fluorescence Using ImageJ. *Anat. Rec.* **2013**, *296*, 378–381. [CrossRef]

57. Rinoldi, C.; Costantini, M.; Kijeńska-Gawrońska, E.; Testa, S.; Fornetti, E.; Heljak, M.; Ćwiklińska, M.; Buda, R.; Baldi, J.; Cannata, S.; et al. Tendon Tissue Engineering: Effects of Mechanical and Biochemical Stimulation on Stem Cell Alignment on Cell-Laden Hydrogel Yarns. *Adv. Healthc. Mater.* **2019**, *8*, 1801218. [CrossRef]

58. Bustin, S.A.; Benes, V.; Garson, J.A.; Hellemans, J.; Huggett, J.; Kubista, M.; Mueller, R.; Nolan, T.; Pfaffl, M.W.; Shipley, G.L.; et al. The MIQE Guidelines: Minimum Information for Publication of Quantitative Real-Time PCR Experiments. *Clin. Chem.* **2009**, *55*, 611–622. [CrossRef]

59. Hellemans, J.; Mortier, G.; De Paepe, A.; Speleman, F.; Vandesompele, J. qBase relative quantification framework and software for management and automated analysis of real-time quantitative PCR data. *Genome Biol.* **2007**, *8*, R19. [CrossRef]

60. Della Porta, G.; Falco, N.; Reverchon, E. Continuous supercritical emulsions extraction: A new technology for biopolymer microparticles production. *Biotechnol. Bioeng.* **2011**, *108*, 676–686. [CrossRef]

61. Della Porta, G.; Falco, N.; Giordano, E.; Reverchon, E. PLGA microspheres by Supercritical Emulsion Extraction: A study on insulin release in myoblast culture. *J. Biomater. Sci. Polym. Ed.* **2013**, *24*, 1831–1847. [CrossRef]

62. Duong, H.; Wu, B.; Tawil, B. Modulation of 3D Fibrin Matrix Stiffness by Intrinsic Fibrinogen–Thrombin Compositions and by Extrinsic Cellular Activity. *Tissue Eng. Part A* **2009**, *15*, 1865–1876. [CrossRef]

63. De Winter, J.C.F. Using the Student's t-test with extremely small sample sizes. *Pract. Assess Res. Eval.* **2013**, *18*, 10.

64. Pajala, A.; Melkko, J.; Leppilahti, J.; Ohtonen, P.; Soini, Y.; Risteli, J. Tenascin-C and type I and III collagen expression in total Achilles tendon rupture. An immunohistochemical study. *Histol. Histopathol.* **2009**, *24*, 1207–1211. [CrossRef] [PubMed]

65. Shukunami, C.; Yoshimoto, Y.; Takimoto, A.; Yamashita, H.; Hiraki, Y. Molecular characterization and function of tenomodulin, a marker of tendons and ligaments that integrate musculoskeletal components. *Jpn. Dent. Sci. Rev.* **2016**, *52*, 84–92. [CrossRef] [PubMed]

66. Wang, T.; Thien, C.; Wang, C.; Ni, M.; Gao, J.; Wang, A.; Jiang, Q.; Tuan, R.S.; Zheng, Q.; Zheng, M.H. 3D uniaxial mechanical stimulation induces tenogenic differentiation of tendon-derived stem cells through a PI3K/AKT signaling pathway. *FASEB J.* **2018**, *32*, 4804–4814. [CrossRef]

67. Wagenhäuser, M.U.; Pietschmann, M.F.; Sievers, B.; Docheva, D.; Schieker, M.; Jansson, V.; Müller, P.E. Collagen type I and decorin expression in tenocytes depend on the cell isolation method. *Bmc Musculoskelet. Disord.* **2012**, *13*, 140. [CrossRef]

68. Alberton, P.; Popov, C.; Prägert, M.; Kohler, J.; Shukunami, C.; Schieker, M.; Docheva, D. Conversion of Human Bone Marrow-Derived Mesenchymal Stem Cells into Tendon Progenitor Cells by Ectopic Expression of Scleraxis. *Stem Cells Dev.* **2012**, *21*, 846–858. [CrossRef]

69. Bagchi, R.A.; Czubryt, M.P. Scleraxis: A New Regulator of Extracellular Matrix Formation. In *Genes and Cardiovascular Function*; Ostadal, B., Nagano, M., Dhalla, N.S., Eds.; Springer: Boston, MA, USA, 2011; pp. 57–65. ISBN 978-1-4419-7206-4.

70. Chen, X.; Yin, Z.; Chen, J.; Shen, W.; Liu, H.; Tang, Q.; Fang, Z.; Lu, L.; Ji, J.; Ouyang, H. Force and scleraxis synergistically promote the commitment of human ES cells derived MSCs to tenocytes. *Sci. Rep.* **2012**, *2*, 977. [CrossRef]

71. Scott, A.; Danielson, P.; Abraham, T.; Fong, G.; Sampaio, A.V.; Underhill, T.M. Mechanical force modulates scleraxis expression in bioartificial tendons. *J. Musculoskelet Neuronal Interact* **2011**, *11*, 124–132.

72. Gao, F.; Chiu, S.M.; Motan, D.A.L.; Zhang, Z.; Chen, L.; Ji, H.-L.; Tse, H.-F.; Fu, Q.-L.; Lian, Q. Mesenchymal stem cells and immunomodulation: Current status and future prospects. *Cell Death Dis.* **2016**, *7*, e2062. [CrossRef]

73. Sugg, K.B.; Lubardic, J.; Gumucio, J.P.; Mendias, C.L. Changes in macrophage phenotype and induction of epithelial-to-mesenchymal transition genes following acute Achilles tenotomy and repair. *J. Orthop. Res.* **2014**, *32*, 944–951. [CrossRef]

74. Millar, N.L.; Akbar, M.; Campbell, A.L.; Reilly, J.H.; Kerr, S.C.; McLean, M.; Frleta-Gilchrist, M.; Fazzi, U.G.; Leach, W.J.; Rooney, B.P.; et al. IL-17A mediates inflammatory and tissue remodelling events in early human tendinopathy. *Sci. Rep.* **2016**, *6*. [CrossRef] [PubMed]

75. Gruber, H.E.; Hoelscher, G.L.; Ingram, J.A.; Bethea, S.; Hanley, E.N. Growth and differentiation factor-5 (GDF-5) in the human intervertebral annulus cells and its modulation by IL-1ß and TNF-α in vitro. *Exp. Mol. Pathol.* **2014**, *96*, 225–229. [CrossRef] [PubMed]

76. Govoni, M.; Lamparelli, E.P.; Ciardulli, M.C.; Santoro, A.; Oliviero, A.; Palazzo, I.; Reverchon, E.; Vivarelli, L.; Maso, A.; Storni, E.; et al. Demineralized Bone Matrix Paste Formulated with biomimetic PLGA microcarriers for the Vancomycin Hydrochloride Controlled Delivery: Release Profile, Citotoxicity and Efficacy against S. aureus. *Int. J. Pharm.* **2020**, *582*, 119322. [CrossRef] [PubMed]

77. Qazi, T.H.; Mooney, D.J.; Duda, G.N.; Geissler, S. Biomaterials that promote cell-cell interactions enhance the paracrine function of MSCs. *Biomaterials* **2017**, *140*, 103–114. [CrossRef] [PubMed]

78. Forcina, L.; Miano, C.; Scicchitano, B.; Musarò, A. Signals from the Niche: Insights into the Role of IGF-1 and IL-6 in Modulating Skeletal Muscle Fibrosis. *Cells* **2019**, *8*, 232. [CrossRef]

Article

Self-Assembling Scaffolds Supported Long-Term Growth of Human Primed Embryonic Stem Cells and Upregulated Core and Naïve Pluripotent Markers

Christina McKee [1,2], Christina Brown [1,2] and G. Rasul Chaudhry [1,2,*]

1   Department of Biological Sciences, Oakland University, Rochester, MI 48309, USA; cmmckee@oakland.edu (C.M.); brown3@oakland.edu (C.B.)
2   OU-WB Institute for Stem Cell and Regenerative Medicine, Rochester, MI 48309, USA
*   Correspondence: chaudhry@oakland.edu; Tel.: +1-248-370-3350

Received: 5 October 2019; Accepted: 14 December 2019; Published: 16 December 2019

Abstract: The maintenance and expansion of human embryonic stem cells (ESCs) in two-dimensional (2-D) culture is technically challenging, requiring routine manipulation and passaging. We developed three-dimensional (3-D) scaffolds to mimic the in vivo microenvironment for stem cell proliferation. The scaffolds were made of two 8-arm polyethylene glycol (PEG) polymers functionalized with thiol (PEG-8-SH) and acrylate (PEG-8-Acr) end groups, which self-assembled via a Michael addition reaction. When primed ESCs (H9 cells) were mixed with PEG polymers, they were encapsulated and grew for an extended period, while maintaining their viability, self-renewal, and differentiation potential both in vitro and in vivo. Three-dimensional (3-D) self-assembling scaffold-grown cells displayed an upregulation of core pluripotency genes, *OCT4*, *NANOG*, and *SOX2*. In addition, the expression of primed markers decreased, while the expression of naïve markers substantially increased. Interestingly, the expression of mechanosensitive genes, *YAP* and *TAZ*, was also upregulated. YAP inhibition by Verteporfin abrogated the increased expression of *YAP/TAZ* as well as core and naïve pluripotent markers. Evidently, the 3-D culture conditions induced the upregulation of makers associated with a naïve state of pluripotency in the primed cells. Overall, our 3-D culture system supported the expansion of a homogenous population of ESCs and should be helpful in advancing their use for cell therapy and regenerative medicine.

Keywords: embryonic stem cells; three-dimensional; self-assembling scaffold; pluripotency; culture conditions; expansion; growth; niche

---

1. Introduction

Pluripotency is defined by the potential of a cell to differentiate into germline cells as well as the cells of all three germ layers [1,2]. During development, pluripotency ranges from the formation of the epiblast until gastrulation and lineage commitment, resulting in a population of cells representing a range of pluripotent states [3,4]. Pluripotency in vitro is determined by the developmental stage from which the cell line is derived and by culture conditions [5–8] with distinct naïve and primed states of pluripotency corresponding to in vivo pre-implantation and post-implantation epiblast cells, respectively [4].

Embryonic stem cells (ESCs) display unlimited self-renewal and differentiation potential in vitro [9], making them an ideal source for the development of cell therapies and regenerative medicine applications. However, the clinical use of ESCs requires a high quality and quantity of cells, which is limited by currently used culturing techniques [10]. ESCs are typically grown as a monolayer in two-dimensional (2-D) plastic culture plates coated with extracellular matrix (ECM) components (such as gelatin from porcine skin [11], matrigel [12], laminin [13], fibronectin [14], vitronectin [15], or collagen [16]) or a mouse embryonic fibroblast (MEF) feeder layer to aid in attachment [17]. Monolayer culture also necessitates routine passaging and the removal of spontaneously differentiated colonies to maintain the self-renewal and potency of cells [18], which pose a major impediment to the large-scale expansion of cells. Generally, 2-D culturing methods often lead to heterogeneous cell populations as well as batch-to-batch variation.

Moreover, 2-D culture conditions lack the intricacies necessary to mimic the ESC niche, dynamic, and specialized three-dimensional (3-D) microenvironments, which are critical for regulating cell fate in vivo. Furthermore, native 3-D niches allow for complex spatial interactions between cells ECM components as well as gradients of nutrients, oxygen, and metabolic waste [19–24]. The microenvironment is important for the self-renewal of ESCs, since cell fate and function is affected by the composition and organization of the ECM [25–27] as well as mechanical forces generated between cells and attachment substrates [25,28–31]. Furthermore, mechanical forces generated by the expansion of the blastocoel have been shown to play an important role in blastocyst lineage formation, stimulating the generation of pluripotent cells [32]. These early morphogenic events in the mammalian embryo indicate a significant interaction between mechanical forces, cell polarity, and the regulation of gene expression in cell fate determination [33]. We hypothesized that 3-D culture would better mimic the in vivo microenvironment, promoting the proliferation and maintenance of human ESCs.

We have previously demonstrated that 3-D self-assembling scaffolds composed of thiolated dextran and 4-arm polyethylene glycol (PEG) functionalized with acrylate groups (Dex-SH/PEG-4-Acr) supported the growth and maintenance of naïve mouse ESCs [34]. However, these 3-D culture conditions failed to support the growth of human primed ESCs. Unlike mouse ESCs, human cells display poor viability and clonogenicity following single cell dissociation [9,10].

In this study, we describe 3-D hydrogel scaffolds that support the long-term growth and maintenance of human primed ESCs (H9). Chemically cross-linked hydrogels were formed by a Michael-type addition reaction by combining two 8-arm PEG polymers functionalized with either thiol (PEG-8-SH) or acrylate end groups (PEG-8-Acr). The PEG-8-SH/PEG-8-Acr scaffold provided a microenvironment that maintained self-renewal and pluripotency for an extended time period. Interestingly, H9 cells displayed an upregulation of core pluripotency markers during 3-D culture. However, the expression of core markers reverted to normal levels when 3-D grown cells were subcultured under 2-D culture conditions. Interestingly, 3-D cultured H9 cells also exhibited a significantly higher expression of naïve pluripotent markers when compared to human naïve ESCs (Elf1) cultured under 2-D conditions. Our results suggest the importance of the 3-D scaffold microenvironment in maintaining the stemness of ESCs.

2. Materials and Methods

2.1. Maintenance of Human ESCs in 2-D culture

H9 cells, derived from a human blastocyst [35], obtained from WiCell (Madison, WI, USA) were maintained in the culture medium containing Knockout/F12 Dulbecco's Modified Eagle Media (DMEM; Life Technologies, Carlsbad, CA, USA) with 20% KnockOut serum replacement (Life Technologies), 0.1 mM 2-mercaptoehtanol (Life Technologies), 1% GlutaMax (Life Technologies), and 1% non-essential amino acid solution (Life Technologies), supplemented with 20 ng/mL of basic fibroblast growth factor (FGF2, Prospec, Ness Ziona, Israel) and 10 µM ROCK inhibitor, Y-27632 (Cayman Chemical, Ann Arbor, MI, USA) and subcultured by manual passaging. For dissociation into single cells, H9 were treated with Accutase (Thermo Fisher Scientific, Waltham, MA, USA) for 5 min, and then cells were centrifuged at 200 g for 5 min. The supernatant was aspirated, and the cell pellet was resuspended in H9 culture media for plating.

Elf1 cells, isolated from a cryopreserved 8-cell human embryo [36], were maintained in the culture medium containing KnockOut/F12 DMEM with GlutaMax (Life Technologies) with 20% KnockOut serum replacement (Life Technologies), 1 mM sodium pyruvate (Life Technologies), 0.1mM

2-mercaptoehtanol (Life Technologies), 1% non-essential amino acids (Life Technologies), and 0.2% penicillin–streptomycin solution (Life Technologies), supplemented with 12 ng/mL of FGF2 (Prospec), 1.5 µM CHIR99021 (Caymen Chemical), 0.4 µM PD03296501 (Caymen Chemical), and 0.01 µg/mL human LIF (Prospec) and cultured according to the published protocol [36].

2.2. Selection and Composition of Scaffolds for 3-D Culture of Human ESCs

Several polymers end functionalized with thiol and acrylate end groups were tested for their ability to self-assemble and form hydrogels scaffolds via a thiol–Michael addition reaction, allowing for the formation of covalent bonds by the addition of a nucleophile to a nucleophile acceptor containing an α,β-unsaturated carbonyl compound [37]. Then, the scaffolds were tested in various compositions to determine the optimal concentration, molar ratio, and degree of modification that would best support the growth of human ESCs. Our preliminary results showed that scaffolds made using 8-arm PEG polymers yielded optimal growth. These scaffolds were prepared using 8-arm PEG-thiol (PEG-8-SH, 20 kDa) and 8-arm PEG-acrylate (PEG-8-Acr, 20 kDa) purchased from JenKem Technology USA (Plano, TX, USA). Functionalized PEG polymers were stored at −20 °C and protected from light.

The preparation of scaffolds for the 3-D culture of ESCs is depicted in Figure 1. Briefly, PEG-8-SH and PEG-8-Acr polymers were dissolved at a concentration of 2.5 *w/v* % (dry weight of polymer per volume of culture medium), combined at a 1:1 molar ratio and mixed with cells. Then, the resulting mixture was transferred to a 1 cc syringe mold for polymerization. After self-assembly, scaffolds were placed in a 24-well culture plate (Fisher Scientific, Pittsburgh, PA, USA), supplemented with culture medium, and maintained in a 5% CO_2 incubator at 37 °C. The medium was changed daily or as needed. Cell growth in the scaffolds was monitored by phase-contrast microscopy.

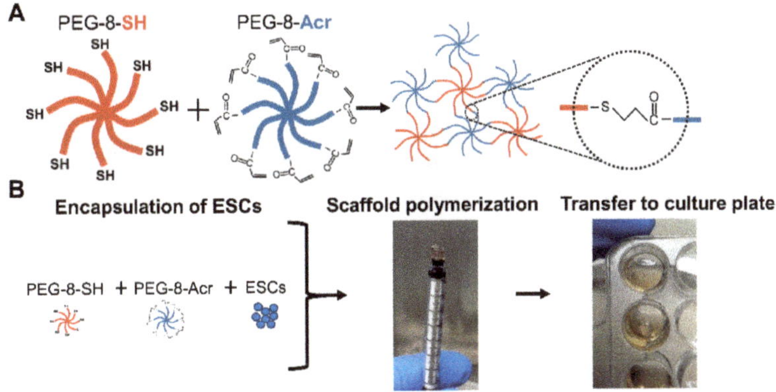

Figure 1. Schematic of self-assembling scaffolds. (**A**) Self-assembly of functionalized polymers, 8-arm polyethylene glycol functionalized with thiol (PEG-8-SH) and acrylate (PEG-8-Acr) via a thiol–Michael addition reaction. (**B**) The encapsulation of H9 cells human embryonic stem cells (ESCs), was achieved upon mixing with the self-assembling polymers in a syringe mold. Following polymerization, the scaffolds were then incubated in culture plates containing medium.

2.3. Cell Proliferation and Viability Assays

The growth rate of cells grown under 2-D and 3-D culture conditions were analyzed at various time intervals using a proliferation assay. Briefly, triplicate samples were treated with 5 mg/mL 3-(4,5-dimethylthiazol-2-yl)-2,5-diphenyltetrazolium bromide (MTT) reagent (Sigma, St. Louis, MO, USA), protected from light, and incubated at 37 °C for 4 h to obtain insoluble formazan, which was then solubilized using 15:1 isopropanol/hydrochloride. Then, the absorbance of the solubilized formazan was measured at 570 nm using an Epoch microplate reader (BioTek, Winooski, VT), and the background absorbance of the cells was subtracted from all measured values. The viability of encapsulated cells was determined by direct microscopic counts and trypan blue exclusion assay. Briefly, cells were counted using a hemocytometer and cells stained blue were considered non-viable.

2.4. Differentiation of Human ESCs

Germ layer differentiation was achieved by the spontaneous formation of embryoid bodies (EBs). ESCs were allowed to spontaneously aggregate for 3 days in non-adherent flat-bottomed 96-well plates in their respective ESC culture medium containing growth factors. Then, the resultant EBs were transferred to 0.1% gelatin-coated wells for adherent growth and grown in high-glucose DMEM supplemented with 10% fetal bovine serum (FBS). Spontaneous differentiation into all three germ layers was assessed by germ layer marker expression by quantitative real time-polymerase chain reaction (qRT-PCR) and immunocytochemistry.

2.5. Teratoma Assay

For teratoma formation, ESCs were harvested following accutase treatment, washed and resuspended in PBS, and mixed with an equal volume of matrigel (BD Biosciences, San Jose, CA, USA). Cells (1×10^6) were subcutaneously injected (20 µL) using a Hamilton syringe into 4-week-old male immune-compromised SCID (severe combined immunodeficient) Beige mice (Fox Chase SCID Beige, Charles River, Wilmington, MA, USA). Animals were monitored daily and humanely euthanized by CO_2 overdose after teratoma formation at 10–12 weeks. Teratomas were explanted, and teratoma tissue was either fixed for histological analysis or flash frozen in liquid nitrogen for RNA isolation. Teratoma assays were performed in triplicate. All the procedures involving animals were reviewed and approved by the Institutional Animal Care and Use Committee of Oakland University (IACUC protocol number: 17031).

2.6. Gene Expression Analysis

Transcriptional analysis was performed by qRT-PCR. Briefly, cells, scaffolds, and teratoma tissue (100–250 mg) were harvested and total cellular mRNA was isolated following the manufacturer's instructions using the GeneJET RNA purification kit (Thermo Fisher Scientific) and RNeasy Midi kit (Qiagen, Germantown, MD, USA), respectively. cDNA was synthesized with the iScript kit (BioRad, Hercules, CA, USA). qRT-PCR was performed using SsoAdvanced SYBR Green Supermix (Bio-Rad) and the CFX90 Real-Time PCR system. The primers (IDT Technologies, Coralville, IA, USA) used in this study are in Table 1. All reactions were prepared in triplicate and normalized to reference genes, *HMBS*, *GAPDH*, and *β-ACTIN*.

2.7. Immunocytochemical Analysis

Protein expression was determined by immunocytochemical staining using selected antibodies. H9 cells grown under 2-D culture conditions on coverslips, H9 cells encapsulated in the self-assembling scaffolds, and harvested teratoma tissue were fixed in 4% paraformaldehyde for 10 min, 30 min, and overnight, respectively. Subsequently, teratoma tissue and scaffolds were embedded and frozen in optimal cutting temperature (O.C.T) compound and cryosectioned into 10 µm sections. For immunochemical analysis, fixed cells and cryosections were permeabilized with 0.5% Triton X-100 (Sigma) for 10 min and blocked with 2% bovine serum albumin (BSA; Sigma) for 1 h at room temperature. Next, samples were incubated with primary antibodies (1:100 dilution) overnight at 4 °C. Primary antibody-treated samples were washed three times with phosphate buffer saline (PBS), stained with secondary antibody at 1:200 dilutions for 2 h at 37 °C, and counterstained with 1 mg/mL 4′,6-diamidino-2-phenylindole (DAPI; Life Technologies). The stained samples were visualized by using confocal microscopy. The antibodies used are listed in Table 2.

Table 1. List of human primer sequences used in qRT-PCR.

Gene	Primer Sequence		Product Length
	Forward (5′-3′)	Reverse (5′-3′)	
ACTIN	AATCTGGCACCACACCTTCTAC	ATAGCACAGCCTGGATAGCAAC	170
BRACHYURY	TGCTTCCCTGAGACCCAGTT	GATCACTTCTTTCCTTTGCATCAAG	121
DNMT3L	CTGCTCCATCTGCTGCTCC	ATCCACACACTCGAAGCAGT	85
DPPA3	AGACCAACAAACAAGGAGCCT	CCCATCCATTAGACACGCAGA	88
ESRRB	GACATTGCCTCTGGCTACCA	CTCCGTTTGGTGATCTCGCA	131
FGF4	CGTGGTGAGCATCTTCGGC	GTAGGACTCGTAGGCGTTGT	145
FOXA2	GGGAGCGGTGAAGATGGA	TCATGTTGCTCACGGAGGAGTA	89
FOXF1	AAGCCGCCCTATTCCTACATC	GCGCTTGGTGGGTGAACT	63
GAPDH	ACAACTTTGGTATCGTGGAAGG	GCCATCACGCCACAGTTTC	101
GATA4	TCCCTCTTCCCTCCTCAAAT	TCAGCGTGTAAAGGCATCTG	194
GATA6	CCCACAACACAACCTACAGC	GCGAGACTGACGCCTATGTA	131
GDF3	GTCTCCCGAGACTTATGCTACG	AGTAGAGGAGGCTTCTGCAGGCA	136
GP130	GGAGTGAAGAAGCAAGTGGGA	AGGCAATGTCTTCCACACGA	128
HMBS	AGGAGTTCAGTGCCATCATCCT	CACAGCATACATGCATTCCTCA	104
ITGA2	TTGCGTGTGGACATCAGTCT	GCTGGTATTTGTCGGACATCT	158
ITGA5	GCCGATTCACATCGCTCTCAA	GTCTTCTCCACAGTCCAGCAA	139
ITGA6	CGAAACCAAGGTTCTGAGCCC	CTTGGATCTCCACTGAGGCAG	151
ITGAV	AGGAGAAGGTGCCTACGAAGC	GCACAGGAAAGTCTTGCTAAGG	105
ITGB1	GGATTCTCCAGAAGGTGGTTT	TGCCACCAAGTTTCCCATCT	143
ITGB3	CATGGATTCCAGCAATGTCCTC	TTGAGGCAGGTGGCATTGAAG	126
ITGB5	GCCTTTCTGTGAGTGCGACAA	CCGATGTAACCTGCATGGCAC	111
KLF17	TCAGGAAGGGACTGGTAGAA	GTACCCGCATATGTCGTCTAAG	206
KLF2	CCAAGAGTTCGCATCTGAAGG	CCGTGTGCTTTCGGTAGTG	132
KLF4	CGAACCCACACAGGTGAGAA	TACGGTAGTGCCTGGTCAGTTC	75
KLF5	ACCCTGGTTGCACAAAAGTT	CAGCCTTCCCAGGTACACTT	100
LATS1	CTCTGCACTGGCTTCAGATG	TCCGCTCTAATGGCTTCAGT	145
LATS2	ACATTCACTGGTGGGGACTC	GTGGGAGTAGGTGCCAAAAA	147
LIFR	CACCTTCCAAAATAGCGAGTATGG	ATGGTTCCGACCGAGACGAGTT	159
MIXL1	CCGAGTCCAGGATCCAGGTA	CTCTGACGCCGAGACTTGG	58
NANOG	AAAGAATCTTCACCTATGCC	GAAGGAAGAGGAGAGACAGT	110
N-CADHERIN	TGTTTGGCCTGGCGTTCTTT	AGGAGACAGAAACGAAGCCA	156
NCAM	AGGAGACAGAAACGAAGCCA	GGTGTTGGAAATGCTCTGGT	161
OCT4	CCCCTGGTGCCGTGAA	GCAAATTGCTCGAGTTCTTTCTG	97
PAX6	CTTTGCTTGGGAAATCCGAG	AGCCAGGTTGCGAAGAACTC	103
PRDM14	CCTTGTGTGGTATGGAGACTGC	CTTTCACATCTGTAGCCTTCTGC	126
RAC1	ATGTCCGTGCAAAGTGGTATC	CTCGGATCGCTTCGTCAAACA	249
RHOA	CATCCGGAAGAAACTGGT	TCCCACAAAGCCAACTC	168
ROCK1	GGTGGTCGGTTGGGGTATTTT	CGCCCTAACCTCACTTCCC	196
SALL2	GGCTTGCCTTATGGTATGTCCG	TGGCACTGAGTGCTGTTGTGGA	115
SOCS3	ATTCGGGACCAGCCCCC	AAACTTGCTGTGGGTGACCA	121
SOX11	CGACGACCTAATGTTCGACC	GACAGGGATAGGTTCCCCG	105
SOX17	CGCACGGAATTTGAACAGTA	GGATCAGGGACCTGTCACAC	182
SOX2	TTGCTGCCTCTTTAAGACTAGGA	CTGGGGCTCAAACTTCTCTC	75
SOX7	ACGCCGAGCTCAGCAAGAT	TCCACGTACGGCCTCTTCTG	73
STAT3	CTTTGAGACCGAGGTGTATCAC	GGTCAGCATGTTGTACCACAG	133
TAZ	GAGGGTGTATGGTGGAGATAAA	CCAACTGTAGCAAACAGGATTAG	86
TBX3	GGACACTGGAAATGGCCGAAG	GCTGCTTGTTCACTGGAGGAC	123
TBX3	CGGACATACTTGTTCCCCGA	GCAGGGTGAGCTGTTTTCTTTT	154
TEAD4	CCAAGCTCTGGATGTTGGAGTTC	GATGTCCACGGCTTCGAGGTA	161
TFCP2L1	TTTGTGGGACCCTGCGAAG	TGCTTAAACGTGTCAATCTGGA	129
TUJ1	GGCCAAGTTCTGGGAAGTCA	CGAGTCGCCCACGTAGTTG	70
YAP	GCTGCCACCAAGCTAGATAA	GTGCATGTGTCTCCTTAGATCC	101
ZFP42	CGCAATCGCTTGTCCTCAGA	GCTCTCAACGAACGCTTTCC	130
ZIC2	CGCTCCGAGAACCTCAAGAT	CCCTCAAACTCACACTGGAA	71

Table 2. List of primary and secondary antibodies used in immunocytochemical staining.

Antibody	Primary	Secondary
BRACHYURY	Rabbit Polyclonal	Anti-Rabbit Alexa Fluor 488
GATA4	Mouse Polyclonal	Anti-Mouse-Cy3
NANOG	Rabbit Polyclonal	Anti-Rabbit Alexa Fluor 488
OCT4	Rabbit Polyclonal	Anti-RabbitAlexa Fluor 488
SOX2	Rabbit polyclonal	Anti-Mouse Alexa Fluor 488
TUJ1	Mouse Polyclonal	Anti-Mouse-Cy3
YAP	Rabbit Polyclonal	Anti-Rabbit Alexa Fluor 488

2.8. Effect of YAP Inhibitor on Human ESCs Grown in 3-D Self-Assembling Scaffolds

For analysis of yes-associated protein (YAP) signaling in 3-D culture, human ESCs were grown in self-assembling scaffolds for 7 days, and then treated with 2 µM YAP inhibitor (YAPi), Verteporfin (VP, R&D Systems, Minneapolis, MN), for an additional 7 days. Cell growth in the scaffolds was monitored by light microscopy. ESCs grown in 3-D self-assembling scaffolds were harvested after 14 days of culture with and without YAPi treatment as a control. Cells were harvested for RNA and immunocytochemical analysis.

2.9. Statistical analysis

Data are presented as mean ± standard error of the mean (SE). One-way ANOVA analysis was performed and analyzed for unequal variances using post hoc tests for multiple comparisons. Results with a p-value less than 0.05 were considered to be significant (* $p < 0.05$ and ** $p < 0.01$). All analyses were performed using SPSS version 26 (SPSS Inc., Chicago, IL., USA).

3. Results

3.1. Growth and Characterization of H9 Cells Grown under 3-D Culture Conditions

H9 cells encapsulated in self-assembling scaffolds composed of PEG-8-SH and PEG-8-Acr polymers grew for extended periods without requiring routine passaging or manipulation. The optimal growth of ESCs was achieved by using a concentration of 2.5 *w/v* % (dry weight of polymer per volume of culture medium) at a 1:1 molar ratio of PEG-8-SH and PEG-8-Acr. The results depicted in Figure 2 show steady cell growth up to 21 days as observed by light microscopy (Figure 2A–D). Cells grew clonally in a time-dependent manner. Visual observations were consistent with the results obtained by MTT assay. A significant and continuous increase in the proliferation of cells was observed during day 7 to 21. However, cells grew more rapidly between day 14 and 21 than between day 1 and 14 (Figure 2E), suggesting that the growth of ESCs required a longer acclimation period in 3-D self-assembling scaffolds compared to 2-D culture conditions. The viability of 3-D cultured ESCs was further validated by direct cell counts, as depicted in Figure 2F. The results showed that while cell proliferation significantly and consistently increased from day 1 to day 21, the number of dead cells remained low.

To confirm whether H9 cells remained pluripotent during 3-D culture, the transcriptional and translational analysis of selected ESC-specific markers was performed using qRT-PCR and immunocytochemical analysis, respectively. The results indicated a successive increase in the expression of *OCT4*, *NANOG*, and *SOX2* throughout the duration of the 3-D culture, which increased 2.85, 2.23, and 3.69-fold, respectively in ESCs grown in 3-D scaffolds for 21 days as compared to cells grown under 2-D culture conditions (Figure 2G). The protein expression of these markers was also increased in cells cultured in the 3-D scaffolds (Figure 2H), which was consistent with the transcriptional upregulation.

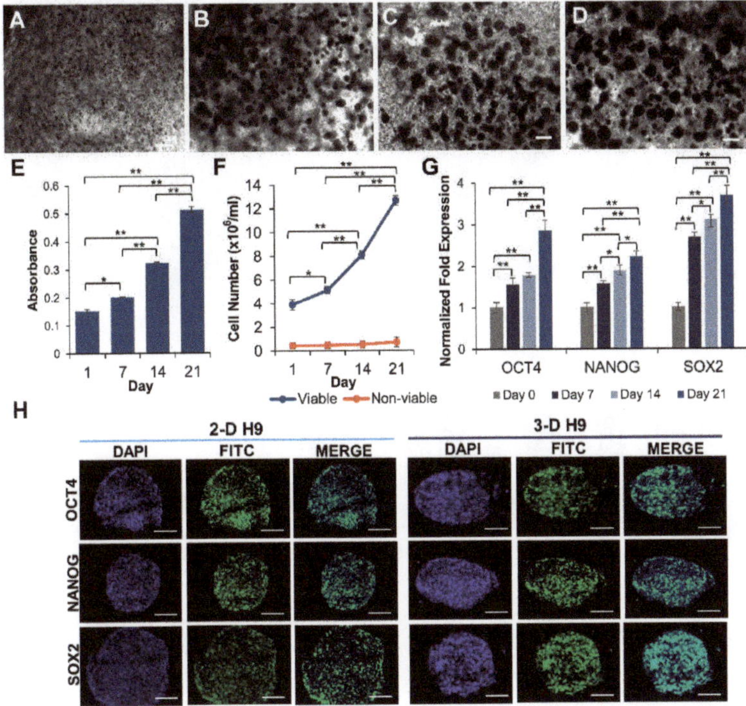

Figure 2. Growth of pluripotent human ESCs in 3-D self-assembling scaffolds. (**A–D**) Clonal growth of ESCs (H9 cells) encapsulated in PEG-8-SH/PEG-8-Acr scaffolds and incubated in culture medium was observed by light microscopy at 0, 7, 14, and 21 days. (**E**) Quantitative determination of cell proliferation by 3-(4,5-dimethylthiazol-2-yl)-2,5-diphenyltetrazolium bromide (MTT) assay using microplate reader. Results were expressed as the absorbance ± standard error (SE) with a significant increase in cell number. (**F**) Growth of ESCs encapsulated in 3-D scaffolds was assayed by direct counts using a hemocytometer, and cell viability was determined by trypan blue exclusion assay at various time intervals. Data presented as cell number ($\times 10^6$ cells/mL) ± SE. (**G**) Expression of selected pluripotency markers, *OCT4*, *NANOG*, and *SOX2*, in ESCs cultured in self-assembling scaffolds for 0, 7, 14, and 21 days as determined by qRT-PCR. The expression of genes at day 0 was set to 1 and results were expressed as the fold expression ± SE normalized to reference genes *HMBS*, *GAPDH*, and *β-ACTIN* (* $p < 0.05$ and ** $p < 0.01$). (**H**) Confocal images (20X) of 2-D and 3-D grown ESCs displaying the expression of pluripotent proteins, OCT4, NANOG, and SOX2. All scale bars represent 100 µm.

3.2. Maintenance of Pluripotency in H9 Cells Grown under 3-D Culture Conditions

To investigate if the 3-D grown H9 cells maintained their pluripotency after 21 days in 3-D culture, cells were subcultured under 2-D culture conditions and analyzed for cell morphology and the expression of pluripotent markers. The results showed that there were no morphological differences between cells that were passaged from 3-D to 2-D culture conditions and the initial ESCs used for encapsulation in the 3-D self-assembling scaffolds (Figure 3A–C). When H9 cells grown in 3-D self-assembling scaffolds for 21 days were subsequently subcultured under 2-D culture conditions, the expression of core pluripotent markers, *OCT4* and *NANOG*, was statistically similar to that of initial 2-D grown cells, while the expression of *SOX2* (1.55 fold) was slightly upregulated (Figure 3D). Apparently, repeated the subculturing of 3-D grown cells in 2-D culture conditions reverted the expression of these markers back to the normal level of expression. Despite the high expression of key pluripotent markers, it was prudent to further investigate the 3-D grown cells for the maintenance of their differentiation potential both in in vitro and in vivo.

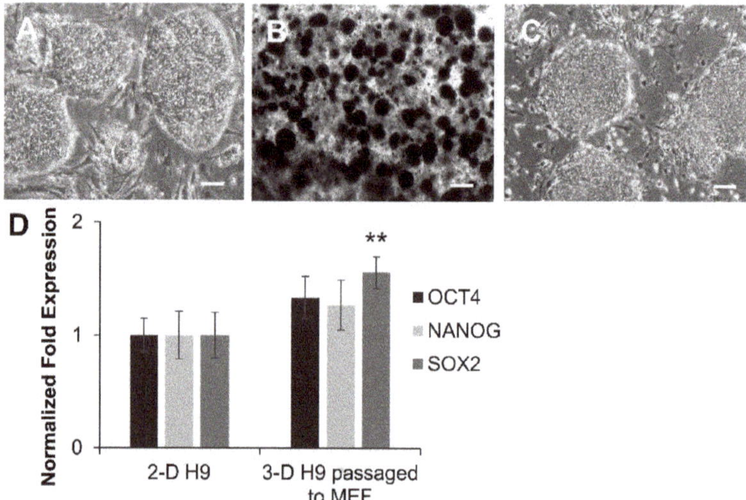

Figure 3. Self-renewal and pluripotency of ESCs were maintained following extended culture in 3-D scaffolds. (**A–C**) Cell morphology of ESCs (H9 cells) grown in 2-D cultures prior to encapsulation, in self-assembling scaffolds for 3-D culture for 21 days, and then subsequently subcultured back to 2-D culture conditions, respectively, as determined by light microscopy. All scale bars represent 100 μm. (**D**) Comparison of expression of *OCT4*, *NANOG*, and *SOX2* in ESCs grown in 2-D conditions and first in 3-D self-assembling scaffolds, and then passaged in 2-D culture conditions as determined by qRT-PCR. Results were expressed as the fold expression ± SE normalized to reference genes *HMBS*, *GAPDH*, and *β-ACTIN* (* $p < 0.05$ and ** $p < 0.01$).

3.3. Differentiation of H9 Cells Grown in 3-D Self-Assembling Scaffolds

The maintenance of the pluripotency of 3-D grown H9 cells was further investigated by the induction of differentiation into three germ layers following EB formation. The results presented in Figure 4 show that EBs from 3-D grown cells spontaneously differentiated into endoderm, mesoderm, and ectoderm germ layers, as evident by the protein expression of specific markers, GATA4, BRACHYURY, and TUJ1, respectively (Figure 4A). Transcriptional analysis also confirmed that differentiated derivatives of 3-D grown cells expressed markers of all three germ layers, including *SOX7* and *GATA6*, *BRACHYURY* and *MIXL1*, and *PAX6* and *NCAM*, representing the endoderm, mesoderm, and ectoderm, respectively (Figure 4B).

Next, the pluripotency of 3-D grown H9 cells was validated in vivo by an analysis of teratoma formation. The results showed that the 3-D grown cells injected into SCID Beige mice formed teratomas (Figure 5). An analysis of teratoma tissue showed the expression of GATA4, BRACHYURY, and TUJ1 proteins, indicating that teratoma tissue had cells representing the endoderm, mesoderm, and ectoderm, respectively (Figure 5C). The transcriptional analysis of teratoma tissues also showed the expression of genes of all three germ layers: the endoderm (*SOX7*, and *SOX17*), mesoderm (*BRACHYURY*, and *MIXL1*), and ectoderm (*PAX6*, and *TUJ1*) (Figure 5D). Taken together, these results indicated that H9 cells grown in 3-D self-assembling scaffolds maintained their pluripotency and differentiation potential.

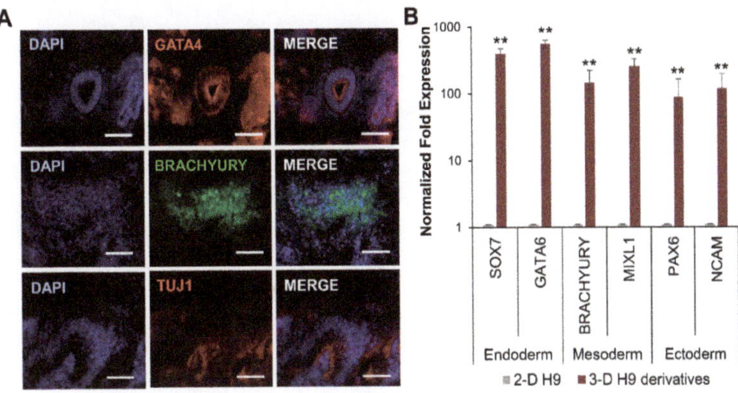

Figure 4. Differentiation potential of human ESCs grown in 3-D self-assembling scaffolds. Embryoid bodies (EBs) derived from 3-D grown ESCs (H9 cells) spontaneously differentiated into three germ layers and expressed specific proteins analyzed by immunocytochemistry. (**A**) Differentiated derivatives of 3-D grown ESCs expressed GATA4, BRACHYURY, and TUJ1 proteins representing the endoderm, mesoderm, and ectoderm, respectively, as shown by confocal images (20X). All scale bars represent 100 μm. (**B**) Differentiated derivatives of 3-D grown ESCs expressed germ layer-specific genes *SOX7* and *GATA6* (endoderm), *BRACHYURY* and *MIXL1* (mesoderm), and *PAX6* and *NCAM* (ectoderm) as determined by qRT-PCR. Results are expressed as the fold expression ± SE normalized to reference genes *HMBS*, *GAPDH*, and *β-ACTIN* (* *p* < 0.05 and ** *p* < 0.01).

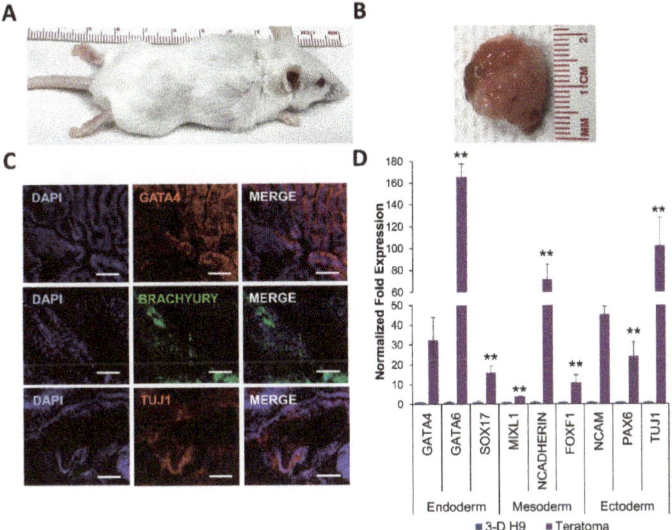

Figure 5. Teratoma formation by 3-D grown human ESCs in severe combined immunodeficient (SCID) Beige mice. (**A**) Tumor growth was observed in all mice (*n* = 3) injected with 3-D grown ESCs (H9 cells). (**B**) Explanted tumor at 4 weeks showed encapsulated, lobular, and well-circumscribed gross morphology consistent with teratoma growth. (**C**) Expression of GATA4, BRACHYURY, and TUJ1 proteins representing the endoderm, mesoderm, and ectoderm, respectively in excised teratomas, as shown by confocal images (20X). All scale bars represent 100 μm. (**D**) Expression of germ layer-specific genes, *SOX7* and *GATA6* (endoderm), *BRACHYURY* and *MIXL1* (mesoderm), and *PAX6* and *NCAM* (ectoderm) in excised teratomas, as determined by qRT-PCR. Results are expressed as the fold expression ± SE normalized to reference genes *HMBS*, *GAPDH*, and *β-ACTIN* (* *p* < 0.05 and ** *p* < 0.01).

3.4. Expression of Naïve Pluripotent Markers in H9 Cells Grown under 3-D Culture Conditions and the Effect of YAP Inhibition

Since a significant upregulation of core pluripotent markers, OCT4, NANOG, and SOX2, was observed in ESCs cultured in the 3-D scaffolds, we also assessed the expression of both primed and naïve pluripotent markers. A comparative transcriptional analysis of H9 cells grown in 3-D scaffolds and 2-D cultured H9 and Elf1 cells (representing primed and naïve ESC lines, respectively) is depicted in Figure 6. Expected primed pluripotent markers (*FOXA2, ZIC2, SALL2,* and *SOX11*) were expressed at higher levels in H9 than Elf1 cells cultured under 2-D conditions. Interestingly, the expression of primed pluripotent markers was significantly decreased in H9 cells grown in 3-D scaffolds to levels comparable to Elf1 cells grown under 2-D conditions. More strikingly, 3-D grown H9 cells expressed significantly higher levels of naïve markers (*KLF2, ESRRB, DNMT3L, KLF17, STAT3, DPPA3, TBX3, PRDM14, KLF5, ZFP42, TFCP2L1, FGF4,* and *GDF3*) in comparison to 2-D cultured Elf1 cells, suggesting that the 3-D scaffold microenvironment modulated gene expression.

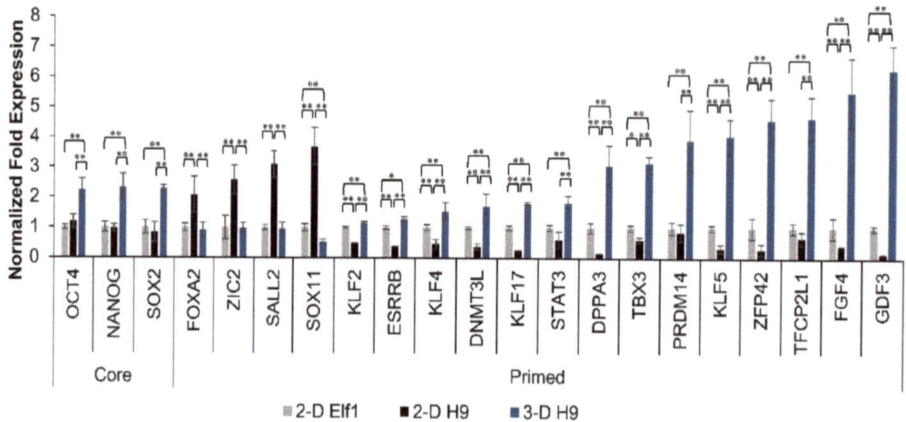

Figure 6. Effect of 3-D culture on the expression of naïve pluripotent markers in primed ESCs. (**A**) Expression of select core (*OCT4, NANOG,* and *SOX2*), primed (*FOXA2, ZIC2, SALL2,* and *SOX11*), and naïve (*KLF2, ESRRB, DNMT3L, KLF17, STAT3, DPPA3, TBX3, PRDM14, KLF5, ZFP42, TFCP2L1, FGF4,* and *GDF3*) pluripotent markers in ESCs cultured in 3-D scaffolds for 21 days and 2-D grown primed ESCs (H9 cells) and naïve ESCs (Elf1 cells, set to control) was analyzed by qRT-PCR. Results were expressed as the fold expression ± SE normalized to reference genes *HMBS, GAPDH,* and *β-ACTIN* (* $p < 0.05$ and ** $p < 0.01$).

Many reports have stated that 3-D scaffolds induce differential gene expression due to mechanical and biological stimuli [10,38,39]. Since the overexpression of YAP has been shown to induce the naïve state of pluripotency in primed ESCs [40], we investigated the effect of YAP in 3-D cultured H9 cells using VP, which is a YAP inhibitor (YAPi). Our results in Figure 7A showed that mechanosensitive genes, *YAP* and *TAZ,* were upregulated in H9 cells grown in 3-D scaffolds in comparison to 2-D cultured cells. In contrast, *YAP* and *TAZ* expression significantly decreased when 3-D grown H9 cells were subjected to YAPi. A similar trend was observed at a translational level, where it appears that the immunofluorescence signal for YAP in 3-D grown H9 cells was brighter than in 2-D cultured cells. Moreover, an increased signal for YAP was observed in the cytoplasm of 3-D grown cells treated with YAPi (Figure 7B).

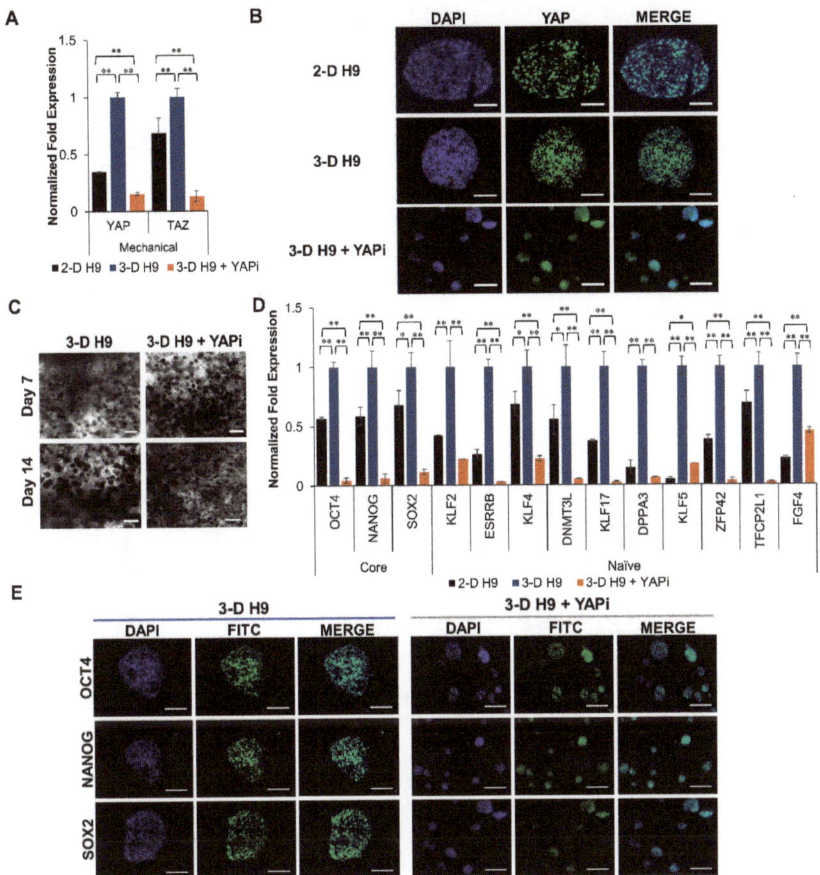

Figure 7. Effect of Yes-associated protein (YAP) inhibition (YAPi) on cell growth and expression of pluripotent markers in human ESCs cultured under 3-D conditions. ESCs (H9 cells) were grown under 3-D culture conditions for 14 days and incubated in the absence (control) or presence of a YAPi. (**A**) The effect of YAPi on the expression of mechanosensitive markers, *YAP* and *TAZ*, in H9 cells grown in 2-D and 3-D culture conditions as determined by qRT-PCR. (**B**) Merged confocal images (20X) showing YAP protein expression in 2-D and 3-D grown H9 cells. (**C**) Light micrographs depicting the cell morphology and colony size of 3-D grown H9 cells incubated in the presence or absence of YAPi. (**D**) Comparison of relative gene expression of core and naïve pluripotent markers in H9 cells grown under 2-D, 3-D, and 3-D + YAPi conditions as determined by qRT-PCR. Results were expressed as the fold expression ± SE normalized to reference genes *HMBS, GAPDH,* and *β-ACTIN* (* $p < 0.05$ and ** $p < 0.01$). (**E**) Merged confocal images (20X) of 3-D H9 cells with and without YAPi treatment for the protein expression of OCT4, NANOG, and SOX2. All scale bars represent 100 μm.

The incubation of YAP also affected the clonal growth of H9 cells encapsulated in 3-D scaffolds, with YAPi-treated H9 cells exhibiting a significantly smaller colony size in comparison to untreated cells (Figure 7C). Interestingly, YAPi treatment also abrogated the upregulation of core (*OCT4, NANOG,* and *SOX2*), and naïve (*ESRRB, KLF4, DNMT3L, KLF17, DPPA3, KLF5, ZFP42, TFCP2L1,* and *FGF4*) pluripotent markers in 3-D cultured cells (Figure 7D). In addition, a concurrent decrease in the expression of core pluripotent proteins, OCT4, NANOG, and SOX2 was observed when compared with the untreated 3-D grown cells (Figure 7E).

3.5. Mechanism of Regulation of Pluripotent Genes in H9 Cells Grown under 3-D Culture Conditions

To determine the basis for the upregulation of pluripotent markers observed in H9 cells grown in 3-D self-assembling scaffolds, we screened multiple signaling pathways with YAP-associated mechanotransduction using transcriptional analysis (Figure 8). The results depicted in Figure 8A show that 3-D cultured cells expressed higher levels of genes encoding integrin subunits, *ITGA5*, *ITGA6*, *ITGAV*, *ITGB1*, and *ITGB3* as well as G-coupled protein receptors (GCPRs), *LPAR1*, *LPAR2*, *S1PR1*, and *S1PR3*. In addition, these cells displayed the upregulation of *RHOA* and *RAC1* (Rho signaling), *YAP*, *TAZ*, and *TEAD4* (Hippo signaling), *LIFR*, *GP130*, *SOCS3*, and *TBX3* (LIF signaling). However, the expression of *ITGA2*, *ROCK1*, *LATS1*, and *LATS2*, which is associated with integrin, Rho, and Hippo signaling, respectively, decreased significantly, while *ITGB5* remained unchanged in 3-D cultured cells.

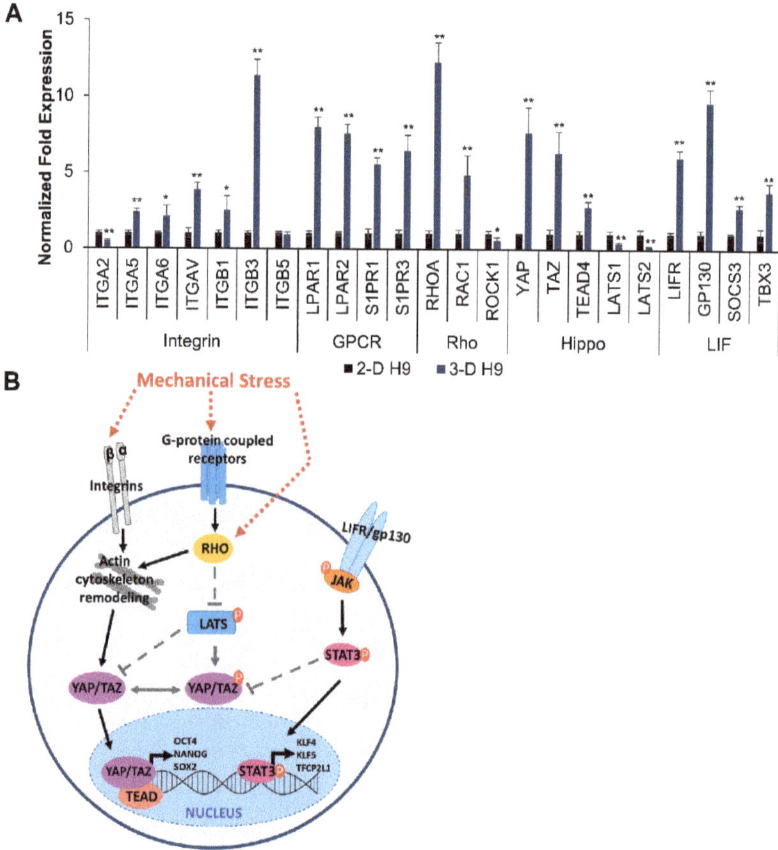

Figure 8. Molecular mechanism of upregulation of pluripotent markers in human ESCs grown in 3-D scaffolds. (**A**) The effect of 3-D culture of ESCs (H9 cells) on the expression of markers associated with integrin (*ITGA2*, *ITGA5*, *ITGA6*, *ITGAV*, *ITGB1*, *ITGB3*, and *ITGB5*) and G-coupled protein receptors (GCPRs; *LPAR1*, *LPAR2*, *S1PR1*, and *S1PR3*), Hippo (*YAP*, *TAZ*, *TEAD4*, *LATS1*, and *LATS2*), Rho (*RHOA*, *RAC1*, and *ROCK1*), and LIF (*LIFR*, *GP130*, *SOCS3*, and *TBX3*) signaling pathways as determined by qRT-PCR. Results were expressed as the fold expression ± SE normalized to reference genes *HMBS*, *GAPDH*, and *β-ACTIN* (* $p < 0.05$ and ** $p < 0.01$). (**B**) Proposed pathway involved in the induction of naïve pluripotency in primed human ESCs (H9 cells) encapsulated in 3-D self-assembling scaffolds.

These results further indicate that 3-D self-assembling scaffolds stimulated mechanosensitive signaling resulting in the upregulation of integrin receptors and GPCRs, thus promoting the activation of Rho signaling, which is associated with actin cytoskeleton remodeling; in turn, this led to the upregulation of mechanosensitive YAP/TAZ signaling. Activated YAP/TAZ act as transcription factors in the nucleus assisted by transcription co-factor, TEAD4, which binds to the DNA, to stimulate the expression of pluripotent genes. The simultaneous upregulation of LIF signaling also contributed to the upregulation of naïve pluripotent markers in H9 cells grown in 3-D self-assembling scaffolds, as proposed in Figure 8B.

4. Discussion

The expansion of human ESCs using traditional 2-D culture techniques is technically challenging and requires routine manipulation and passaging by dissecting colonies via enzymatic digestion or non-enzymatic methods [41–43]. These manipulations can result in poor viability, large batch-to-batch variation, and spontaneous differentiation. To address these problems, we hypothesized that 3-D culture may better mimic the in vivo environment from which ESCs are derived, which would improve both the long-term growth and maintenance of these cells.

Our study investigated the effect of the microenvironment in stemness by developing 3-D scaffolds made of two functionalized PEG polymers that self-assembled via a Michael addition reaction. When ESCs were included in the polymer mixtures, they were encapsulated upon self-assembly of the scaffolds. It has been previously reported that fully hydrated hydrogels mimic the 3-D native microenvironment, which allow nutrient diffusion and promote the growth of cells [44]. We have previously shown that soft 3-D scaffolds composed of Dex-SH and PEG-4-Acr self-assembling polymers were capable of supporting mouse ESC pluripotency for over 6 weeks [34]. Several other studies have also reported that mouse ESC self-renewal could be maintained on soft and low stiffness substrates in 2-D culture [45–47]. In addition, mechanically stiffer prefabricated scaffolds have been shown to promote the differentiation of mouse ESCs [48], while softer 3-D scaffolds supported the differentiation of human adult stem cells [49]. The culture of H9 cells in soft scaffolds did not support the viability of encapsulated cells, and growth was severely curtailed. This is consistent with previous studies, which showed that stiffer substrates promoted the maintenance of human ESCs [22,50,51]. Taken together, this led us to develop stiffer scaffolds made of functionalized PEG polymers.

To optimize scaffold polymerization as well as the encapsulation and growth of H9 cells, we tested several self-assembling polymers at various polymer concentrations in our preliminary studies (unpublished data). These studies showed that scaffolds made of PEG-8-SH and PEG-8-Acr prepared at 2.5% *w/v* improved the clonal growth of H9 cells in comparison to scaffolds with lower cross-linking densities and higher swelling ratios (i.e., Dex-SH/PEG-4-Acr and PEG-4-SH/PEG-4-Acr). These results were in line with other the reported studies performed using scaffolds made of multi-arm PEG functionalized with vinyl sulfone (VS) [52]. In these studies, H9 cells grew upon encapsulation in scaffolds made of VS functionalized 4-arm and 8-arm but not 3-arm PEG hydrogels [52]. However, this study was performed using clumps of H9 cells and not single cells for encapsulation. Whereas, in our study, PEG-8-SH/PEG-8-Acr scaffolds supported cell viability, allowing for even dispersal and clonal growth of H9 cells encapsulated as single cells. This has important implications because the growth of single cell inoculations and the generation of homologous populations of pluripotent cells is necessary for cell-based therapeutics [53]. Furthermore, the maintenance and growth of H9 cells was achieved using PEG-8-SH/PEG-8-Acr self-assembling scaffolds for long-term 3-D culture without passaging or manipulation. The pluripotency of the 3-D grown H9 cells was further demonstrated by their ability to differentiate into three germ layers and teratoma formation in vitro and in vivo, respectively. Additionally, core pluripotent markers, *OCT4*, *NANOG*, and *SOX2*, were upregulated during growth in the self-assembling scaffolds, showing significantly higher expression on day 21 of 3-D culture as compared to cells grown in 2-D culture. When 3-D grown H9 cells were passaged back to 2-D culture conditions, they exhibited undifferentiated morphology, and the expression of pluripotent markers decreased to levels similar to the 2-D cultured cells, suggesting that the encapsulated cells cultured under 3-D conditions were not altered.

Changes in the expression of core and naïve pluripotent markers in H9 cells cultured in 3-D self-assembling scaffolds can be attributed to multiple factors, including matrix dimensionality, stiffness, and/or bioinductive signaling [10]. While the incorporation of natural biomaterials has been shown to increase biological signaling, synthetic biomaterials lack biological activities minimizing batch-to-batch variation, but still allow for biophysical modifications, including pore size and mechanical stiffness [54]. Scaffolds composed of natural polymers including hyaluronic acid [55], chitosan, and alginate [56] have been shown to support human ESC self-renewal without a significant change in pluripotent marker expression. Whereas thermoresponsive synthetic hydrogels composed of PEG functionalized with poly-N-isopropylacrylamide allowed for the continuous 3-D culture of cells for 60 passages but only yielded cells 95% positive for *OCT4* [57]. In contrast, we observed the upregulation of not only *OCT4* but also *NANOG* and *SOX2* during the maintenance of human ESCs in 3-D scaffolds.

Previous studies with mouse ESCs reported that the differential upregulation of pluripotent markers in 3-D culture was dependent on scaffold composition and stiffness [34,58]. A recent study using single cell inoculation and the expansion of human ESCs in large-scale bioreactors resulted in the maintenance of *OCT4* levels but the upregulation of *SOX2* in 3-D grown ESC aggregates [59]. In another study, a 3-D culture of H9 clumps in VS functionalized PEG scaffolds resulted in upregulated gene expression of *SOX2* and *KLF4* but not *OCT4* and *NANOG* when compared to 2-D cultured cells [52]. Interestingly, we observed that primed ESCs, H9 cells, encapsulated in the 3-D self-assembling scaffolds exhibited a decreased expression of primed pluripotency markers, *FOXA2*, *ZIC2*, *SALL2*, and *SOX11*, which is characteristic of post-implantation epiblast cells [60,61], to levels comparable to naïve Elf1 cells. Furthermore, a 3-D culture of H9 cells resulted in an increase in the expression of naïve pluripotent markers, (*KLF2*, *ESRRB*, *DNMT3L*, *KLF17*, *STAT3*, *DPPA3*, *TBX3*, *PRDM14*, *KLF5*, *ZFP42*, *TFCP2L1*, *FGF4*, and *GDF3*), which was associated with naïve pre-implantation epiblast cells [7,60–66]. Recent studies have shown that traditionally derived primed human ESC lines can be reprogrammed to naïve pluripotency using the ectopic expression of select genes and specific media conditions containing inhibitor cocktails [67,68]. In one study, a transcriptomic comparison of primed and reprogrammed naïve human demonstrated the differential expression of certain genes such as *KLF2*, *KLF4*, *GDF3*, *SOCS3*, *STAT3*, and *TBX3* expressed at higher levels in naïve than primed cells [63]. However, the expression of core pluripotent markers *OCT4* and *NANOG* remained unchanged, while *SOX2* levels decreased upon reversion to naïve pluripotency [63]. Here, we observed an increase in core and naïve pluripotent markers during 3-D culture, which was presumably influenced by the 3-D scaffold microenvironment.

It has been well established that the transduction of biophysical signals, including substrate stiffness, cell–cell interactions, and mechanical forces can influence human ESC fate and self-renewal in 2-D culture conditions [39]. The interplay between external and internal mechanical stresses of cells and their ECM play important roles in tensional homeostasis of tissues in vivo [69]. However, the effect of 3-D culture on the regulation of the pluripotency of ESCs has not been fully explored.

Physical interactions with cells or the ECM can be transduced into biological signals and influence actin dynamics via mechanosensitive receptors, such as integrin receptors and GPCRs [70]. Evidence has shown that integrin heterodimers, α5β1, α6β1, and αvβ3, mediate interactions between ESCs and various ECM substrates [13,71–73]; they also play an important role in the maintenance of pluripotency [74], and matrix stiffness regulates integrin binding [75]. Whereas GPCRs, including lysophosphatidic acid receptors (LPARs) and sphingosine-1-phosphate receptors (S1PRs), play a role in the YAP signaling axis. In response to LPA and S1P ligand binding, the dephosphorylation of YAP and TAZ allows them to enter the nucleus to activate transcription [76]. Treatment with exogenous LPA has been shown to aid in the reversion of primed pluripotency to naïve pluripotency in ESCs due to the role that both YAP and LPA serve in suppressing differentiation associated with GSK3 inhibition [40]. However, the transduction of mechanical signals has also been shown to activate G-protein signaling independent of ligand binding [77]. We observed the upregulation of numerous integrin subunits as well as LPARs and S1PRs in 3-D grown H9 cells. This suggests that the 3-D scaffolds promoted mechanical signaling pathways in H9 cells; however, future studies are required to determine the mechanism by which these receptors are activated.

Mechanical signals such as substrate stiffness have been shown to lead to the activation of Rho signaling, which regulates actin cytoskeleton organization, leading to increased F-actin and nuclear localization of YAP and TAZ [78]. YAP is a mechanosensitive transcription factor in the Hippo signaling pathway, which plays a crucial role in cancer, regeneration, and the regulation of organ size [79]. In addition, Hippo signaling regulates the first cell fate specification to the trophoectoderm and the inner cell mass (ICM) of early blastocysts via mechanically sensitive pathways [80]. YAP is mostly retained in the cytoplasm in the ICM of early blastocysts, but becomes active during epiblast specification with a strong nuclear signal [64]. YAP and its transcription cofactor, TAZ, act as major downstream effectors in the Rho signaling pathway and have been shown to control the pluripotent state, allowing for the long-term survival and expansion of human ESCs in vitro [81].

Furthermore, cell culture on stiffer substrates has also been shown to increase the nuclear function of YAP/TAZ in human ESCs [82]. TAZ itself is required for the maintenance of self-renewal marker expression in ESCs, and the loss of TAZ leads to the inhibition of transforming growth factor beta (TGFβ) signaling and differentiation into a neuroectoderm lineage [83]. Likewise, a knockdown of YAP results in a loss of pluripotency in mouse ESCs, whereas the ectopic expression of YAP prevents differentiation [84].

In the nucleus, YAP mediates transcriptional enhanced associate domain (TEAD) transcription, and YAP/TAZ complexes associate with TEAD to regulate pluripotency by activating *OCT4* expression [85]. Pluripotency is determined by an autoregulatory core transcriptional circuitry comprised of *OCT4*, *NANOG*, and *SOX2*, which inhibits lineage specification genes [86]. The phosphorylation of YAP by LATS kinases prevents interaction with TEAD and results in cytoplasmic retention and the inactivation of YAP [87]. In contrast, YAP overexpression has been shown to promote the reversion of primed pluripotency into naïve pluripotency [40].

When 3-D grown H9 cells were cultured with a YAPi (VP), the upregulation of core and naïve pluripotent markers was abrogated. VP acts to inhibit the interaction of YAP and TEAD, disrupting downstream transcriptional activation, and sequestering YAP in the cytoplasm for inactivation [88]. Therefore, we postulate that mechanical forces in 3-D scaffolds stimulated the upregulation of mechanosensitive receptors, including integrins and GPCRs, leading to enhanced Rho signaling and higher levels of YAP/TAZ. In addition, the mechanical forces generated by the expansion of cells may also have contributed to the inhibition of phosphorylated LATS kinases [87], which in turn allowed YAP/TAZ to enter the nucleus, activating the transcription of pluripotent genes. This was concurrent with the observed increase in LIF signaling receptors and downstream naïve pluripotent target genes including *KLF4*, *KLF5*, and *TFCP2L1* [89]. While LIF/STAT3 signaling fails to maintain the self-renewal of primed human ESCs [90], naïve ESCs are dependent on LIF signaling [63]. Although these results are interesting and suggest a role of mechanical signaling in the regulation of cell fate in vitro, further analysis is needed to confirm the mechanism of upregulation of pluripotency markers and the genetic stability of ESCs grown in 3-D culture conditions.

Overall, the 3-D scaffolds made of PEG-8-SH/PEG-8-Acr support the clonal growth of primed ESCs as well as the enhanced expression of both core and naïve pluripotency markers, suggesting that the scaffold provided a permissive microenvironment for the induction of a naïve-like state of pluripotency. Our 3-D culture method is robust, simple, and less labor-intensive for the long-term amplification of homogenous populations of ESCs, which could promote their use in basic science and therapeutic applications.

Author Contributions: Conceptualization, G.R.C.; methodology, C.M.; validation, C.M., and C.B.; data curation, C.M., and C.B.; formal analysis, G.R.C.; writing—original draft preparation, C.M.; writing—review and editing, C.B, and G.R.C.; supervision, G.R.C.

Funding: This research received no external funding.

Acknowledgments: The study was supported by the OU-WB Institute for Stem Cell and Regenerative Medicine (ISCRM), Oakland University, and Michigan Head and Spine Institute. C. McKee received the Provost Graduate Research Award from Oakland University for this project. We are thankful for C. Govind and S. Dinda for reviewing the manuscript.

Conflicts of Interest: The authors declare no conflict of interest.

References

1. Reubinoff, B.E.; Pera, M.F.; Fong, C.Y.; Trounson, A.; Bongso, A. Embryonic stem cell lines from human blastocysts: Somatic differentiation in vitro. *Nat. Biotech.* **2000**, *18*, 399–404. [CrossRef]
2. Takahashi, K.; Yamanaka, S. Induction of pluripotent stem cells from mouse embryonic and adult fibroblast cultures by defined factors. *Cell* **2006**, *126*, 663–676. [CrossRef]
3. Nakamura, T.; Okamoto, I.; Sasaki, K.; Yabuta, Y.; Iwatani, C.; Tsuchiya, H.; Seita, Y.; Nakamura, S.; Yamamoto, T.; Saitou, M. A developmental coordinate of pluripotency among mice, monkeys and humans. *Nature* **2016**, *537*, 57–62. [CrossRef]
4. Nichols, J.; Smith, A. Naive and primed pluripotent states. *Cell Stem Cell* **2009**, *4*, 487–492. [CrossRef]
5. Gafni, O.; Weinberger, L.; Mansour, A.A.; Manor, Y.S.; Chomsky, E.; Ben-Yosef, D.; Kalma, Y.; Viukov, S.; Maza, I.; Zviran, A.; et al. Derivation of novel human ground state naive pluripotent stem cells. *Nature* **2013**, *504*, 282–286. [CrossRef]
6. Morgani, S.; Nichols, J.; Hadjantonakis, A.-K. The many faces of pluripotency: In vitro adaptations of a continuum of in vivo states. *BMC Dev. Biol.* **2017**, *17*, 7. [CrossRef] [PubMed]
7. Theunissen, T.W.; Powell, B.E.; Wang, H.; Mitalipova, M.; Faddah, D.A.; Reddy, J.; Fan, Z.P.; Maetzel, D.; Ganz, K.; Shi, L.; et al. Systematic identification of culture conditions for induction and maintenance of naive human pluripotency. *Cell Stem Cell* **2014**, *15*, 471–487. [CrossRef] [PubMed]
8. Ying, Q.L.; Wray, J.; Nichols, J.; Batlle-Morera, L.; Doble, B.; Woodgett, J.; Cohen, P.; Smith, A. The ground state of embryonic stem cell self-renewal. *Nature* **2008**, *453*, 519–523. [CrossRef] [PubMed]
9. Wu, J.; Yamauchi, T.; Izpisua Belmonte, J.C. An overview of mammalian pluripotency. *Development* **2016**, *143*, 1644–1648. [CrossRef] [PubMed]
10. McKee, C.; Chaudhry, G.R. Advances and challenges in stem cell culture. *Colloids Surf. B Biointerfaces* **2017**, *159*, 62–77. [CrossRef] [PubMed]
11. Li, Y.; Lin, C.; Wang, L.; Liu, Y.; Mu, X.; Ma, Y.; Li, L. Maintenance of human embryonic stem cells on gelatin. *Chin. Sci. Bull.* **2009**, *54*, 4214. [CrossRef]
12. Xu, C.; Inokuma, M.S.; Denham, J.; Golds, K.; Kundu, P.; Gold, J.D.; Carpenter, M.K. Feeder-free growth of undifferentiated human embryonic stem cells. *Nat. Biotechnol.* **2001**, *19*, 971–974. [CrossRef] [PubMed]
13. Rodin, S.; Domogatskaya, A.; Strom, S.; Hansson, E.M.; Chien, K.R.; Inzunza, J.; Hovatta, O.; Tryggvason, K. Long-term self-renewal of human pluripotent stem cells on human recombinant laminin-511. *Nat. Biotechnol.* **2010**, *28*, 611–615. [CrossRef] [PubMed]
14. Kalaskar, D.M.; Downes, J.E.; Murray, P.; Edgar, D.H.; Williams, R.L. Characterization of the interface between adsorbed fibronectin and human embryonic stem cells. *J. R. Soc. Interface* **2013**, *10*, 20130139. [CrossRef]
15. Braam, S.R.; Zeinstra, L.; Litjens, S.; Ward-van Oostwaard, D.; van den Brink, S.; van Laake, L.; Lebrin, F.; Kats, P.; Hochstenbach, R.; Passier, R.; et al. Recombinant vitronectin is a functionally defined substrate that supports human embryonic stem cell self-renewal via alphavbeta5 integrin. *Stem Cells* **2008**, *26*, 2257–2265. [CrossRef]
16. Richards, M.; Fong, C.Y.; Chan, W.K.; Wong, P.C.; Bongso, A. Human feeders support prolonged undifferentiated growth of human inner cell masses and embryonic stem cells. *Nat. Biotechnol.* **2002**, *20*, 933–936. [CrossRef]
17. Jozefczuk, J.; Drews, K.; Adjaye, J. Preparation of mouse embryonic fibroblast cells suitable for culturing human embryonic and induced pluripotent stem cells. *J. Vis. Exp.* **2012**. [CrossRef]
18. Kent, L. Culture and maintenance of human embryonic stem cells. *J. Vis. Exp.* **2009**. [CrossRef]
19. Akhmanova, M.; Osidak, E.; Domogatsky, S.; Rodin, S.; Domogatskaya, A. Physical, spatial, and molecular aspects of extracellular matrix of in vivo niches and artificial scaffolds relevant to stem cells research. *Stem Cells Int.* **2015**, *2015*, 35. [CrossRef]

20. Cosson, S.; Lutolf, M.P. Chapter 7—Microfluidic patterning of protein gradients on biomimetic hydrogel substrates. In *Methods in Cell Biology*; Piel, M., Théry, M., Eds.; Academic Press: Cambridge, MA, USA, 2014; Volume 121, pp. 91–102.

21. Infantes, E.C.; Prados, A.B.H.; Contreras, I.D.; Cahuana, G.M.; Hmadcha, A.; Bermudo, F.M.; Soria, B.; Huamán, J.R.T.; Bergua, F.J.B. Nitric oxide and hypoxia response in pluripotent stem cells. *Redox Biol.* **2015**, *5*, 417–418. [CrossRef]

22. Musah, S.; Morin, S.A.; Wrighton, P.J.; Zwick, D.B.; Jin, S.; Kiessling, L.L. Glycosaminoglycan-binding hydrogels enable mechanical control of human pluripotent stem cell self-renewal. *ACS Nano* **2012**, *6*, 10168–10177. [CrossRef] [PubMed]

23. Peerani, R.; Rao, B.M.; Bauwens, C.; Yin, T.; Wood, G.A.; Nagy, A.; Kumacheva, E.; Zandstra, P.W. Niche-mediated control of human embryonic stem cell self-renewal and differentiation. *EMBO J.* **2007**, *26*, 4744–4755. [CrossRef] [PubMed]

24. Szablowska-Gadomska, I.; Zayat, V.; Buzanska, L. Influence of low oxygen tensions on expression of pluripotency genes in stem cells. *Acta Neurobiol. Exp.* **2011**, *71*, 86–93.

25. Burdick, J.A.; Vunjak-Novakovic, G. Engineered microenvironments for controlled stem cell differentiation. *Tissue Eng. Part A* **2009**, *15*, 205–219. [CrossRef]

26. Janmey, P.A.; Miller, R.T. Mechanisms of mechanical signaling in development and disease. *J. Cell Sci.* **2011**, *124*, 9–18. [CrossRef]

27. Provenzano, P.P.; Keely, P.J. Mechanical signaling through the cytoskeleton regulates cell proliferation by coordinated focal adhesion and rho gtpase signaling. *J. Cell Sci.* **2011**, *124*, 1195–1205. [CrossRef]

28. Baker, B.M.; Chen, C.S. Deconstructing the third dimension: How 3d culture microenvironments alter cellular cues. *J. Cell Sci.* **2012**, *125*, 3015–3024. [CrossRef]

29. Lund, A.W.; Yener, B.; Stegemann, J.P.; Plopper, G.E. The natural and engineered 3d microenvironment as a regulatory cue during stem cell fate determination. *Tissue Eng. Part B Rev.* **2009**, *15*, 371–380. [CrossRef]

30. Lutolf, M.P.; Gilbert, P.M.; Blau, H.M. Designing materials to direct stem-cell fate. *Nature* **2009**, *462*, 433–441. [CrossRef]

31. Saha, K.; Pollock, J.F.; Schaffer, D.V.; Healy, K.E. Designing synthetic materials to control stem cell phenotype. *Curr. Opin. Chem. Biol.* **2007**, *11*, 381–387. [CrossRef]

32. Mammoto, A.; Mammoto, T.; Ingber, D.E. Mechanosensitive mechanisms in transcriptional regulation. *J. Cell Sci.* **2012**, *125*, 3061–3073. [CrossRef] [PubMed]

33. Krupinski, P.; Chickarmane, V.; Peterson, C. Simulating the mammalian blastocyst—Molecular and mechanical interactions pattern the embryo. *PLOS Comput. Biol.* **2011**, *7*, e1001128. [CrossRef] [PubMed]

34. McKee, C.; Perez-Cruet, M.; Chavez, F.; Chaudhry, G.R. Simplified three-dimensional culture system for long-term expansion of embryonic stem cells. *World J. Stem Cells* **2015**, *7*, 1064–1077. [PubMed]

35. Thomson, J.A.; Itskovitz-Eldor, J.; Shapiro, S.S.; Waknitz, M.A.; Swiergiel, J.J.; Marshall, V.S.; Jones, J.M. Embryonic stem cell lines derived from human blastocysts. *Science* **1998**, *282*, 1145–1147. [CrossRef]

36. Ware, C.B.; Nelson, A.M.; Mecham, B.; Hesson, J.; Zhou, W.; Jonlin, E.C.; Jimenez-Caliani, A.J.; Deng, X.; Cavanaugh, C.; Cook, S.; et al. Derivation of naive human embryonic stem cells. *Proc. Natl. Acad. Sci. USA* **2014**, *111*, 4484–4489. [CrossRef]

37. Nair, D.P.; Podgórski, M.; Chatani, S.; Gong, T.; Xi, W.; Fenoli, C.R.; Bowman, C.N. The thiol-michael addition click reaction: A powerful and widely used tool in materials chemistry. *Chem. Mater.* **2014**, *26*, 724–744. [CrossRef]

38. Antoni, D.; Burckel, H.; Josset, E.; Noel, G. Three-dimensional cell culture: A breakthrough in vivo. *Int. J. Mol. Sci.* **2015**, *16*, 5517–5527. [CrossRef]

39. Kraehenbuehl, T.P.; Langer, R.; Ferreira, L.S. Three-dimensional biomaterials for the study of human pluripotent stem cells. *Nat. Methods* **2011**, *8*, 731–736. [CrossRef]

40. Qin, H.; Hejna, M.; Liu, Y.; Percharde, M.; Wossidlo, M.; Blouin, L.; Durruthy-Durruthy, J.; Wong, P.; Qi, Z.; Yu, J.; et al. Yap induces human naive pluripotency. *Cell Rep.* **2016**, *14*, 2301–2312. [CrossRef]

41. Beers, J.; Gulbranson, D.R.; George, N.; Siniscalchi, L.I.; Jones, J.; Thomson, J.A.; Chen, G. Passaging and colony expansion of human pluripotent stem cells by enzyme-free dissociation in chemically defined culture conditions. *Nat. Protoc.* **2012**, *7*, 2029–2040. [CrossRef]

42. Chen, G.; Gulbranson, D.R.; Hou, Z.; Bolin, J.M.; Ruotti, V.; Probasco, M.D.; Smuga-Otto, K.; Howden, S.E.; Diol, N.R.; Propson, N.E.; et al. Chemically defined conditions for human ipsc derivation and culture. *Nat. Methods* **2011**, *8*, 424–429. [CrossRef] [PubMed]

43. Nie, Y.; Walsh, P.; Clarke, D.L.; Rowley, J.A.; Fellner, T. Scalable passaging of adherent human pluripotent stem cells. *PLoS ONE* **2014**, *9*, e88012. [CrossRef] [PubMed]

44. Lee, J.; Cuddihy, M.J.; Kotov, N.A. Three-dimensional cell culture matrices: State of the art. *Tissue Eng. Part B Rev.* **2008**, *14*, 61–86. [CrossRef] [PubMed]

45. Chowdhury, F.; Li, Y.; Poh, Y.C.; Yokohama-Tamaki, T.; Wang, N.; Tanaka, T.S. Soft substrates promote homogeneous self-renewal of embryonic stem cells via downregulating cell-matrix tractions. *PLoS ONE* **2010**, *5*, e15655. [CrossRef]

46. Chowdhury, F.; Na, S.; Li, D.; Poh, Y.C.; Tanaka, T.S.; Wang, F.; Wang, N. Material properties of the cell dictate stress-induced spreading and differentiation in embryonic stem cells. *Nat. Mater.* **2010**, *9*, 82–88. [CrossRef]

47. Higuchi, S.; Watanabe, T.M.; Kawauchi, K.; Ichimura, T.; Fujita, H. Culturing of mouse and human cells on soft substrates promote the expression of stem cell markers. *J. Biosci. Bioeng.* **2014**, *117*, 749–755. [CrossRef]

48. McKee, C.; Hong, Y.; Yao, D.; Chaudhry, G.R. Compression induced chondrogenic differentiation of embryonic stem cells in three-dimensional polydimethylsiloxane scaffolds. *Tissue Eng. Part A* **2017**, *23*, 426–435. [CrossRef]

49. McKee, C.; Beeravolu, N.; Brown, C.; Perez-Cruet, M.; Chaudhry, G.R. Mesenchymal stem cells transplanted with self-assembling scaffolds differentiated to regenerate nucleus pulposus in an ex vivo model of degenerative disc disease. *Appl. Mater. Today* **2019**. [CrossRef]

50. Keung, A.J.; Asuri, P.; Kumar, S.; Schaffer, D.V. Soft microenvironments promote the early neurogenic differentiation but not self-renewal of human pluripotent stem cells. *Integr. Biol.* **2012**, *4*, 1049–1058. [CrossRef]

51. Sun, Y.; Villa-Diaz, L.G.; Lam, R.H.W.; Chen, W.; Krebsbach, P.H.; Fu, J. Mechanics regulates fate decisions of human embryonic stem cells. *PLoS ONE* **2012**, *7*, e37178. [CrossRef]

52. Jang, M.; Lee, S.T.; Kim, J.W.; Yang, J.H.; Yoon, J.K.; Park, J.C.; Ryoo, H.M.; van der Vlies, A.J.; Ahn, J.Y.; Hubbell, J.A.; et al. A feeder-free, defined three-dimensional polyethylene glycol-based extracellular matrix niche for culture of human embryonic stem cells. *Biomaterials* **2013**, *34*, 3571–3580. [CrossRef] [PubMed]

53. Chen, K.G.; Mallon, B.S.; McKay, R.D.G.; Robey, P.G. Human pluripotent stem cell culture: Considerations for maintenance, expansion, and therapeutics. *Cell Stem Cell* **2014**, *14*, 13–26. [CrossRef] [PubMed]

54. Kharkar, P.M.; Kiick, K.L.; Kloxin, A.M. Designing degradable hydrogels for orthogonal control of cell microenvironments. *Chem. Soc. Rev.* **2013**, *42*, 7335–7372. [CrossRef] [PubMed]

55. Gerecht, S.; Burdick, J.A.; Ferreira, L.S.; Townsend, S.A.; Langer, R.; Vunjak-Novakovic, G. Hyaluronic acid hydrogel for controlled self-renewal and differentiation of human embryonic stem cells. *Proc. Natl. Acad. Sci. USA* **2007**, *104*, 11298–11303. [CrossRef] [PubMed]

56. Li, Z.; Leung, M.; Hopper, R.; Ellenbogen, R.; Zhang, M. Feeder-free self-renewal of human embryonic stem cells in 3d porous natural polymer scaffolds. *Biomaterials* **2010**, *31*, 404–412. [CrossRef]

57. Lei, Y.; Schaffer, D.V. A fully defined and scalable 3d culture system for human pluripotent stem cell expansion and differentiation. *Proc. Natl. Acad. Sci. USA* **2013**, *110*, E5039–E5048. [CrossRef]

58. Wei, J.; Han, J.; Zhao, Y.; Cui, Y.; Wang, B.; Xiao, Z.; Chen, B.; Dai, J. The importance of three-dimensional scaffold structure on stemness maintenance of mouse embryonic stem cells. *Biomaterials* **2014**, *35*, 7724–7733. [CrossRef]

59. Li, X.; Ma, R.; Gu, Q.; Liang, L.; Wang, L.; Zhang, Y.; Wang, X.; Liu, X.; Li, Z.; Fang, J.; et al. A fully defined static suspension culture system for large-scale human embryonic stem cell production. *Cell Death Dis.* **2018**, *9*, 892. [CrossRef]

60. Collier, A.J.; Panula, S.P.; Schell, J.P.; Chovanec, P.; Plaza Reyes, A.; Petropoulos, S.; Corcoran, A.E.; Walker, R.; Douagi, I.; Lanner, F.; et al. Comprehensive cell surface protein profiling identifies specific markers of human naive and primed pluripotent states. *Cell Stem Cell* **2017**, *20*, 874–890. [CrossRef]

61. Rostovskaya, M.; Stirparo, G.G.; Smith, A. Capacitation of human naïve pluripotent stem cells for multi-lineage differentiation. *Development* **2019**, *146*, dev172916. [CrossRef]

62. Guo, G.; von Meyenn, F.; Santos, F.; Chen, Y.; Reik, W.; Bertone, P.; Smith, A.; Nichols, J. Naive pluripotent stem cells derived directly from isolated cells of the human inner cell mass. *Stem Cell Rep.* **2016**, *6*, 437–446. [CrossRef] [PubMed]

63. Hanna, J.; Cheng, A.W.; Saha, K.; Kim, J.; Lengner, C.J.; Soldner, F.; Cassady, J.P.; Muffat, J.; Carey, B.W.; Jaenisch, R. Human embryonic stem cells with biological and epigenetic characteristics similar to those of mouse escs. *Proc. Natl. Acad. Sci. USA* **2010**, *107*, 9222–9227. [CrossRef] [PubMed]

64. Hashimoto, M.; Sasaki, H. Epiblast formation by tead-yap-dependent expression of pluripotency factors and competitive elimination of unspecified cells. *Dev. Cell* **2019**, *50*, 139–154. [CrossRef]

65. Kilens, S.; Meistermann, D.; Moreno, D.; Chariau, C.; Gaignerie, A.; Reignier, A.; Lelièvre, Y.; Casanova, M.; Vallot, C.; Nedellec, S.; et al. Parallel derivation of isogenic human primed and naive induced pluripotent stem cells. *Nat. Commun.* **2018**, *9*, 360. [CrossRef] [PubMed]

66. Takashima, Y.; Guo, G.; Loos, R.; Nichols, J.; Ficz, G.; Krueger, F.; Oxley, D.; Santos, F.; Clarke, J.; Mansfield, W.; et al. Resetting transcription factor control circuitry toward ground-state pluripotency in human. *Cell* **2014**, *158*, 1254–1269. [CrossRef]

67. Sim, Y.J.; Kim, M.S.; Nayfeh, A.; Yun, Y.J.; Kim, S.J.; Park, K.T.; Kim, C.H.; Kim, K.S. 2i maintains a naive ground state in escs through two distinct epigenetic mechanisms. *Stem Cell Rep.* **2017**, *8*, 1312–1328. [CrossRef]

68. Van der Jeught, M.; Taelman, J.; Duggal, G.; Ghimire, S.; Lierman, S.; Chuva de Sousa Lopes, S.M.; Deforce, D.; Deroo, T.; De Sutter, P.; Heindryckx, B. Application of small molecules favoring naive pluripotency during human embryonic stem cell derivation. *Cell. Reprogramming* **2015**, *17*, 170–180. [CrossRef]

69. Zhang, Y.; Liao, K.; Li, C.; Lai, A.C.K.; Foo, J.J.; Chan, V. Progress in integrative biomaterial systems to approach three-dimensional cell mechanotransduction. *Bioengineering* **2017**, *4*, 72. [CrossRef]

70. Finch-Edmondson, M.; Sudol, M. Framework to function: Mechanosensitive regulators of gene transcription. *Cell. Mol. Biol. Lett.* **2016**, *21*, 28. [CrossRef]

71. Baxter, M.A.; Camarasa, M.V.; Bates, N.; Small, F.; Murray, P.; Edgar, D.; Kimber, S.J. Analysis of the distinct functions of growth factors and tissue culture substrates necessary for the long-term self-renewal of human embryonic stem cell lines. *Stem Cell Res.* **2009**, *3*, 28–38. [CrossRef]

72. Mei, Y.; Saha, K.; Bogatyrev, S.R.; Yang, J.; Hook, A.L.; Kalcioglu, Z.I.; Cho, S.W.; Mitalipova, M.; Pyzocha, N.; Rojas, F.; et al. Combinatorial development of biomaterials for clonal growth of human pluripotent stem cells. *Nat. Mater.* **2010**, *9*, 768–778. [CrossRef] [PubMed]

73. Vitillo, L.; Kimber, S.J. Integrin and fak regulation of human pluripotent stem cells. *Curr. Stem Cell Rep.* **2017**, *3*, 358–365. [CrossRef] [PubMed]

74. Lee, S.T.; Yun, J.I.; Jo, Y.S.; Mochizuki, M.; van der Vlies, A.J.; Kontos, S.; Ihm, J.E.; Lim, J.M.; Hubbell, J.A. Engineering integrin signaling for promoting embryonic stem cell self-renewal in a precisely defined niche. *Biomaterials* **2010**, *31*, 1219–1226. [CrossRef] [PubMed]

75. Sun, Y.; Chen, C.S.; Fu, J. Forcing stem cells to behave: A biophysical perspective of the cellular microenvironment. *Annu. Rev. Biophys.* **2012**, *41*, 519–542. [CrossRef]

76. Yu, F.X.; Zhao, B.; Panupinthu, N.; Jewell, J.L.; Lian, I.; Wang, L.H.; Zhao, J.; Yuan, H.; Tumaneng, K.; Li, H.; et al. Regulation of the hippo-yap pathway by g-protein-coupled receptor signaling. *Cell* **2012**, *150*, 780–791. [CrossRef]

77. Lam, R.M.; Chesler, A.T. Shear elegance: A novel screen uncovers a mechanosensitive gpcr. *J. Gen. Physiol.* **2018**, *150*, 907. [CrossRef]

78. Varelas, X. The hippo pathway effectors taz and yap in development, homeostasis and disease. *Development* **2014**, *141*, 1614–1626. [CrossRef]

79. Elosegui-Artola, A.; Andreu, I.; Beedle, A.E.M.; Lezamiz, A.; Uroz, M.; Kosmalska, A.J.; Oria, R.; Kechagia, J.Z.; Rico-Lastres, P.; Le Roux, A.L.; et al. Force triggers yap nuclear entry by regulating transport across nuclear pores. *Cell* **2017**, *171*, 1397–1410. [CrossRef]

80. Sasaki, H. Roles and regulations of hippo signaling during preimplantation mouse development. *Dev. Growth Differ.* **2017**, *59*, 12–20. [CrossRef]

81. Ohgushi, M.; Minaguchi, M.; Sasai, Y. Rho-signaling-directed yap/taz activity underlies the long-term survival and expansion of human embryonic stem cells. *Cell Stem Cell* **2015**, *17*, 448–461. [CrossRef]

82. Hsiao, C.; Lampe, M.; Nillasithanukroh, S.; Han, W.; Lian, X.; Palecek, S.P. Human pluripotent stem cell culture density modulates yap signaling. *Biotechnol. J.* **2016**, *11*, 662–675. [CrossRef] [PubMed]

83. Varelas, X.; Sakuma, R.; Samavarchi-Tehrani, P.; Peerani, R.; Rao, B.M.; Dembowy, J.; Yaffe, M.B.; Zandstra, P.W.; Wrana, J.L. Taz controls smad nucleocytoplasmic shuttling and regulates human embryonic stem-cell self-renewal. *Nat. Cell Biol.* **2008**, *10*, 837. [CrossRef] [PubMed]

84. Lian, I.; Kim, J.; Okazawa, H.; Zhao, J.; Zhao, B.; Yu, J.; Chinnaiyan, A.; Israel, M.A.; Goldstein, L.S.; Abujarour, R.; et al. The role of yap transcription coactivator in regulating stem cell self-renewal and differentiation. *Genes Dev.* **2010**, *24*, 1106–1118. [CrossRef] [PubMed]

85. Papaspyropoulos, A.; Bradley, L.; Thapa, A.; Leung, C.Y.; Toskas, K.; Koennig, D.; Pefani, D.E.; Raso, C.; Grou, C.; Hamilton, G.; et al. Rassf1a uncouples wnt from hippo signalling and promotes yap mediated differentiation via p73. *Nat. Commun.* **2018**, *9*, 424. [CrossRef]

86. Young, R.A. Control of the embryonic stem cell state. *Cell* **2011**, *144*, 940–954. [CrossRef]

87. Meng, Z.; Moroishi, T.; Guan, K.L. Mechanisms of hippo pathway regulation. *Genes Dev.* **2016**, *30*, 1–17. [CrossRef]

88. Wang, C.; Zhu, X.; Feng, W.; Yu, Y.; Jeong, K.; Guo, W.; Lu, Y.; Mills, G.B. Verteporfin inhibits yap function through up-regulating 14-3-3σ sequestering yap in the cytoplasm. *Am. J. Cancer Res.* **2015**, *6*, 27–37.

89. Onishi, K.; Zandstra, P.W. Lif signaling in stem cells and development. *Development* **2015**, *142*, 2230. [CrossRef]

90. Daheron, L.; Opitz, S.L.; Zaehres, H.; Lensch, M.W.; Andrews, P.W.; Itskovitz-Eldor, J.; Daley, G.Q. Lif/stat3 signaling fails to maintain self-renewal of human embryonic stem cells. *Stem Cells* **2004**, *22*, 770–778. [CrossRef]

Article

Generation of Differentiating and Long-Living Intestinal Organoids Reflecting the Cellular Diversity of Canine Intestine

Nina Kramer [1,*], Barbara Pratscher [1,‡], Andre M. C. Meneses [1,2,†,‡], Waltraud Tschulenk [3], Ingrid Walter [3], Alexander Swoboda [1], Hedwig S. Kruitwagen [2], Kerstin Schneeberger [2], Louis C. Penning [2], Bart Spee [2], Matthias Kieslinger [1], Sabine Brandt [4] and Iwan A. Burgener [1]

1 Division of Small Animal Internal Medicine, Department for Small Animals and Horses, University of Veterinary Medicine, 1210 Vienna, Austria
2 Department of Clinical Sciences, Faculty of Veterinary Medicine, Utrecht University, 3584 Utrecht, The Netherlands
3 Institute of Pathology, Department for Pathobiology, University of Veterinary Medicine, 1210 Vienna, Austria
4 Research Group Oncology, Equine Surgery, Department of Small Animals and Horses, University of Veterinary Medicine, 1210 Vienna, Austria
* Correspondence: nina.kramer@vetmeduni.ac.at
† Current address: Institute of Animal Health and Production, Veterinary Hospital, Amazon Rural Federal University, Belém 66.077-830, Brazil.
‡ These authors contributed equally to this work.

Received: 17 February 2020; Accepted: 26 March 2020; Published: 28 March 2020

Abstract: Functional intestinal disorders constitute major, potentially lethal health problems in humans. Consequently, research focuses on elucidating the underlying pathobiological mechanisms and establishing therapeutic strategies. In this context, intestinal organoids have emerged as a potent in vitro model as they faithfully recapitulate the structure and function of the intestinal segment they represent. Interestingly, human-like intestinal diseases also affect dogs, making canine intestinal organoids a promising tool for canine and comparative research. Therefore, we generated organoids from canine duodenum, jejunum and colon, and focused on simultaneous long-term expansion and cell differentiation to maximize applicability. Following their establishment, canine intestinal organoids were grown under various culture conditions and then analyzed with respect to cell viability/apoptosis and multi-lineage differentiation by transcription profiling, proliferation assay, cell staining, and transmission electron microscopy. Standard expansion medium supported long-term expansion of organoids irrespective of their origin, but inhibited cell differentiation. Conversely, transfer of organoids to differentiation medium promoted goblet cell and enteroendocrine cell development, but simultaneously induced apoptosis. Unimpeded stem cell renewal and concurrent differentiation was achieved by culturing organoids in the presence of tyrosine kinase ligands. Our findings unambiguously highlight the characteristic cellular diversity of canine duodenum, jejunum and colon as fundamental prerequisite for accurate in vitro modelling.

Keywords: intestinal organoids; canine intestine; differentiation; organoid culture

---

1. Introduction

Gastrointestinal (GI) disorders such as inflammatory bowel disease (IBD), infection-induced GI disorders and cancer have a major negative impact on human health and impose a high financial burden on healthcare systems. Infectious diarrhea is the second leading cause of death among young children [1] and GI cancers are the third most common cancer type worldwide, with one million newly diagnosed cases per year [2]. Potentially lethal GI diseases also affect livestock and companion animals, with enterotoxic bacteria and enteropathogenic viruses being frequently involved in disease onset and progression [3]. Although considerable efforts are made to establish new therapies for human and veterinary GI diseases, mortality rates remain high, because translation of biomedical research into clinical practice is hampered by the lack of epithelial models that reliably mimic the organ and recapitulate disease in patients [4–6].

Early attempts to bring the intestinal epithelium to in vitro culture were mostly short-lived, as exemplified by explanted biopsies and epithelial cells, which disintegrated after 72 h and two weeks, respectively [7–9]. To establish culture systems for long-lived homogenous cell populations tissue explants were either cultivated in collagen gels or on 3T3 feeder layers on an air-liquid interface [10,11]. However, growth factor-providing (myo)fibroblasts, which are prerequisites for prolonged cultivation in these systems, made them susceptible to inconsistencies from one experiment to another. A major step to overcome these limitations was the establishment of three-dimensional (3D) murine intestinal organoids derived from adult intestinal stem cells [12]. These organoids contained long-lived stem cells that differentiated into the main cell types of murine small intestine such as Paneth cells, enterocytes, enteroendocrine cells and goblet cells. Establishment of this innovative in vitro model was achieved by using a laminin-rich extracellular matrix (Matrigel) and growth media supplemented with epithelial growth factor (EGF), Noggin, and R-spondin as reviewed in detail by Date and Sato [13]. For the cultivation of human intestinal organoids, growth medium had to be supplemented with Wnt3a, gastrin, p38-MAPK inhibitor, nicotinamide and ALK4/5/7 inhibitor, thereby preventing the differentiation into goblet and enteroendocrine cells and retaining enterocytes in a premature state [14]. Withdrawing nicotinamide and p38-MAPK inhibitor induced de novo differentiation, thus shortening the lifespan of the intestinal stem cell pool and rendering this system into an endpoint assay. Further improvement of human 3D intestinal organoid models was achieved by microscaffolds that mimic the size and distance between crypts in transwell assays [15]. Differentiation was promoted by using a growth factor gradient system based on (i) intestinal stem cell-supporting expansion medium in lower wells, and (ii) addition of differentiation medium to upper wells, thereby inducing differentiation of migrating cells to goblet cells, enteroendocrine cells and enterocytes along the modeled crypt/villus axis. However, this technique is unsuitable for, e.g., high-throughput screenings because it is very labor-intensive, time-consuming and hampers down-stream analysis. An approach characterizing receptor tyrosine kinase signaling between the crypt base and its niche managed to omit nicotinamide and p38-MAPK inhibitor usage by a combination of two ligands, i.e., insulin-like growth factor 1 (IGF1) and basic fibroblast growth factor (FGF2). This combination sustains stem cell growth and allows simultaneous differentiation into enteroendocrine and goblet cells, mimicking corresponding natural epithelia more authentically with minimal effort.

In recent years, organoid technology has also been introduced into veterinary research, albeit to a lesser extent [16–22]. Animal organoid systems that are already available today mainly consist of stem and undifferentiated cells requiring tailor-made species-specific media to allow for differentiation [17]. Interestingly, dogs and their owners share similar environments, food and carcinogenic load, and develop similar GI diseases including GI cancer, infectious disease and IBD [23]. Consequently, canine patients represent an interesting natural model for human GI disorders, even more so since dogs show reduced genetic variation within the majority of breeds [24]. To exploit the full potential of canine intestinal organoids to mimic the corresponding organ accurately, a balance between self-renewal and differentiation of stem cells must be achieved to design a physiologically relevant system.

Herein, we report the establishment of a novel culture system for canine intestinal organoids that is based on recent findings by Fujii et al. [25] and reliably supports the sustained proliferation of duodenum-, jejunum- and colon-derived stem cells, while concomitantly allowing their differentiation into secretory lineage cells.

2. Materials and Methods

2.1. Isolation and Cultivation of Canine Intestinal Organoids

Duodenal, jejunal and colonic samples were obtained from three dogs (two males and one female) that were euthanized due to non-intestinal disease. In addition, duodenal biopsies were taken from one dog undergoing routine gastroduodenal endoscopy. Tissue sampling was approved by the institutional ethics committee in accordance with Good Scientific Practice guidelines and Austrian legislation. Based on the guidelines of the institutional ethics committee, use of tissue material collected during therapeutic excision or post mortem is included in the University's "owner's consent for treatment", which was signed by all patient owners. Intestinal crypts were isolated from tissue samples and biopsies according to established protocols [12]. In short, tissue was incubated with 5 mM EDTA (Sigma-Aldrich, St. Louis, MO, USA) in order to dissociate crypts for 30 min to one hour depending on bowel segment, followed by vigorous shaking until crypts were released. After two washing steps with PBS and Advanced Dulbecco's modified Eagle's medium/F12 (DMEM/F12, Invitrogen, Thermo Fisher Scientific, Waltham, MA, USA) 500 crypts were resuspended in 50 µL Matrigel (BD Biosciences, Franklin Lakes, NJ, USA) and seeded per well of a 24-well plate. Following Matrigel polymerization expansion medium was added. Canine intestinal growth media were prepared as indicated in Table 1. Penicillin/streptomycin, HEPES, Glutamax, B27 (with Vitamin A) and N2 supplement were obtained from Invitrogen, Thermo Fisher Scientific. N-acetylcysteine, gastrin and nicotinamide were purchased from Sigma-Aldrich. EGF was provided by Thermo Fisher Scientific. ALK5 kinase inhibitor (A83-01) was obtained from Tocris Bioscience, Bristol, UK. p38 MAPK inhibitor (SB202190) was purchased from Selleck Chemicals, Houston, TX, USA. Human hepatocyte growth factor (HGF), human Noggin, human IGF1 and human FGF2 were provided by PeproTech, Rocky Hill, NJ, USA. Cultrex R-spondin1 cells were obtained from Trevigen, Gaithersburg, MD, USA. Expansion medium was supplemented with 10 µM Rock inhibitor (Y-27632, Selleck Chemicals) for the first two days after isolation, then medium was changed to not-supplemented expansion medium. For isolations using refined medium, EGF and 10 µM Y-27632 were additionally added only for the first two days. Growth medium was changed every two to three days. For weekly passaging at 1:4 to 1:8 split ratios, organoids were harvested, mechanically disrupted using a flame-polished Pasteur pipette. Depending on the splitting ratio, the corresponding quantity of organoid fragments were then embedded in 50 µL fresh Matrigel, seeded per well of a 24-well plate and cultured with either expansion or refined growth medium. To induce differentiation, organoids cultivated in expansion medium were transferred to differentiation medium and analyzed after 4 days. For assessing derivation efficiency in expansion and refined medium, organoids were counted 10 days after initial isolation and projected area was calculated using ImageJ64 (NIH).

Table 1. List of media compositions used in this study. Basal medium consists of Advanced DMEM/F12, 2 mM GlutaMAX, 10 mM HEPES, 1x B27, 1 mM N-acetylcystein. Final concentrations are given.

Reagent	Expansion Medium	Differentiation Medium	Refined Medium
basal medium	+	+	+
N2	1x	1x	−
EGF	50 ng/mL	50 ng/mL	− [a]
Noggin	100 ng/mL	100 ng/mL	100 ng/mL
Rspo1	10% *v/v*	10% *v/v*	10% *v/v*
Wnt3a	43% *v/v*	43% *v/v*	50% *v/v*
Nicotinamide	10 mM	−	−
Gastrin	10 nM	10 nM	10 nM
A83-01	500 nM	500 nM	500 nM
SB202190	10 µM	−	−
HGF	50 ng/mL	50 ng/mL	50 ng/mL
IGF1	−	−	100 ng/mL
FGF2	−	−	50 ng/mL

[a] in refined medium 50 ng/mL EGF is only added for the first two days after isolation or medium adaption.

2.2. Gene Expression Analysis

Total RNA was purified from duodenal, jejunal and colonic organoids in expansion, differentiation and refined medium and from corresponding intestinal epithelium that was harvested during crypt isolation using ReliaPrep™ RNA Tissue Miniprep System (Promega, Madison, WI, USA) according to instructions of the manufacturer. Individual RNA concentrations were determined spectrophotometrically (Nanodrop One C, Thermo Fisher Scientific) and RNA integrity was checked using Agilent Tape Station 4200 (Agilent, Santa Clara, CA, USA). 500 ng RNA was reverse-transcribed to cDNA with oligo-dT and random hexamer primers according to recommendations of the manufacturer (GoScript™ Reverse Transcription System, Promega). Subsequently, qPCR reactions were carried out using GoTaq® qPCR Master Mix (Promega). Primer sequences are given in Table S1. Amplification conditions were as follows: 2 min of initial denaturation at 95 °C, 40 cycles of 15 s of denaturation at 95 °C, 60 s of annealing/extension at 60 °C and a read step, followed by 10 s of dissociation at 95 °C and a melting curve from 65 °C to 95 °C in 5 s per 0.5 °C increments. Data analysis was performed according to Pfaffl et al. [26] taking PCR efficiency into account. Relative quantification was achieved by normalization of values to the stably expressed canine housekeeping gene hypoxanthine phosphoribosyl transferase (HPRT). Heat maps were generated using Excel conditional formatting of log2 fold changes.

2.3. Periodic Acid-Schiff Reaction of Organoid and Tissue Sections

For the generation of formalin-fixed paraffin-embedded (FFPE) samples, organoids were fixed using 2% *v/v* paraformaldehyde (PFA, Merck, Darmstadt, Germany) for 15 min at room temperature, cast in 1.5% *w/v* agarose and dehydrated prior to embedding in paraffin (Sigma-Aldrich). 2.5 μm FFPE sections of organoids derived from duodenum, jejunum and colon and corresponding tissue were stained using PAS-Reaction staining kit and counterstained with Haematoxylin acidic after Mayer according to manufacturer's instructions (Morphisto, Frankfurt am Main, Germany). Images were acquired with a DMi8 microscope and LASX software (Leica, Wetzlar, Germany).

2.4. Transmission Electron Microscopy of Organoids

Samples were fixed in 3% *v/v* buffered glutaraldehyde (pH 7.4, Merck, Darmstadt, Germany), pre-embedded in 1.5% *w/v* agarose (Invitrogen, Thermo Fisher Scientific), washed three times in 0.1 M phosphate buffer (Soerensen, pH 7.4) afterwards and post-fixed in 1% *v/v* osmium tetroxide (Electron Microscopy Sciences, Hatfield, USA) for 2 h at room temperature. Dehydration was performed in a series of graded ethanol solutions (70%, 80%, 96% and 100%), subsequently infiltrated with propylene oxide (Sigma-Aldrich), followed by increasing ratios of epoxy resin-propylene oxide and finally pure resin (Serva, Mannheim, Germany). After an additional change, the resin was polymerized at 60 °C for 48 h. Semi-thin sections were cut at 0.8 μm and stained with toluidine blue, ultra-thin sections were cut at 70 nm, mounted on copper grids (Gröpl, Tulln, Austria) and stained with uranyl acetate (Fluka Chemie AG, Buchs, CH) and lead citrate (Merck, Darmstadt, Germany). Transmission electron micrographs were made with an EM900 (Zeiss, Oberkochen, Germany).

2.5. Proliferation Assay

In order to assess cell proliferation, the Click-iT® EdU Imaging Kit (Invitrogen, Thermo Fisher Scientific) was used. Canine intestinal organoids cultivated in expansion and refined medium were incubated with 5-ethynyl-2'deoxyuridine (EdU) at a final concentration of 10 μM for one hour at 37 °C and were then fixed with 2% *v/v* PFA for 15 min at room temperature. Staining procedure was carried out according to manufacturer's instructions. DNA was counterstained using 4',6-Diamidine-2'-phenylindole dihydrochloride (DAPI, Sigma-Aldrich). Confocal images were taken using an LSM 880 (Zeiss). For 3D reconstruction of z-stack images, the Arivis 3D plugin of Zen 2.3 lite (Zeiss) was used.

2.6. Viability and Apoptosis Assays

Viability and apoptosis of organoids in the three media compositions were assessed using the RealTime-Glo MT Cell Viability Assay (Promega) and RealTime-Glo Annexin V Apoptosis Assay (Promega). Organoids were trypsinized for 2 min and seeded in 24-well plates with expansion or refined medium, both supplemented with Y-27632 inhibitor. After 2 days in culture, organoids were harvested, counted and 50 organoids were seeded in six wells of white 96-well plate for each condition and assay. Detection reagents were prepared according to manufacturer and added to the different culture media before overlaying the organoids with either expansion, differentiation or refined medium. Luminescence was measured 0 h, 24 h, 48 h and 72 h after adding the substrates using a GloMax plate reader (Promega). Six wells without organoids, but with medium, served as background controls.

2.7. Statistical Evaluation

Data were analyzed and plotted in Prism5 (GraphPad, San Diego, CA, USA). Using a t-test for independent samples (two tailed, unpaired) p-values ≤ 0.05 were considered statistically significant. qPCR data were analyzed by two-way ANOVA (factors are growth medium and individual dogs) and Tukey multiple comparison test using R (version 3.3.1). Statistical results are shown in Tables S2–S7. *** $p < 0.001$ ** $p < 0.01$ * $p < 0.05$. $p < 0.1$.

3. Results

3.1. Expansion Medium Does Not Support Expression of Secretory Lineage Differentiation Markers

Canine intestinal organoids derived from duodenal, jejunal and colonic tissue were generated and cultivated in expansion medium. Initial cell isolates representing the intestinal epithelium after separation from the underlying tissue, as well as low- (≤ 5) and high-passage (≥ 10) organoids derived therefrom, were quantitatively assessed for expression of selected cell markers using RT-qPCR.

The stem cell marker *LGR5* displayed similar expression levels in low- and high-passage organoids, except high-passage duodenal organoids, which exhibited significant up-regulation of *LGR5* in comparison to their low-passage and initial cell isolate counterparts (Figure 1A). The enterocyte marker villin 1 (*VIL1*) exhibited significantly reduced expression in duodenal, jejunal and colonic organoids in low- and high-passage organoids compared to the initial cell isolates. Enteroendocrine cell marker chromogranin A (*CHGA*) [27] and goblet cell marker mucin 2 (*MUC2*) [28,29] transcripts were barely detectable in organoids when kept in expansion medium throughout passages (Figure 1A).

3.2. Canine Intestinal Organoids Grow Efficiently in Refined Medium

To best reflect the in vivo situation and thus allow for downstream applications including drug and toxicity assessment, microbiome studies and modelling of infectious disease, intestinal organoids should harbor stem cells, undifferentiated transit-amplifying cells and differentiated cells. Three different media were compared to identify the ideal medium composition supporting such a scenario: (i) an expansion medium for the culture of canine organoids; (ii) a differentiation medium characterized by omission of nicotinamide and p38-MAPK inhibitor [14]; and (iii) a novel refined medium based on canine expansion medium without nicotinamide, p38-MAPK inhibitor, N2 and EGF, but supplemented with IGF1 and FGF2 as suggested for human organoid culture [25]. After several passages, duodenal, jejunal and colonic canine organoids grown in expansion medium were partly transferred to refined medium for long-term cultivation. For experiments, expansion medium-grown organoids where partly transferred to differentiation medium and after four days, organoids in all three media were harvested (Figure S1A). Growth characteristics and gene expression were assessed.

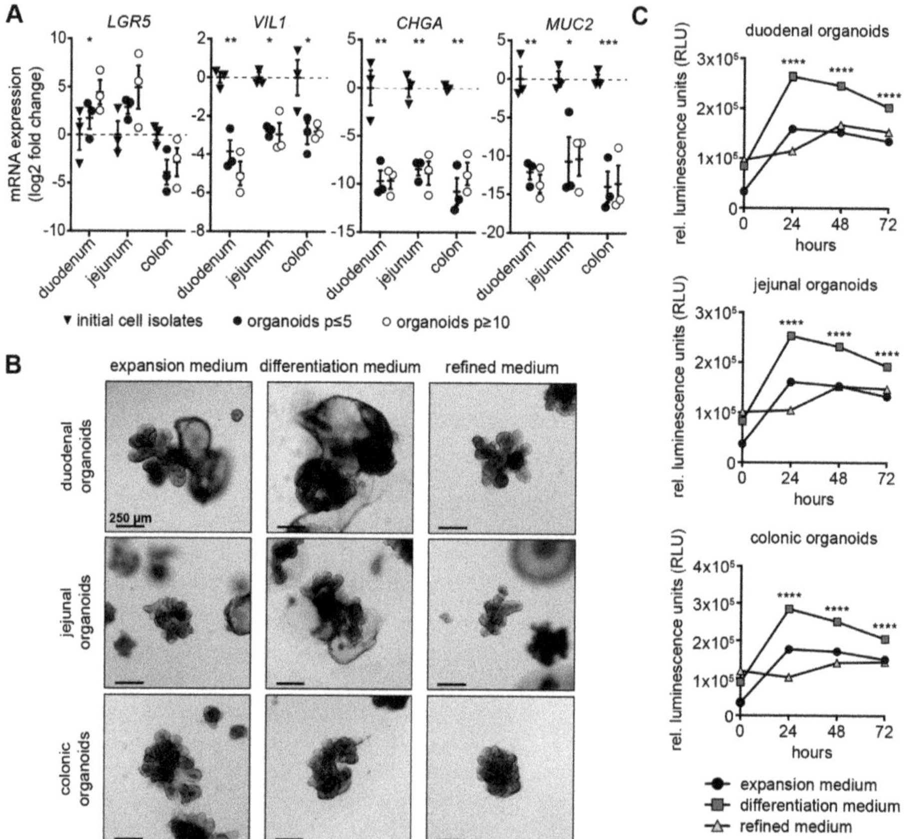

Figure 1. Establishment of new culture conditions for canine intestinal organoids. (**A**) RT-qPCR analysis of initial cell isolates (triangles), organoids with up to five passages (dark circles) and organoids with 10 or more passages (light circles) isolated from duodenum, jejunum or colon using *LGR5*, *VIL1*, *CHGA* and *MUC2* primer. log2 fold changes normalized to expression of initial cell isolates are shown with scatter dot plots; mean is shown; whiskers present SEM; * $p < 0.05$, ** $p < 0.01$ and *** $p < 0.001$, statistical analysis given in detail in Supplementary Tables S2–S4; n = 3 individual dogs. (**B**) Light microscopic images of organoids derived from duodenum, jejunum and colon cultivated in expansion, differentiation and refined medium four days after seeding; scale bars represent 250 μm. (**C**) Mean activity of Annexin V-induced luciferase is shown for 0, 24, 48 and 72 h after addition of substrate to duodenal, jejunal and colonic organoids in expansion (circle), differentiation (square) and refined medium (triangle); whiskers represent SEM; **** $p < 0.0001$; 50 organoids seeded per replicate; n = 6 replicates.

Bright field images of duodenal, jejunal and colonic organoids grown in expansion and refined medium displayed intact, budding structures (Figure 1B). In contrast, all organoid types grown in differentiation medium displayed dark and necrotic areas (Figure 1B, Figure S1B). In accordance with these observations, luciferase-based Annexin V assay revealed low and comparable apoptosis in organoids grown in expansion and refined medium, but pronounced apoptosis in organoids grown in differentiation medium (Figure 1C). Organoids in expansion and refined medium could be cultivated for prolonged period of 25 passages after adaptation of growth medium (Figure S1C). Taken together, only expansion and refined medium supported organoid growth.

3.3. Refined Medium Promotes Expression of Cell Differentiation Markers in Canine Intestinal Organoids

We next aimed at defining the cellular composition of organoids with respect to the different culture conditions. To this end, total RNA isolated from initial cell isolates and from the three different organoid types cultured in the three different media were subjected to gene expression analysis of stem cell and differentiation markers using RT-qPCR.

In general, a heat map representing log2 fold changes of organoids to respective initial cell isolates showed that transcription of stem cell markers was up-regulated in intestinal organoids grown in all three media. Conversely, transcription of differentiation markers was exclusively up-regulated in intestinal organoids grown in differentiation or refined medium (Figure 2A).

In more detail, duodenal organoids displayed significant up-regulation of stem cell marker *LGR5* [30,31] gene expression irrespective of culture conditions (Figure 2B). Culture of duodenal organoids in differentiation and refined medium enhanced mRNA expression of the differentiation markers *NEUROG3* (enteroendocrine progenitor cells) [32,33], *CHGA* (enteroendocrine cells) and *MUC2* (Goblet cells). Transcription of stem cell markers *ASCL2*, *EPHB2*, *BMI1*, *NOTCH1* and *PROM1* [30,31] did not vary significantly throughout duodenal organoids analyzed, while enterocyte marker *VIL1* mRNA was significantly down-regulated in duodenal organoids irrespective of culture conditions (Figure S2A).

Jejunal organoids grown in the three different media exhibited no significant difference in *LGR5*, *ASCL2*, *EPHB2*, *BMI1*, *NOTCH1* or *PROM1* transcription compared to corresponding initial cell isolates (Figure 2C, Figure S2B). *NEUROG3*, *CHGA* and *MUC2* gene expression was significantly down-regulated in jejunal organoids grown in expansion medium. Transcription of differentiation markers *NEUROG3*, *CHGA*, and *MUC2* mRNA was up-regulated when culturing jejunal organoids in refined medium. Expression of *VIL1* was slightly, but significantly, reduced throughout jejunal organoids (Figure S2B).

In colonic organoids, medium compositions had no significant impact on *LGR5* expression, while *NEUROG3*, *CHGA* and *MUC2* transcription was significantly down-regulated in organoids kept in expansion medium (Figure 2D). Culturing colonic organoids in differentiation and refined medium induced an increase of *CHGA* and *MUC2* expression almost to levels displayed by initial cell isolates. In contrast to duodenal and jejunal organoids, colonic organoids showed aberrant *ASCL2*, *EPHB2*, *BMI1* and *PROM1* transcription in comparison to initial cell isolates, whereas *NOTCH1* expression remained grossly unchanged (Figure S2C). *VIL1* transcription was generally slightly lower expressed in colonic organoids irrespective of culture conditions.

Marker transcription levels in organoids varied to some extent with respect to the individual dog they were derived from. These variations were less pronounced when cultivating organoids—notably jejunal and colonic organoids—in refined medium (Figure 2, Figure S2). Despite these individual variations the presented data indicate that differentiation and refined medium supported cell differentiation to secretory lineage cells within organoids.

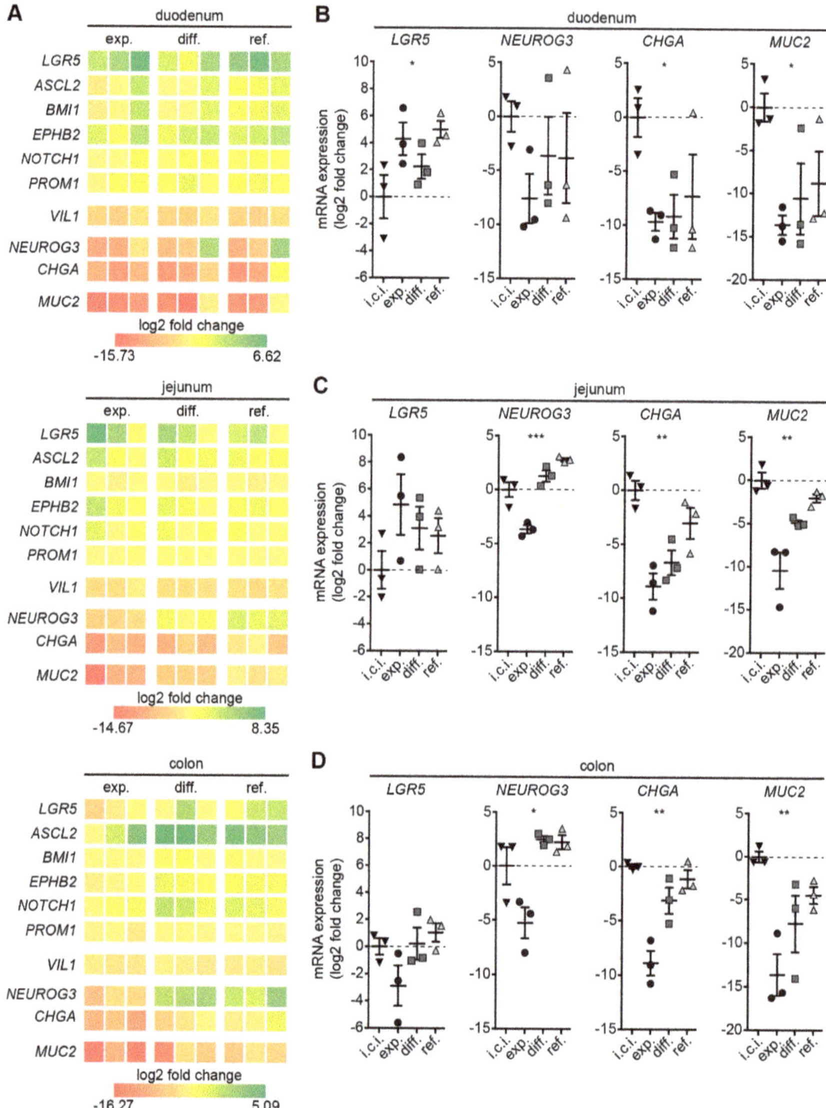

Figure 2. Refined medium and differentiation medium show elevated marker expression of enteroendocrine and goblet cells. (**A**) Heat map of log2 fold changes derived from RT-qPCR data of duodenal, jejunal and colonic organoids cultivated four days in expansion, differentiation and refined medium normalized to initial cell isolates (i.c.i. = 0); data from three individuals are shown in side-by-side columns. (**B–D**) Individual scatter dot plots of gene expression data from (**A**) shown for stem cell marker *LGR5*, secretory lineage precursor marker *NEUROG3*, enteroendocrine cell marker *CHGA* and goblet cell marker *MUC2* for organoids derived from duodenum (**B**), jejunum (**C**) and colon (**D**) four days after seeding; mean is shown, whiskers are SEM; * $p < 0.05$, ** $p < 0.01$, *** $p < 0.001$, statistical analysis given in detail in Supplementary Tables S5–S7; n = 3 dogs; i.c.i., initial cell isolates; exp., expansion medium; diff., differentiation medium; ref., refined medium.

3.4. Refined and Differentiation Medium Both Support Organoid Differentiation

The detection of markers for goblet and enteroendocrine cells by RT-qPCR prompted us to confirm and localize these cells within organoids. Therefore, periodic-acid Schiff (PAS) reaction and transmission electron microscopy (TEM) were used. PAS reaction of mucin as revealed by a purple signal demonstrated the presence of goblet cells that were evenly distributed in duodenal and jejunal tissue sections along the crypt-villus axis, and at higher frequency around the crypt base in colon tissue sections (Figure 3A). In organoid culture, expansion medium failed to induce differentiation into goblet cells, while growth in differentiation and refined medium gave rise to goblet cells irrespective of the intestinal segment. These observations were in accordance with *MUC2* transcription data (see Figure 2) and further substantiated by TEM, which revealed the presence of goblet cells with apical large mucin granules and enteroendocrine cells characterized by small dark basal granules in organoids grown in differentiation and refined medium (Figure 3B and Figure S3B). In contrast, TEM of organoids cultivated in expansion medium revealed uniformly appearing cells harboring microvilli (Figure S3A). These cells were also present in all other conditions suggesting the presence of enterocytes, which is correlating with *VIL1* expression data (see Figure 2 and Figure S2). Interestingly, cell-to-cell interactions such as tight junctions, adherens junctions and desmosomes could be visualized by organoid TEM throughout culture conditions (Figure S3C). These data demonstrate the presence of secretory lineage cells in canine organoids upon cultivation in refined and differentiation medium.

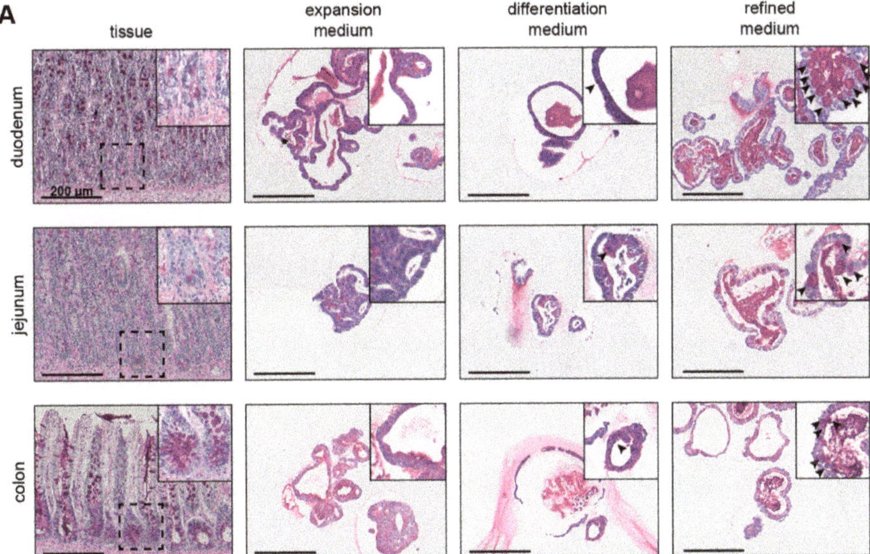

Figure 3. *Cont.*

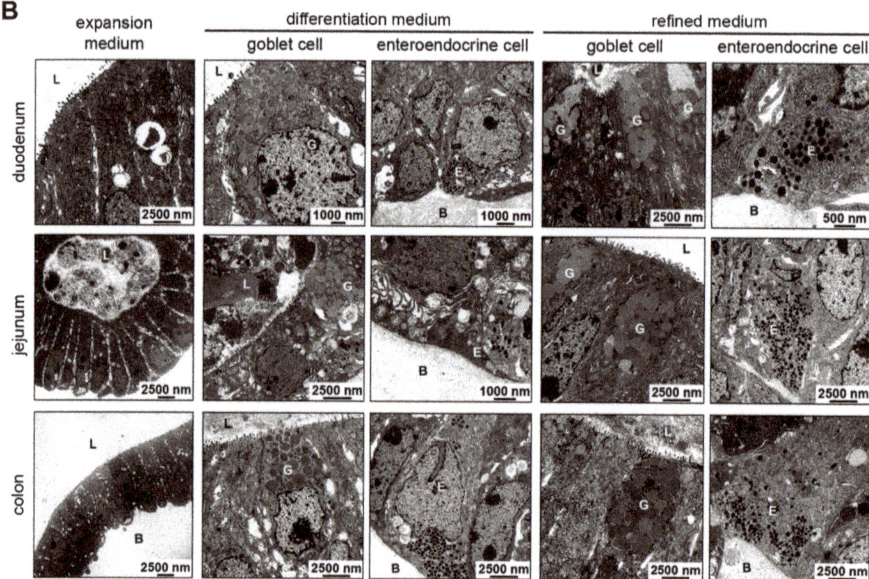

Figure 3. Goblet and enteroendocrine cells are present in organoids upon cultivation in differentiation and refined medium, but not in expansion medium. (**A**) PAS reaction in FFPE sections of canine intestinal tissue or organoids in expansion, differentiation and refined medium derived from duodenum, jejunum and colon; counterstained with Hematoxylin, images of duodenal and jejunal tissue were cropped in order to visualize crypts; scale bars 200 μm, inset depicts higher magnification; arrow heads indicate PAS positive goblet cells; one representative image is shown. (**B**) TEM images of duodenal, jejunal and colonic organoids in expansion, differentiation and refined medium; scale bars are indicated; L, lumen; B, basal lamina; G, goblet cell; E, enteroendocrine cell.

3.5. Refined Medium Supports Continuous Growth and Derivation of Organoids

So far, we have been able to verify that organoids in refined medium were differentiating similarly or even better than organoids in differentiation medium (see Figures 2 and 3) without concomitant induction of apoptosis (see Figure 1B,C). Therefore, we conclude that a refined medium composition is beneficial over the use of a separate differentiation medium. In a next step we assessed the effectiveness of refined versus expansion medium on the continuous growth of organoids.

Importantly, all organoid types showed a similar proliferation pattern, as revealed by a scattered distribution of single, EdU-positive cells within organoids when exposed to the two different medium conditions (Figure 4A). However, viability of duodenal and jejunal organoids in expansion medium was reduced compared to colonic organoids, which could be significantly enhanced by using refined medium (Figure 4B).

To assess the suitability of refined medium for the establishment of new organoid lines, duodenal biopsies were taken during routine gastroduodenoscopy. Following isolation, starting cell material was cultured either with expansion medium containing Rock inhibitor, or with EGF- and Rock inhibitor-supplemented refined medium for the first two days, then changed to unsupplemented expansion and refined medium. After 10 days in culture, an almost 4-fold higher number of organoids had grown in refined medium compared to in expansion medium (Figure 4C). Organoids exposed to refined medium also displayed a higher projected area compared to organoids formed in expansion medium, indicating that refined medium allows for easier adaption of canine intestinal cells to in vitro culture conditions.

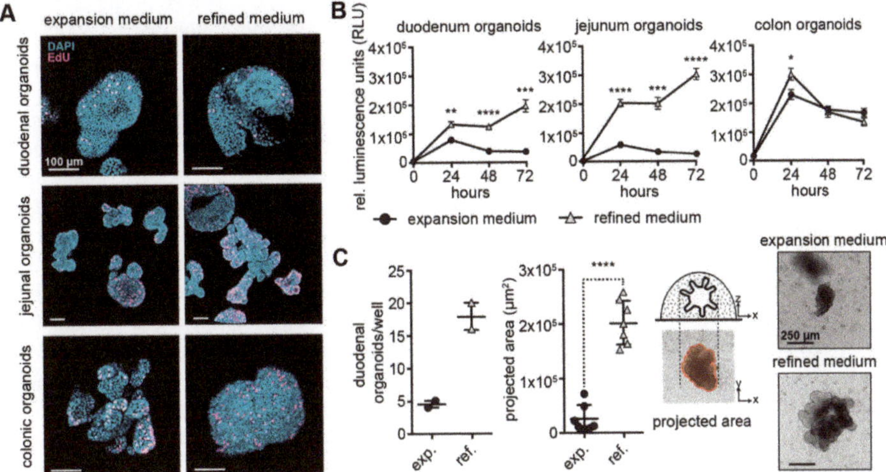

Figure 4. Derivation of organoids is more efficient using refined medium conditions. (**A**) 3D reconstruction of confocal z-stack images from duodenal, jejunal and colonic organoids in expansion and refined medium; staining for EdU after one hour labeling (magenta); organoids counterstained using DAPI (blue); scale bar represents 100 μm. (**B**) Relative luciferase activity of cell viability assay is shown for serial measurements at 0, 24, 48 and 72 h after addition of substrate to duodenal, jejunal and colonic organoids in expansion (circle) and refined medium (triangle); mean is shown; whiskers represent SEM; * $p < 0.05$, ** $p < 0.01$, *** $p < 0.001$ and **** $p < 0.0001$; 50 organoids seeded per replicate; n = 6 replicates. (**C**) Organoid number is shown for freshly isolated duodenal biopsies seeded in expansion and refined medium, in scatter dot plots with mean and range indicated by whiskers, n = 2 wells; projected area of organoids is depicted with mean and SEM defined by whiskers, **** $p < 0.0001$, n = 7 organoids; bright field images of representative organoids are shown, scale bar represents 250 μm.

4. Discussion

Since the first reports on the derivation of organoids from murine and subsequently from human intestine, the establishment of canine intestinal organoids was a matter of time. Particularly as dogs develop GI diseases like cancer, infectious disease and IBD naturally [23], thereby representing a faithful model system for the corresponding human diseases. In early 2017 our group [17], Kingsbury et al. [34], and Powel and Behnke [16] reported on the successful derivation of canine intestinal organoids. We then focused on advancing organoid cultivation to bring them a step closer to the in vivo situation.

In the presented study, we found that expansion medium does not support stem cell commitment in the secretory lineage using transcription profiling. This finding is supported by reports on human intestinal organoids, which usually harbor only a limited number of differentiated cell types, because they are grown in expansion media that allow for the maintenance of stem cells but do not support secretory lineage differentiation [14,17]. This absence of differentiated cells in expansion medium could not be overcome by prolonged cultivation: low- (≤5) and high-passage (≥10) canine intestinal organoids all tested negative for differentiation marker expression. Recently, Chandra et al. published a well-written paper on the derivation of canine intestinal organoids and the presence of goblet and enteroendocrine cells [35]. The formulation of the expansion medium differs in a 5-fold higher concentration of p38-MAPK inhibitor and the addition of 8% fetal bovine serum FBS (Chandra) compared to the medium formulation used herein. p38-MAPK inhibitor in human organoid culture is used to suppress secretory lineage commitment of intestinal stem cells to prevent their depletion and maintain long-term culture [14,36,37]. The addition of 8% FBS is a likely explanation for the presence of secretory cells, potentially via subsequent activation of p38-MAPK signaling. To create a culture system with defined components, the culture medium used in the study presented herein corresponds to a widely used human organoid medium without FBS [14] that was only supplemented with HGF to allow for the cultivation of over 50 passages. These variations in medium additives may explain the differences between both studies.

To further develop canine intestinal organoid culture, we focused on optimizing the growth medium to obtain a more in vivo-like cell composition with stem cells, enterocytes, goblet and enteroendocrine cells. Recently, a newly composed medium was shown to enable stem cells of human intestinal organoids to undergo differentiation, while concomitantly preserving the stem cell pool [25]. To overcome the intrinsic limitations of classical expansion medium, we adapted our canine intestinal expansion medium based on the suggestions of Fujii et al. [25] by (i) withdrawal of nicotinamide, p38-MAPK inhibitor, N2 and EGF; and (ii) addition of IGF1 and FGF2. This medium termed "refined medium" promoted stem cell growth and differentiation of canine intestinal organoids simultaneously for over 6 months of cultivation (data not shown). To assess the benefits and limitations of this refined medium in regard to organoid growth and lineage commitment, canine intestinal organoids grown in expansion, differentiation or refined medium were comparatively characterized.

While duodenal, jejunal and colonic organoids displayed a viable morphology upon culture in expansion and refined medium, differentiation medium induced apoptosis in all three intestinal segments. Interestingly, apoptosis peaked as early as 24 h after transfer to differentiation medium, indicating that programmed cell death was initiated by differentiation-unrelated factors. This assumption is supported by previous reports on the promoting effect of nicotinamide on cell survival in human pluripotent stem cells [38], on proliferation of aged murine organoids [39] and on inducing apoptosis via reduced intracellular NAD concentrations upon its withdrawal [40]. Taken together, differentiation medium induced apoptosis, thus not supporting long-term culture of canine intestinal organoids. This induction of cell death by the medium formulation is unfavorable when differentiation, proliferation or even apoptosis are investigated upon compound treatment for drug development. In addition, differentiation of organoids in expansion medium must be repeated for every individual experiment making them susceptible to fluctuations between media batches, illustrating the need for a better culture system.

Refined medium induced similar levels of differentiation or even outperformed differentiation medium in this regard. Duodenal, jejunal and colonic organoids grown in refined medium revealed enhanced expression of markers indicating the presence of secretory lineage precursors, enteroendocrine and goblet cells. Interestingly, organoids in refined medium expressed comparable or even higher levels of stem cell marker compared to organoids in expansion medium. Organoids irrespective of the intestinal segment or media composition expressed the enterocyte marker *VIL1*. These data confirm findings reported for human intestinal organoids grown in niche-inspired culture media that supported expression of enteroendocrine, goblet and stem cell markers [25] and data reported for villin1 positive enterocytes in murine intestinal organoids [12]. However, expression of the differentiation markers *CHGA*, *MUC2* and *VIL1* in differentiation and refined medium reached the levels of the i.c.i. only to a certain extent. While an in vitro culture model cannot fully reproduce the complex in vivo situation, our data show that growth conditions in refined medium are more similar compared to those in expansion medium.

Since only a limited number of specific antibodies were available for reliable characterization of canine cells, we used two well-established alternative methods, PAS reaction and TEM, for the detection of goblet cells and enteroendocrine cells. These results supported our RT-qPCR data, substantiating that only organoids grown in differentiation and refined medium harbored enteroendocrine and goblet cells. These findings are in accordance with data obtained for human intestinal organoids grown in differentiation [14] or niche-inspired medium [25]. Furthermore, the presence of enterocytes in organoids cultivated in expansion, differentiation and refined medium of all segments was supported by microvilli bearing cells as reported for murine intestinal organoids [12], indicating that enterocytes are present irrespective of culture condition.

So far, our data provided evidence that cultivation of canine intestinal organoids in refined medium is preferable to the use of differentiation medium, since it promotes stem cell growth and differentiation simultaneously, allows long-term growth without fluctuations between media batches and does not induce apoptosis. Therefore, further analysis focused on comparison of organoid growth in expansion and refined medium.

The comparison of proliferative behavior differences between canine intestinal organoids grown in expansion versus refined medium by EdU incorporation assays revealed similar proliferation levels for all organoid types irrespective of the medium used. Yet, viability data of duodenal and jejunal organoids indicate that expansion medium does not support small intestinal organoid cultivation as effective as refined medium. The colonic epithelium is highly populated with various bacteria and has less digestive and absorptive function than small intestinal epithelium possibly resulting in more robustness with respect to cultivation. Therefore, colonic organoids can be expanded in both expansion and refined medium efficiently.

Importantly, generation of organoids from duodenal biopsies was more successful in refined than expansion medium. Not only the number of organoids per well was enhanced, but also their projected area. This feature has an important impact on canine intestinal disease modelling, especially when donor tissue stems from GI enteropathies with severely affected epithelium. The high suitability of refined medium for canine organoid establishment also matches the higher plating efficacy shown for human single-cell dissociated organoids grown in niche-inspired media [25].

5. Conclusions

Taken together, we strongly recommend the use of refined medium for establishment and long-term culture of canine intestinal organoids. We could clearly show that this medium sustains stem cell growth, while simultaneously promoting differentiation of stem cells/immature cells into enterocytes, enteroendocrine and goblet cells. Canine intestinal organoids cultivated in refined medium bear the advantage of an easy to handle, reproducible and stable culture system, thereby representing a physiologically superior in vitro system for disease modelling, drug development, toxicity studies and personalized medicine. Compared to conventional expansion and differentiation media, the refined medium presented herein allows for more accurate assessment of genetic and epigenetic impacts on canine intestinal cell differentiation, bringing organoids a step closer to the in vivo situation.

Supplementary Materials: The following are available online at http://www.mdpi.com/2073-4409/9/4/822/s1, Figure S1: Cultivation scheme of organoids and images of organoids in differentiation medium and after prolonged cultivation in expansion and refined medium; Figure S2: Gene expression of stem cell marker is increased in colonic organoids; Figure S3: TEM images of microvilli and cell-cell junctions in all conditions and goblet cells in organoids cultivated in differentiation and refined medium; Table S1: qPCR primer specifications used in this study; Table S2: Statistical analysis of duodenal tissue vs. duodenal organoids; Table S3: Statistical analysis of jejunal tissue vs. jejunal organoids; Table S4: Statistical analysis of colonic tissue vs. colonic organoids; Table S5: Statistical analysis of duodenal organoids in different media; Table S6: Statistical analysis of jejunal organoids in different media; Table S7: Statistical analysis of colonic organoids in different media.

Author Contributions: Conceptualization, N.K., B.P., A.M.C.M., A.S., H.S.K., K.S., L.C.P., B.S., M.K., S.B. and I.A.B.; Formal analysis, N.K.; Funding acquisition, L.C.P., B.S., S.B. and I.A.B.; Investigation, N.K., B.P., A.M.C.M., W.T. and I.W.; Methodology, N.K., B.P., A.M.C.M., H.S.K. and K.S.; Project administration, B.P. and I.A.B.; Resources, I.A.B.; Supervision, M.K., S.B. and I.A.B.; Validation, N.K., W.T. and I.W.; Visualization, N.K.; Writing—original draft, N.K. and S.B.; Writing—review & editing, B.P., A.M.C.M., W.T., I.W., A.S., H.S.K., K.S., L.C.P., B.S., M.K. and I.A.B. All authors have read and agreed to the published version of the manuscript.

Funding: The Austrian Research Association FFG, project № 13610974, funded this work. The Dutch Research Council NWO ZON/MW (116004121) funded parts of this work.

Acknowledgments: This research was supported using resources of the VetCore Facility (Imaging, VetBioBank and Genomics) of the University of Veterinary Medicine Vienna. Open Access Funding by the University of Veterinary Medicine Vienna.

Conflicts of Interest: The authors declare no conflict of interest. The funders had no role in the design of the study; in the collection, analyses, or interpretation of data; in the writing of the manuscript, or in the decision to publish the results.

References

1. Kim, H.P.; Crockett, S.; Shaheen, N.J. The Burden of Gastrointestinal and Liver Disease around the World. In *GI Epidemiology*; Wiley: Oxford, UK, 2014; pp. 1–13.

2. Bray, F.; Ferlay, J.; Soerjomataram, I.; Siegel, R.L.; Torre, L.A.; Jemal, A. Global cancer statistics 2018: GLOBOCAN estimates of incidence and mortality worldwide for 36 cancers in 185 countries. *CA Cancer J. Clin.* **2018**, *68*, 394–424. [CrossRef] [PubMed]

3. Rubin, S.I. Infectious Diseases of the GI Tract. Available online: https://www.msdvetmanual.com/digestive-system/digestive-system-introduction/infectious-diseases-of-the-gi-tract (accessed on 19 September 2019).

4. Albani, S.; Colomb, J.; Prakken, B. Translational Medicine 2.0: From Clinical Diagnosis–Based to Molecular-Targeted Therapies in the Era of Globalization. *Clin. Pharmacol. Ther.* **2010**, *87*, 642–645. [CrossRef]

5. Albani, S.; Prakken, B. The advancement of translational medicine—From regional challenges to global solutions. *Nat. Med.* **2009**, *15*, 1006–1009. [CrossRef]

6. Scherzer, M.; Kramer, N.; Unger, C.; Walzl, A.; Walter, S.; Stadler, M.; Hengstschläger, M.; Dolznig, H. Preclinical Cancer Models with the Potential to Predict Clinical Response. In *Drug Discovery in Cancer Epigenetics*; Elsevier BV: Oxford, UK, 2016; pp. 97–122.

7. Browning, T.H.; Trier, J.S. Organ culture of mucosal biopsies of human small intestine. *J. Clin. Investig.* **1969**, *48*, 1423–1432. [CrossRef] [PubMed]

8. Kedinger, M.; Haffen, K.; Simon-Assmann, P. Intestinal tissue and cell cultures. *Differentiation* **1987**, *36*, 71–85. [CrossRef] [PubMed]

9. Evans, G.S.; Flint, N.; Somers, A.S.; Eyden, B.; Potten, C.S. The development of a method for the preparation of rat intestinal epithelial cell primary cultures. *J. Cell Sci.* **1992**, *101*.

10. Ootani, A.; Li, X.; Sangiorgi, E.; Ho, Q.T.; Ueno, H.; Toda, S.; Sugihara, H.; Fujimoto, K.; Weissman, I.L.; Capecchi, M.R.; et al. Sustained in vitro intestinal epithelial culture within a Wnt-dependent stem cell niche. *Nat. Med.* **2009**, *15*, 701–706. [CrossRef]

11. Wang, X.; Yamamoto, Y.; Wilson, L.H.; Zhang, T.; Howitt, B.; Farrow, M.A.; Kern, F.; Ning, G.; Hong, Y.; Khor, C.C.; et al. Cloning and variation of ground state intestinal stem cells. *Nature* **2015**, *522*, 173–178. [CrossRef]

12. Sato, T.; Vries, R.G.; Snippert, H.J.; Van De Wetering, M.; Barker, N.; Stange, D.; Van Es, J.H.; Abo, A.; Kujala, P.; Peters, P.J.; et al. Single Lgr5 stem cells build crypt-villus structures in vitro without a mesenchymal niche. *Nature* **2009**, *459*, 262–265. [CrossRef]

13. Date, S.; Sato, T. Mini-Gut Organoids: Reconstitution of the Stem Cell Niche. *Annu. Rev. Cell Dev. Boil.* **2015**, *31*, 269–289. [CrossRef]

14. Sato, T.; Stange, D.; Ferrante, M.; Vries, R.G.; Van Es, J.H.; Brink, S.V.D.; Van Houdt, W.J.; Pronk, A.; Van Gorp, J.M.; Siersema, P.D.; et al. Long-term Expansion of Epithelial Organoids From Human Colon, Adenoma, Adenocarcinoma, and Barrett's Epithelium. *Gastroenterology* **2011**, *141*, 1762–1772. [CrossRef] [PubMed]

15. Wang, Y.; Kim, R.; Gunasekara, D.B.; Reed, M.I.; Disalvo, M.; Nguyen, D.L.; Bultman, S.J.; Sims, C.E.; Magness, S.T.; Allbritton, N.L. Formation of Human Colonic Crypt Array by Application of Chemical Gradients Across a Shaped Epithelial Monolayer. *Cell. Mol. Gastroenterol. Hepatol.* **2017**, *5*, 113–130. [CrossRef] [PubMed]

16. Powell, R.H.; Behnke, M. WRN conditioned media is sufficient for in vitro propagation of intestinal organoids from large farm and small companion animals. *Boil. Open* **2017**, *6*, 698–705. [CrossRef] [PubMed]

17. 2017 ACVIM Forum Research Abstract Program. *J. Veter- Intern. Med.* **2017**, *31*, 1225–1361. [CrossRef]

18. Nantasanti, S.; Spee, B.; Kruitwagen, H.S.; Chen, C.; Geijsen, N.; Oosterhoff, L.A.; Van Wolferen, M.E.; Peláez, N.; Fieten, H.; Wubbolts, R.W.; et al. Disease Modeling and Gene Therapy of Copper Storage Disease in Canine Hepatic Organoids. *Stem Cell Rep.* **2015**, *5*, 895–907. [CrossRef]

19. Kruitwagen, H.S.; Oosterhoff, L.A.; Vernooij, I.G.; Schrall, I.M.; Van Wolferen, M.E.; Bannink, F.; Roesch, C.; Van Uden, L.; Molenaar, M.; Helms, J.B.; et al. Long-Term Adult Feline Liver Organoid Cultures for Disease Modeling of Hepatic Steatosis. *Stem Cell Rep.* **2017**, *8*, 822–830. [CrossRef]

20. Derricott, H.; Luu, L.; Fong, W.Y.; Hartley, C.S.; Johnston, L.; Armstrong, S.D.; Randle, N.; Duckworth, C.A.; Campbell, B.J.; Wastling, J.M.; et al. Developing a 3D intestinal epithelium model for livestock species. *Cell Tissue Res.* **2018**, *375*, 409–424. [CrossRef]

21. Hamilton, C.A.; Young, R.; Jayaraman, S.; Sehgal, A.; Paxton, E.; Thomson, S.; Katzer, F.; Hope, J.; Innes, E.A.; Morrison, L.; et al. Development of in vitro enteroids derived from bovine small intestinal crypts. *Veter- Res.* **2018**, *49*, 54. [CrossRef]

22. Pierzchalska, M.; Grabacka, M.; Michalik, M.; Zyla, K.; Pierzchalski, P. Prostaglandin E2 supports growth of chicken embryo intestinal organoids in Matrigel matrix. *Biotechniques* **2012**, *52*, 307–315. [CrossRef]

23. Cerquetella, M.; Spaterna, A.; Laus, F.; Tesei, B.; Rossi, G.; Antonelli, E.; Villanacci, V.; Bassotti, G. Inflammatory bowel disease in the dog: Differences and similarities with humans. *World J. Gastroenterol.* **2010**, *16*, 1050–1056. [CrossRef] [PubMed]

24. Shearin, A.L.; Ostrander, E.A. Leading the way: Canine models of genomics and disease. *Dis. Model. Mech.* **2010**, *3*, 27–34. [CrossRef] [PubMed]

25. Fujii, M.; Matano, M.; Toshimitsu, K.; Takano, A.; Mikami, Y.; Nishikori, S.; Sugimoto, S.; Sato, T. Human Intestinal Organoids Maintain Self-Renewal Capacity and Cellular Diversity in Niche-Inspired Culture Condition. *Cell Stem Cell* **2018**, *23*, 787–793.e6. [CrossRef] [PubMed]

26. Pfaffl, M.W. A new mathematical model for relative quantification in real-time RT-PCR. *Nucleic Acids Res.* **2001**, *29*, 45. [CrossRef] [PubMed]

27. Grün, D.; Muraro, M.J.; Boisset, J.-C.; Wiebrands, K.; Lyubimova, A.; Dharmadhikari, G.; Born, M.V.D.; Van Es, J.; Jansen, E.; Clevers, H.; et al. De Novo Prediction of Stem Cell Identity using Single-Cell Transcriptome Data. *Cell Stem Cell* **2016**, *19*, 266–277. [CrossRef]

28. Audie, J.P.; Janin, A.; Porchet, N.; Copin, M.C.; Gosselin, B.; Aubert, J.P. Expression of human mucin genes in respiratory, digestive, and reproductive tracts ascertained by in situ hybridization. *J. Histochem. Cytochem.* **1993**, *41*, 1479–1485. [CrossRef]

29. Kim, Y.S.; Ho, S.B. Intestinal Goblet Cells and Mucins in Health and Disease: Recent Insights and Progress. *Curr. Gastroenterol. Rep.* **2010**, *12*, 319–330. [CrossRef]

30. Barker, N.; Van Es, J.H.; Kuipers, J.; Kujala, P.; Born, M.V.D.; Cozijnsen, M.; Haegebarth, A.; Korving, J.; Begthel, H.; Peters, P.J.; et al. Identification of stem cells in small intestine and colon by marker gene Lgr5. *Nature* **2007**, *449*, 1003–1007. [CrossRef]

31. Munoz, J.; Stange, D.; Schepers, A.G.; Van De Wetering, M.; Koo, B.-K.; Itzkovitz, S.; Volckmann, R.; Kung, K.S.; Koster, J.; Radulescu, S.; et al. The Lgr5 intestinal stem cell signature: Robust expression of proposed quiescent '+4' cell markers. *EMBO J.* **2012**, *31*, 3079–3091. [CrossRef]

32. Jenny, M.; Uhl, C.; Roche, C.; Duluc, I.; Guillermin, V.; Guillemot, F.; Jensen, J.; Kedinger, M.; Gradwohl, G. Neurogenin3 is differentially required for endocrine cell fate specification in the intestinal and gastric epithelium. *EMBO J.* **2002**, *21*, 6338–6347. [CrossRef]

33. Lopez-Diaz, L.; Jain, R.N.; Keeley, T.M.; VanDussen, K.L.; Brunkan, C.S.; Gumucio, D.L.; Samuelson, L.C. Intestinal Neurogenin 3 directs differentiation of a bipotential secretory progenitor to endocrine cell rather than goblet cell fate. *Dev. Boil.* **2007**, *309*, 298–305. [CrossRef]

34. Kingsbury, D.; Sun, L.; Qi, Y.; Fredericks, J.; Wang, Q.; Wannemuehler, M.; Jergens, A.; Allenspach, K. DDW Abstract: Optimizing the Development and Characterization of Canine Small Intestinal Crypt Organoids as a Research Model. *Gastroenterology* **2017**, *152* (Suppl. 1). [CrossRef]

35. Chandra, L.; Borcherding, D.C.; Kingsbury, D.; Atherly, T.; Ambrosini, Y.M.; Bourgois-Mochel, A.; Yuan, W.; Kimber, M.; Qi, Y.; Wang, Q.; et al. Derivation of adult canine intestinal organoids for translational research in gastroenterology. *BMC Boil.* **2019**, *17*, 33. [CrossRef] [PubMed]

36. Kozuka, K.; He, Y.; Koo-McCoy, S.; Kumaraswamy, P.; Nie, B.; Shaw, K.; Chan, P.; Leadbetter, M.; He, L.; Lewis, J.G.; et al. Development and Characterization of a Human and Mouse Intestinal Epithelial Cell Monolayer Platform. *Stem Cell Rep.* **2017**, *9*, 1976–1990. [CrossRef] [PubMed]

37. Wallach, T.E.; Bayrer, J.R. Intestinal Organoids. *J. Pediatr. Gastroenterol. Nutr.* **2017**, *64*, 180–185. [CrossRef]

38. Meng, Y.; Ren, Z.; Xu, F.; Zhou, X.; Song, C.; Wang, V.Y.-F.; Liu, W.; Lu, L.; Thomson, J.A.; Chen, G. Nicotinamide Promotes Cell Survival and Differentiation as Kinase Inhibitor in Human Pluripotent Stem Cells. *Stem Cell Rep.* **2018**, *11*, 1347–1356. [CrossRef]

39. Uchida, R.; Saito, Y.; Nogami, K.; Kajiyama, Y.; Suzuki, Y.; Kawase, Y.; Nakaoka, T.; Muramatsu, T.; Kimura, M.; Saito, H. Epigenetic silencing of Lgr5 induces senescence of intestinal epithelial organoids during the process of aging. *NPJ Aging Mech. Dis.* **2018**, *4*, 1. [CrossRef]
40. Heikal, A.A. Intracellular coenzymes as natural biomarkers for metabolic activities and mitochondrial anomalies. *Biomark. Med.* **2010**, *4*, 241–263. [CrossRef]

Article

PP2A Deficiency Enhances Carcinogenesis of Lgr5$^+$ Intestinal Stem Cells Both in Organoids and In Vivo

Yu-Ting Yen [1,2], May Chien [1,2], Yung-Chih Lai [2], Dao-Peng Chen [3], Cheng-Ming Chuong [2,4], Mien-Chie Hung [5,6] and Shih-Chieh Hung [1,2,7,*]

[1] Drug Development Center, Institute of New Drug Development, China Medical University, Taichung 40402, Taiwan; d92449001@ntu.edu.tw (Y.-T.Y.); maymaychien2k12@gmail.com (M.C.)
[2] Integrative Stem Cell Center, China Medical University Hospital, Taichung 40402, Taiwan; yungchihlai@gmail.com (Y.-C.L.); cmchuong@med.usc.edu (C.-M.C.)
[3] Kim Forest Enterprise Co., Ltd., Taipei 22175, Taiwan; D96743@mail.cmuh.org.tw
[4] Department of Pathology, Keck School of Medicine, University of Southern California, Los Angeles, CA 90033, USA
[5] Center for Molecular Medicine and Graduate Institute of Cancer Biology, China Medical University, Taichung 40402, Taiwan; mhung77030@gmail.com
[6] Cancer Biology Program, The University of Texas Graduate School of Biomedical Sciences, The University of Texas MD Anderson Cancer Center, Houston, TX 77030, USA
[7] Department of Orthopaedics, China Medical University Hospital, Taichung 40402, Taiwan
* Correspondence: hung3340@gmail.com

Received: 16 November 2019; Accepted: 28 December 2019; Published: 30 December 2019

Abstract: In most cancers, cellular origin and the contribution of intrinsic and extrinsic factors toward transformation remain elusive. Cell specific carcinogenesis models are currently unavailable. To investigate cellular origin in carcinogenesis, we developed a tumorigenesis model based on a combination of carcinogenesis and genetically engineered mouse models. We show in organoids that treatment of any of three carcinogens, DMBA, MNU, or PhIP, with protein phosphatase 2A (PP2A) knockout induced tumorigenesis in Lgr5$^+$ intestinal lineage, but not in differentiated cells. These transformed cells increased in stem cell signature, were upregulated in EMT markers, and acquired tumorigenecity. A mechanistic approach demonstrated that tumorigenesis was dependent on Wnt, PI3K, and RAS-MAPK activation. In vivo combination with carcinogen and PP2A depletion also led to tumor formation. Using whole-exome sequencing, we demonstrate that these intestinal tumors display mutation landscape and core driver pathways resembling human intestinal tumor in The Cancer Genome Atlas (TCGA). These data provide a basis for understanding the interplay between extrinsic carcinogen and intrinsic genetic modification and suggest that PP2A functions as a tumor suppressor in intestine carcinogenesis.

Keywords: carcinogen; protein phosphatase 2A (PP2A); intestinal tumor; intestinal organoid; Lgr5$^+$ crypt stem cell

---

1. Introduction

The cells of origin in most cancers have remained unknown. Chemical carcinogenesis mouse models recapitulating most of human cancers that are induced by exposure to environmental carcinogens [1], however, is difficult to be achieved in a cell-specific manner. Therefore, the current strategies to investigate the cellular origins of cancers are using genetically engineered mouse models (GEMMs), with either transgenic or conditionally targeted gene technologies to induce tumor in different cellular contexts [2]. Moreover, both models take a long time to develop cancer, limiting progress in the cancer research field.

The most applied animal model for studying intestinal tumorigenesis is based on activating mutations in the Wnt pathway, which relies on adenomatous polyposis coli (Apc) depletion [3] and beta-catenin (CTNNB1) activation [4], leading to beta-catenin stabilization and constitutive transcription of its down-stream genes. Recent progress in the understanding of the cell of origin of intestinal tumor was made using this model, although several inconsistencies were observed. After in vivo Apc depletion in leucine-rich-repeat containing G-protein-coupled receptor 5 (Lgr5)$^+$ crypt stem cells, tumor formation occurred within 3–5 weeks [3]. However, Apc depletion or being combined with Kras^{G12D} mutation in progenitor and differentiated cells did not induce tumor formation [5,6]. However, tumor-initiating mutations can occur in Lgr5$^+$ crypt stem cells and in differentiated Lgr5$^-$ cells [4], indicating that the two hypotheses are not mutually exclusive. Ablation of Lgr5$^+$ cells in orthotopically transplanted tumors, generated by genetic modification in differentiated villus cells, suppressed tumor growth [7]. Interestingly, Lgr5$^+$ cells reappeared and tumors recurred when ablation was terminated 4 days later. The generation of Lgr5$^+$ cells from Lgr5$^-$ cells after Lgr5$^+$ ablation was also observed in the xenograft mouse model of human colon cancer stem cells (CSCs) [8]. However, the specific mechanism of Lgr5$^+$ cell generation from remaining Lgr5$^-$ cells remains unclear.

Aberrant activation of signal transduction pathway, a dynamic process involving an 'on/off' switch, can transform a normal cell to be malignant or further render cancer cells with the capacities for therapy resistance. Activating mutations in genes encoding kinases or signaling molecules, such as RAS and PI3K, switch on the signaling, continuously activating a survival and/or proliferation pathway, while activations of phosphatases, such as the serine/threonine phosphatase PP2A family, switch off the signaling [9]. Previous efforts through high-throughput screens of tyrosine kinome and tyrosine phosphatome have identified several driver or passenger mutations in a spectrum of malignancies, including intestinal tumor [10–12]. However, there are few if any studies focusing on the altered signalings driven by serine/threonine kinase mutations [13]. Human intestinal tumors contain active mutations in genes encoding proteins involved in the WNT, MAPK, TGF-beta, and PI3K pathways [14]. Ingenuity pathway analysis (IPA) of TCGA-COAD revealed PP2A complex and its subunits, such as PPP2R1A and PPP2CA, are intercalated among several driver mutation pathways (Figure S1A–C). Moreover, endogenous PP2A inhibitors, SET and CIP2A [15], are highly expressed in intestinal tumors in comparison to their matched normal tissue samples (Figure S1D). We showed that PP2A was suppressed in intestinal tumor stem cells (CSCs), thereby activating its substrate kinases to enhance survival under hypoxia and serum depletion [16], thereby increasing resistance to anti-angiogenesis therapy [17]. Our recent studies also demonstrated that reduced PP2A activity in colorectal and lung CSCs enhances suspension survival and induces tumor initiation [18], revealing the tumor suppressive role of PP2A [19]. Although higher numbers of Apc, p53, Kras^{G12D}, and Smad4 driver mutations may be required for human colorectal tumorigenesis, there are some intestinal tumors carry only one or no alteration in these driver mutations [20]. For example, gene fusions involving R-spondin 1 occurring in 10% of intestinal tumor are mutually exclusive with active Wnt signaling caused by *APC* or *CTNNB1* mutations [20]. The emerging novel intestinal tumorigenesis animal models should allow for elucidating the molecular mechanisms of these cancers.

Given that cancer is the product of complex interactions between the genetic and environmental predisposition factors, the combined use of chemical carcinogens that switch on kinases and GEMM with phosphatase deficiency is a logical approach for examining the complex interplay between genetic susceptibility and environmental exposure [21]. To investigate the cell origin of intestinal tumor, we first combined treatment with carcinogen 7,12-dimethylbenzanthracene (DMBA) that has previously been known to induce rodent s in the presence of 1,2-dimethyl-hydrazine [22] and PP2A inhibition via okadaic acid (OA) treatment or genetic deficiency. DMBA not only activates multiple mutations in different codons of ras [23] but also induces activation in other pathways, such as Notch [24], providing a screening approach for identifying key kinases or molecules. Besides DMBA, we also investigated the effects of *N*-methyl-*N*-nitrosourea (MNU) and 2-amino-1-methyl-6-phenylimidazo-[4,5-b]pyridine (PhIP) on intestinal carcinogenesis. MNU is one of the direct alkylating agents, which does not require metabolic activation for initiating carcinogenesis [25]. PhIP has received considerable attention because it has multi-organ targets and it was developed upon broiling of fish and meat [26]. Moreover, we established primary intestinal organoid models that recapitulate the rodent intestinal tumorigenesis paradigm [27]. These rodent intestinal tumorigenesis models are useful in the development of new strategies for targeting rodent intestinal CSCs and treatment of intestinal tumor.

2. Materials and Methods

2.1. Mouse Colonies

Ppp2r1a^{flox/flox} mice, carrying conditional alleles with loxP sites flanking exon 5–6 of *Ppp2r1a*, were purchased from the Jackson Laboratory and crossed to *Lgr5-EGFP-CreERT2* or *Villin-Cre* mice to generate *Lgr5-EGFP-CreERT2; Ppp2r1a^{flox/flox}* or *Villin-Cre; Ppp2r1a^{flox/flox}* mice. NOD/SCID mice were purchased from Lasco Co., Ltd. (Taiwan). All animal studies and care of live animals were approved and performed following the guidelines made by the China Medical University Institutional Animal Care and Use Committee 2016-398-1; 2017-239.

2.2. Mouse Intestinal Organoid Cell Isolation, Culture, and Passage

Organoid culture was preformed according to a protocol modified from previously described methods [28]. In brief, the intestines were dissected, opened longitudinally and cut into small (2 mm) pieces. The tissues were rocked in dissociation reagent and incubated at room temperature (15–25 °C) for 15 min. The tissues were then mixed and filtered through a 70 µm sterile cell strainer. The crypts were collected by centrifugation at 140× *g* for 5 min at 4 °C. Approximately 500 crypts were suspended in 50 µL growth factor reduced phenol-free Matrigel (BD Biosciences, San Jose, CA, USA). Next, a 50 µL droplet of Matrigel/crypt mix was placed and polymerized in the center well of a 48-well plate. The basic culture medium (Dulbecco's modified Eagle's medium/F12 supplemented with penicillin/streptomycin), was supplemented with 50 ng/mL murine recombinant epidermal growth factor (EGF; Peprotech, Hamburg, Germany), Noggin (5% final volume) and R-spondin 1 (5% final volume) called "ENR" medium. Medium change was performed every 3–4 days. Each condition was examined in triplicate with multiple (>15) organoids in each sample. Each experiment was repeated twice.

2.3. Dysplasia Index

Histologic changes were scored blindly on the levels of four histological characteristics as previously described [27]: nuclear grade (enlarged nuclei with diffuse membrane irregularities and prominent nucleoli); stratification; mitoses and invasion (>2 foci). The dysplasia index was evaluated by all microscopic fields containing viable organoids with 5 fields per sample (*n*).

2.4. Primary Organoid Transplantation

For transplantation, cells from passage 7; day 50 organoid cultures were collected. Dissociated cells were pelleted by centrifugation and resuspended with Matrigel (50% Matrigel (BD), in a total volume of 100 µL containing indicated cell numbers and injected s.c. into NOD-SCID mice.

2.5. Immunofluorescence and Immunohistochemistry

Freshly isolated intestines were prepared according to a protocol modified from previously described methods [27]. The intestines were then applied for immunostaining. For immunostaining, the organoid cells were rinsed three times in ice-cold PBS. The organoid cells were spun down at 900 rpm for 10 min at 4 °C. Sections were deparaffinized and stained with H&E for the initial histology analysis. The immunofluorescence was performed on paraffin-embedded sections (5 µm). The permeabilized organoid cell samples were incubated with primary antibodies overnight at 4 °C. The samples were incubated with anti-PPP2R1A (GTX102206; GeneTex, Hsinchu City, Taiwan), anti-CK20 (GTX110600 Genetex), anti-Lgr5 (GTX50839 Genetex), anti-SMA (Abcam, ab5694, Cambridge, MA, USA), anti-beta-catenin (BD Transduction Labs, San Jose, CA, USA; 610154); the secondary antibodies used were DyLight® 650 Conjugated goat anti-rabbit (cat no. A120-101D5; Bethyl Laboratories Inc., Montgomery, TX, USA) and Goat anti-Rabbit IgG Antibody-FITC (Bethyl cat no. A120-101F) and DAPI (Molecular Probes) for 1 h at room temperature. The slides were mounted with SlowFade (SlowFade® AntiFade Kit, Molecular Probes, Waltham, MA, USA) followed by covering with a coverslip, and the edges were sealed to prevent drying. The specimens were examined with a Zeiss 710 Laser Scanning confocal microscope (Zeiss, Oberkochen, Germany).

Intestinal tissue was fixed and processed into paraffin blocks according to standard procedures. beta-catenin immunohistochemistry was performed as previously described [3]. Immunohistochemistry protocol hold as following: freshly isolated intestines were flushed with 10% formalin in PBS and fixed by incubation in a 10-fold excess of formalin overnight at room temperature. The formalin was removed and the intestines washed twice in PBS at room temperature. The intestines were then transferred to a tissue cassette and dehydrated by serial immersion in 20-fold volumes of 70, 96 and 100% EtOH for 2 h each at 4 °C. Excess ethanol was removed by incubation in xylene for 1.5 h room temperature and the cassettes then immersed in liquid paraffin (56 °C) overnight. Paraffin blocks were prepared using standard methods and 4µm tissue sections generated. These sections were de-waxed by immersion in xylene (2 × 5 min) and hydrated by serial immersion in 100% EtOH (2 × 1 min), 96% EtOH (2 × 1 min), 70% EtOH (2 × 1 min) and distilled water (2 × 1 min). Endogenous peroxidase activity was blocked by immersing the slides in peroxidase blocking buffer (0.040 M citric acid, 0.121 M disodium hydrogen phosphate, 0.030 M sodium azide, 1.5% hydrogen peroxide) for 15 min at room temperature. For beta-catenin, antigen retrieval involved 20 min boiling in Tris-EDTA pH 9.0, and blocking buffer (1% BSA in PBS) added to the slides for 30 min at room temperature. For beta-catenin (BD Transduction Labs, 610154), staining involved 1/100 dilution in blocking buffer (0.05% BSA in PBS) for 2 h at room temperature.

The slides were then rinsed in PBS and secondary antibody added (polymer HRP-labeled anti-mouse/rabbit, Envision) for 30 min at room temperature (Dako, Trappes, France). Slides were again washed in PBS and bound peroxidase detected by adding DAB substrate for 10 min at room temperature. Slides were then washed 2× in PBS and nuclei counterstained with Mayer's hematoxylin for 2 min, followed by two rinses in distilled water. Sections were dehydrated by serial immersion for 1 min each in 50 and 60% EtOH, followed by 2 min each in 70, 96, and 100% EtOH and xylene. Slides were mounted in mounting medium and a coverslip placed over the tissue section.

For immunohistochemistry (IHC) analysis, nuclei expressing beta-catenin after IHC staining were counted under 200× magnification. The Histological score (H-score) was determined based on the intensity and percentage of nucleus staining at each intensity [29], and calculated as follows: H-score = (nucleus showing highly beta-catenin expression) × 3 + (nucleus showing beta-catenin expression) × 2 + (nucleus showing weak beta-catenin expression) × 1.

2.6. Viral Infection of Organoids

For in vitro deficiency of the *Ppp2r1a*, organoid cultures containing floxed *Ppp2r1a* alleles were infected with adenovirus-encoding Cre recombinase (Ad-Cre) (Vector Biolabs, Philadelphia, PA, USA) at a titer of 100 multiplicity of infection (MOI) [27].

2.7. Tamoxifen Induction

Mice aged 6–8 weeks were injected intraperitoneally with a single 200 µL dose of tamoxifen in sunflower oil at 10 mg/mL.

2.8. Organoid Disaggregation, FACS, and Immunoblotting

Organoid cultures were recovered and dissociated from collagen gel by collagenase IV incubation, followed by incubation with 0.05% trypsin and EDTA. After extensive washing with 10% FBS, cells were filtered with 40-μm cell strainers (BD Falcon) Pellets were resuspended with FACS staining solution (5% FCS in PBS). Stringent wash was applied using ice-cold PBS, followed by isolation of Lgr5$^-$EGFP$^+$ cells using an FACSAria II (BD) [30]. For immunoblotting, the organoid cells were lysed in lysis buffer (1% Triton X-100, 150 mmol/L NaCl, 10 mmol/L Tris pH 7.4, 1 mmol/L EDTA pH 8.0, protease inhibitor cocktail) and then sonicated. The protein concentration was then measured. Next, equal amounts of protein (20 μg/well) were separated by SDS-polyacrylamide gel electrophoresis, transferred to nitrocellulose, and immunoblotted with primary antibodies. The membranes were blocked with CISblock buffer purchased from Cis-biotechnology, Taiwan. The following antibodies were used: anti-phospho-AKT (Ser-473), anti-AKT; anti-phospho-ERK1/2 (Thr-202/Tyr-204), anti-ERK, and anti-PP2A from Cell Signaling; anti-Lgr5 and anti-alkaline phosphate intestinal (Alpi) (Genetex, epitope C-terminus), beta-catenin (BD Transduction Labs, 610154), anti-Villin (Santa Cruz, Dallas, TX, USA), and anti-beta-actin and GAPDH (Sigma-Aldrich, St. Louis, MO, USA). Following the primary antibody incubation, the nitrocellulose membranes were incubated with secondary antibodies and visualized by ECL.

2.9. Antibody Arrays of Mouse AKT Pathway Phosphorylation

The RayBio™ Mouse AKT Pathway Phosphorylation Array Kit (cat. no. AAH-AKT1-2) was purchased and preformed according to a protocol modified from RayBiotech Inc. (Norcross, GA, USA). Briefly, the array membranes were blocked with blocking buffer for 30 min at room temperature. The membranes were then incubated with 2 mL of lysate prepared from organoid cultures with different treatments after normalization with equal amounts of protein. After extensive washing with wash buffer I (3 washings of 5 min each), and wash buffer II (3 washings of 5 min each) to remove unbound materials, the membranes were incubated with the Detection Antibody Cocktail for 1.5 to 2 h at room temperature, followed by wash with wash buffer I and II. Then the membranes were incubated with HRP-Anti-Rabbit IgG for 2 h at room temperature. The unbound HRP antibody was washed out with wash buffer I and II. Finally, each array membrane was exposed to X-ray film using a chemiluminescence detection system (Perkin Elmer, Wellesley, MA, USA).

2.10. Transcriptome Analysis

RNA was extracted from organoid culture using an RNeasy Kit (Qiagen, Hilden, Germany). RNA integrity was assessed using the RNA Nano6000 assay kit (Agilent Technologies, Santa Clara, CA, USA). For RNA-seq, library preparation and sequencing were performed by Novogene Technology. The output data (FASTQ files) were mapped to the target genome (TopHat v2.0.12), which can generate a database of splice junctions based on the gene model annotation file. HTSeq v0.6.1 was used to count the reads numbers mapped to each gene. Then the FPKM of each gene was calculated based on the length of the gene and reads count mapped to this gene. Differential expression heatmap results and biological variability were analyzed by ClustVis free web server [31] and gene set enrichment analysis (GSEA) [32], respectively. Data were submitted and approved by Gene Expression Omnibus (GEO; accession number GSE120241).

2.11. Whole-Exome Sequencing, Alignment, and Annotation

Exome sequences were captured with SureSelect^{XT} Mouse All Exon Kit (G7550E-001, Agilent, CA, USA) following the standard protocols. The products of exome capture should pass criteria: the length of fragments: 300 ± 30 bp and total amount: >600 ng. After exome capturing, the index-tagged samples were pooled and sequenced on Illumina HiSeq 2000. Burrows-Wheeler Alignment (v0.7.12) was employed to align reads to the reference genome (mm10) with default parameters. Aligned reads were sorted by picard-tools (v1.8). The duplicated reads were marked by picard-tools. Indel Realignment were performed with GenomeAnalysisTK (v3.5) using mm10 dbsnp database as known sites. Base quality score recalibration was also performed with GenomeAnalysisTK (v3.5) using mm10 dbsnp database. SNPs and indels were called by GenomeAnalysisTK HaplotypeCaller (v3.5), which used default parameters. Whole exome sequencing raw data was submitted to SRA database (SRA; http://trace.ncbi.nlm.nih.gov/Traces/sra/, accession number SRP162613)

2.12. Statistics

The *p*-values were determined using two-tailed Student's *t*-test (t groups) and One-way ANOVA (>2 groups). A *p*-value less than 0.05 was considered significant.

3. Results

3.1. Combination of DMBA and OA Treatment Induces Dysplasia and Oncogenic Transformation in Organoid Culture

We chose an organoid culture system supported by epidermal growth factor (EGF), Noggin and R-spondin 1 (ENR) medium to investigate whether DMBA or/and OA could induce oncogenic transformation. As previously described [27], small intestine or colon organoids predominantly exhibited a well-organized, stereotyped epithelial single-layer organization at 7 days of culture, and maintained the similar morphology over a 50-day period of culture (Figure 1). At day 7, DMBA did not affect colony (organoid)-forming efficiency. OA induced a slight increase in both colony-forming efficiency, while a combination of DMBA and OA induced a large and significant increase in colony-forming efficiency (Figure 1A). At day 50, DMBA did not affect organoid morphology, OA induced mild enlargement in part of the epithelial layer with crowded nuclei, while a combination of DMBA and OA induced a very large malformation involving the entire epithelium with a confluent sheet of nuclear pleomorphism (Figure 1B), similar to that observed only when combining *Apc*, *p53*, *Kras*^{G12D}, and *Smad4* mutations in differentiated villus cells [27]. Histological analysis revealed that organoids treated with DMBA alone had a single-layer epithelium, similar to the control. OA-treated organoids showed multi-cell-layer-changes in only a small part of the epithelium, while those treated with DMBA in combination with OA showed multi-cell-layer-changes with loss of the cell border in nearly the entire epithelium (Figure 1C), similar to the histology achieved only by quadruple mutants, *Apc/Kras*^{G12D}*/p53/Smad4* [27]. A dysplasia index quantification of proliferation, nuclear atypia, invasion, and cellular stratification in organoids indicated that DMBA did not induce dysplasia compared to the control, OA induced a marginal increase in dysplasia, while the combination of DMBA and OA induced a large and significant increase in dysplasia (Figure 1D). Furthermore, the combination of DMBA and OA, but not DMBA or OA alone, endowed organoids with robust in vivo tumorigenicity, forming alpha-smooth muscle actin (SMA)$^{+}$ and CK20^{+} intestinal tumor after subcutaneous transplantation (Figure 1E–G).

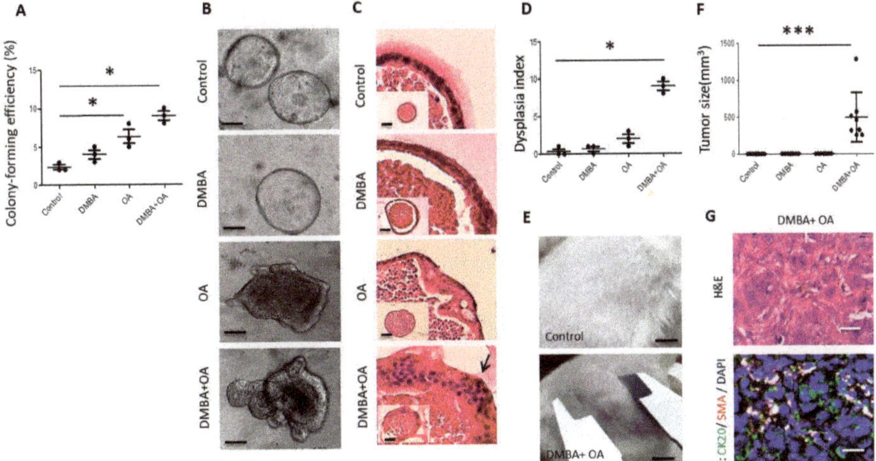

Figure 1. Combination of DMBA and okadaic acid (OA) induces dysplasia and oncogenic transformation in wild-type intestinal organoid culture. In vitro culture of wild type intestinal organoids without (control) or with DMBA or/and tamoxifen (TAM) in the presence of EGF, Noggin, and R-spondin 1 (500 single cells/well). (**A**) Colony (organoid)-forming efficiency was calculated at day 7. Experiment has been carried out in triplicate and each time 100 organoids were counted in each group. (**B**) Bright-field of organoid culture at day 50. Scale bar, 100 μm. (**C**) H&E staining and histologic characterization of cystic stratified epithelium with nuclear pleomorphism (arrow). Scale bar, 50 μm. (**D**) Dysplasia index at day 50 (experiment were repeated twice with $n = 3$ microscopic fields containing viable organoids). (**E**) Dissociated cells in Matrigel (500,000 cells/100 μL) were injected s.c. into NOD-SCID mice. In vivo tumor formation 45 days later (For those without tumor formation, observation was extended for up to 3 months, experiment were repeated twice with $n = 3$). (**F**) Tumor size 45 days after s.c. implantation ($n = 8$). (**G**) H&E staining and immunofluorescence of CK20 and SMA for tumor sections. Scale bar, 100 μm. *, $p < 0.05$; ***, $p < 0.0001$ as determined with one-way ANOVA.

3.2. Combination of DMBA Treatment and PP2A Deficiency Induces Dysplasia and Oncogenic Transformation in Organoid Culture

We further characterized the transformation effect of DMBA or/and adenovirus carrying recombinase (Ad-Cre-GFP)-mediated-PP2A deficiency in *Ppp2r1a^{flox/flox}* mice-derived organoids. Similarly, DMBA had no significant effects, Ad-Cre-GFP mediated-PP2A deficiency had mild effects, and the combination of DMBA and Ad-Cre-GFP mediated-PP2A deficiency had large effects on early organoid forming efficiency, late organoid morphology, histological changes, and dysplasia (Figure 2A–D). Furthermore, only the combination of DMBA and Ad-Cre-GFP-mediated-PP2A deficiency fully endowed organoids with robust in vivo tumorigenicity (Figure 2E–G).

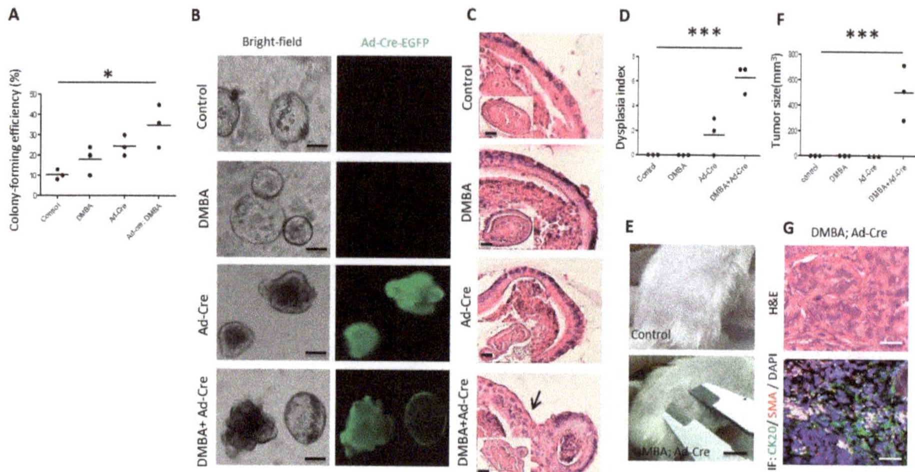

Figure 2. Combination of DMBA and Ad-Cre induces dysplasia and oncogenic transformation in Ppp2r1aflox/flox intestinal organoid culture. In vitro culture of Ppp2r1aflox/flox intestinal organoids without (control) or with DMBA or/and Ad-Cre-GFP (Ad-Cre) infection in the presence of EGF, Noggin and R-spondin 1 (500 single cells/well). (**A**) Colony (organoid)-forming efficiency was calculated at day 7. At least 100 organoids were counted in each group. (**B**) Bright-field and fluorescence images of organoid culture at day 50. Scale bar, 100 μm. (**C**) H&E staining and histologic characterization of cystic stratified epithelium with nuclear pleomorphism (arrow indicated). Scale bar, 50 μm. (**D**) Dysplasia index at day 50 (experiments were repeated twice with $n = 3$ microscopic fields containing viable organoids). (**E**) Dissociated cells in Matrigel (500,000 cells/100 μL) were injected s.c. into NOD-SCID mice. In vivo tumor formation 45 days later (for those without tumor formation, observation was extended for up to 3 months, experiments were repeated twice with $n = 3$). (**F**) Tumor size 45 days after s.c. implantation ($n = 3$). (**G**) H&E staining and immunofluorescence of CK20 and SMA for tumor sections. Scale bar, 100 μm. *, $p < 0.05$; ***, $p < 0.0001$ as determined with one-way ANOVA.

3.3. Combination of DMBA Treatment and PP2A Deficiency in Lgr5+ Rather than in Differentiated Villus Cells Induces Dysplasia and Oncogenic Transformation in Organoid Culture

To investigate whether Lgr5+ crypt stem cells or differentiated villus cells serve as the cell of origin of tumors, we treated organoids from *Lgr5-EGFP-CreERT2; Ppp2r1a^flox/flox^* mice with DMBA or/and tamoxifen. Similarly, DMBA did not have significant effects, tamoxifen mediated-PP2A deficiency had mild effects, and the combination of DMBA and tamoxifen-mediated-PP2A deficiency had great effects on early organoid forming efficiency, late organoid morphology, histological changes, and dysplasia (Figure 3A–E). Furthermore, only the combination of DMBA and tamoxifen-mediated-PP2A deficiency endowed organoids with in vivo tumorigenicity (Figure 3F,G). Interestingly, the combination of DMBA and PP2A deficiency in organoids derived from *Villin-Cre; Ppp2r1a^flox/flox^* mice did not affect early organoid forming efficiency, late organoid morphology, histological changes, and dysplasia (Figure S2A–E), and failed to induce in vivo tumorigenicity (Figure S2F). Of note, the recombinase activity in Villin-Cre mice is gradually reduced from villus to crypt [33], nevertheless, Ppp2r1a protein is only deleted in sorted villus cells but not in sorted Lgr5+ cells (Figure S2A). These data suggest that Lgr5+ crypt stem cells but not differentiated villus cells serve as the cell of origin of intestinal tumor.

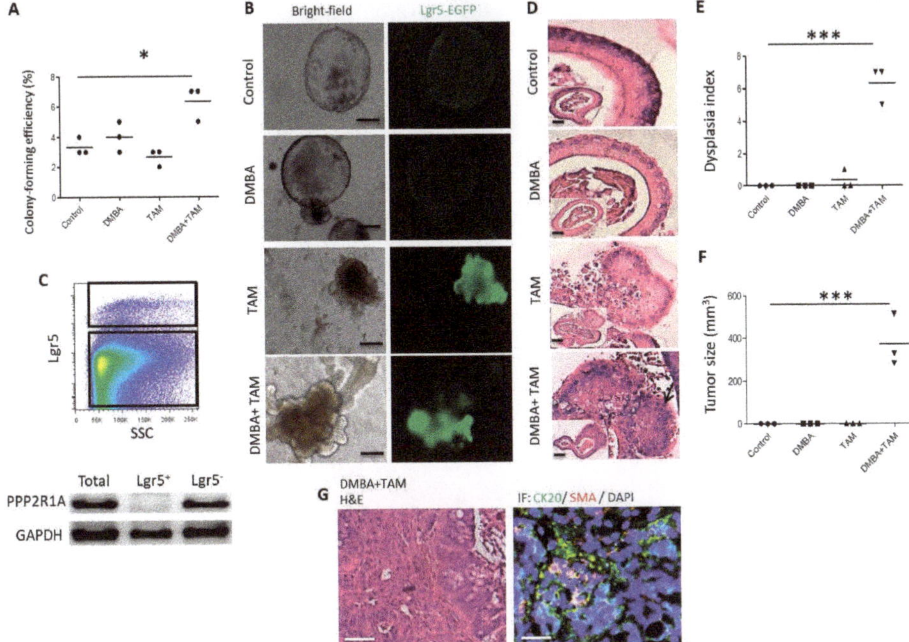

Figure 3. Combination of DMBA and TAM induces dysplasia and oncogenic transformation in *Lgr5-EGFP-CreERT2; Ppp2r1a^{flox/flox}* intestinal organoid culture. In vitro culture of *Lgr5-EGFP-CreERT2; Ppp2r1a^{flox/flox}* intestinal organoids without (control) or with DMBA or/and tamoxifen (TAM) in the presence of EGF, Noggin and R-spondin 1 (500 single cells/well). (**A**) Colony (organoid)-forming efficiency was calculated at day 7. At least 100 organoids were counted in each group. (**B**) Bright-field and fluorescence images of organoid culture at day 50. Scale bar, 100 μm. (**C**) Fluorescence-activated cell sorting (FACS) isolation of Lgr5^+ and Lgr5^− populations. After FACS, PPP2R1A protein levels were detected by western blot. (**D**) H&E staining and histologic characterization of cystic stratified epithelium with nuclear pleomorphism (arrow). Scale bar, 50 μm. (**E**) Dysplasia index at day 50 (experiments were repeated twice with *n* = 3 microscopic fields containing viable organoids). (**F**) Dissociated cells in Matrigel (500,000 cells/100 μL) were injected s.c. into NOD-SCID mice. In vivo tumor formed 45 days later (for those without tumor formation, observation was extended for up to 3 months, experiments were repeated twice with *n* = 3). (**G**) H&E staining and immunofluorescence of CK20 and SMA for tumor sections. Scale bar, 100 μm. *, $p < 0.05$; ***, $p < 0.0001$ as determined with one-way ANOVA.

3.4. Combination of DMBA Treatment and PP2A Deficiency in Lgr5^+ Cells Induces Upregulation in Stem Cell and EMT Markers, Downregulation in Differentiated Markers, and Tumorigenicity in Organoid Culture

Flow cytometric analysis (Figure 4A) revealed that DMBA did not increase the Lgr5^+ cell ratio (as assayed by Lgr5-EGFP). PP2A deficiency induced a marginal increase, while the combination of DMBA and PP2A deficiency induced a significant increase in the Lgr5^+ cell ratio. Comparative gene expression analysis of RNA samples isolated from organoid culture of *Lgr5-EGFP-CreERT2; Ppp2r1a^{flox/flox}* mice 50 days after DMBA and tamoxifen administration revealed marked upregulation of stem cell genes, such as *Lgr5, CD44, Ephb3, Egr2, Notch1*, and *Sox4*; as well as EMT markers, such as *Snail1, Snail2, Twist1, fibronectin*, and *vimentin*; and a marked downregulation of genes associated with differentiated cells, such as Paneth, enterocyte, goblet, and secretory cells compared to other treatment groups (Figure 4B). Furthermore, a small number (10^3) of Lgr5^+ but not Lgr5^− cells isolated from organoid culture treated with DMBA and tamoxifen possessed in vivo tumorigenicity (Figure 4C).

Collectively, these data suggest that the combination of DMBA and PP2A deficiency converted Lgr5$^+$ crypt stem cells into CSCs.

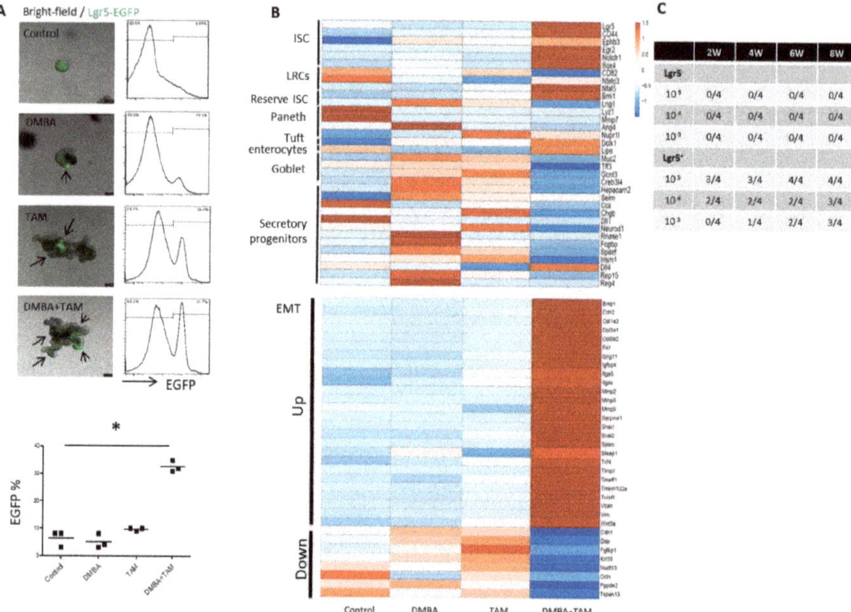

Figure 4. Combination of DMBA and PP2A deficient induces upregulation in stem cell and EMT markers and downregulation in differentiated markers in intestinal organoid. (**A**) In vitro culture of *Lgr5-EGFP-CreERT2; Ppp2r1a^flox/flox* intestinal organoids. Merged bright-field and fluorescence images of organoid culture treated without (control) or with DMBA or/and tamoxifen (TAM) for 50 days in the presence of EGF, Noggin and R-spondin 1. *Lgr5-EGFP* were denoted as arrows. Scale bar, 100 μm. Flow cytometric analysis and quantification of GFP expression (bottom panel). No fluorescence organoid culture was serve as negative control to decide the threshold. *, $p < 0.05$ as determined with one-way ANOVA. (**B**) RNA-*sequencing* (seq) analysis of transcriptomes for 50-day organoids. The upper heat map shows clustering to previously reported RNA-seq data of sorted ISC (Intestinal Stem Cell); reserve ISC; paneth cell; LRCs (label retaining cells); tuft; enterocytes; goblet and secretory progenitor cells. The lower heat map shows clustering to EMT (epithelial–mesenchymal transition). (**C**) Tumor incidence in limiting dilution assay. Tumorigenic potential characterization of indicated numbers of Lgr5$^+$ and Lgr5$^-$ cells from the 50-day organoid culture. Dissociated cells in Matrigel (100 μL) were injected s.c. into NOD-SCID mice. Incidence of tumor formation was calculated more than 2 months.

3.5. Combination of DMBA Treatment and PP2A Deficiency Generates CSCs through PI3K, ERK, and Wnt Activation

To understand the molecular characteristics of the genes and pathways involved in organoid culture of *Lgr5-EGFP-CreERT2; Ppp2r1a^flox/flox* mice after DMBA and tamoxifen administration, RNA-seq and gene set enrichment analysis (GSEA) were performed. Similar to previous findings that intestinal tumor begins with specific molecular alterations in Wnt-beta-catenin pathway [13,23], Wnt signaling was upregulated upon oncogenic transformation of organoid culture (Figure 5A). Western blotting of nuclear proteins (Figure 5B) and immunofluorescence (Figure 5C) revealed that nuclear accumulation of beta-catenin was predominantly observed in organoids with both of DMBA treatment and PP2A deficiency. Screening with a serine/threonine phosphorylation protein array (Figure S3) followed by confirmation with western blotting further revealed that PI3K/AKT/GSK-3beta and Raf/ERK were activated in organoids with both DMBA treatment and PP2A depletion compared to other treatment groups (Figure 5D,E). Interestingly, treatment with the PI3K inhibitor LY294002, MEK inhibitor PD98059, and Wnt inhibitor DKK1 reduced the formation of malformed organoids, inhibited dysplasia (Figure 5F,G), and completely blocked in vivo tumorigenicity (Figure 5H). Notably, not only the WNT signaling was the most, but also the PI3K and RAS-MAPK (ERK) signalings were common altered pathways in human intestinal tumor, as revealed by The Cancer Genome Atlas (TCGA) Project [14]. These data suggest that CSC generation by DMBA treatment and PP2A deficiency depends on the activation of PI3K, ERK, and Wnt signals.

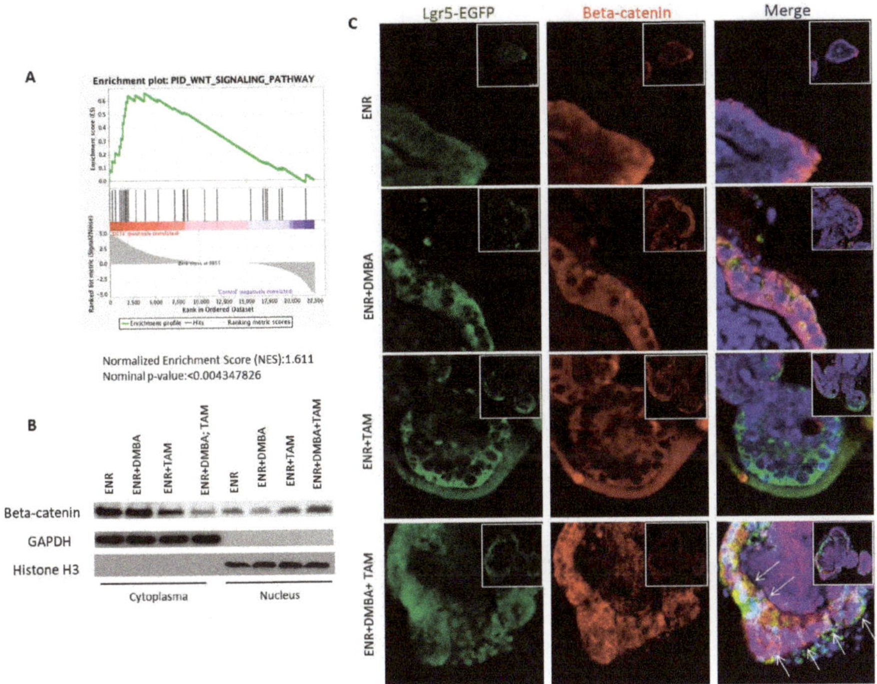

Figure 5. *Cont.*

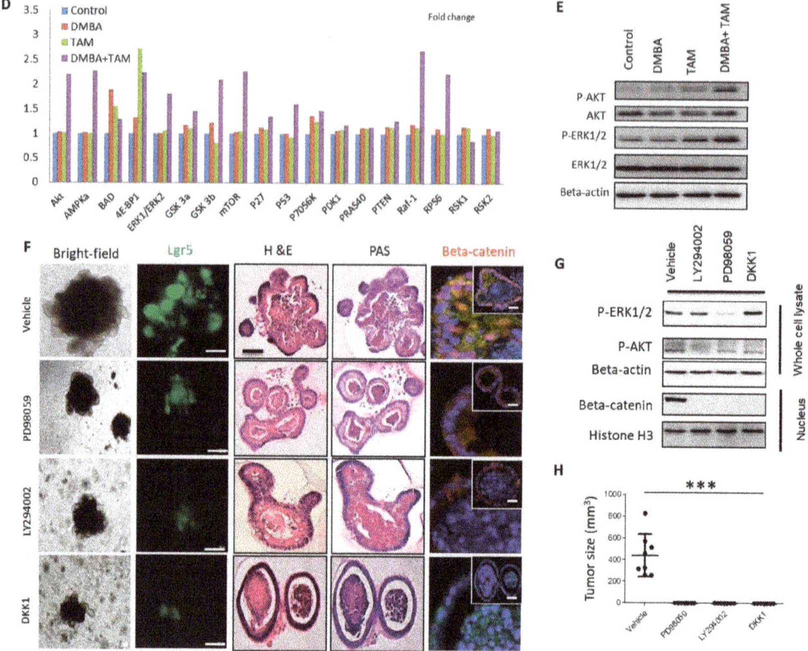

Figure 5. Beta-catenin activation caused by PI3K and ERK mediates dysplasia and oncogenic transformation in organoid culture. *Lgr5-EGFP-CreERT2; Ppp2r1a$^{flox/flox}$* intestinal organoids were treated without (control) or with DMBA or/and tamoxifen (TAM) for 50 days in the presence of EGF, Noggin and R-spondin 1. (**A**) Gene set enrichment analysis (GSEA) shows "WNT SIGNALING PATHWAY" for the organoids treated with DMBA and TAM group versus control group, $p < 0.05$. (**B**) Western blotting of nuclear and cytoplasmic fractions. GAPDH and Histone H3 were used as protein loading controls for the cytoplasmic and nuclear fractions, respectively. (**C**) Immunofluorescence of Lgr5-EGFP and beta-catenin expression in organoid (denoted by arrows). Scale bar, 100 μm. (**D**) Graphs of mouse AKT pathway phosphorylation protein expression array (original data in Extended Data Figure 3) and densitometric analyses, $n = 1$. (**E**) Western blotting. (**F**) Bright-field, fluorescence images, H&E, PAS staining and beta-catenin immunofluorescence images. (**G**) Western blot analysis of whole cell lysate and nuclear fraction from organoid cultures treated with EGF, Noggin, R-spondin 1 (ENR), DMBA, and TAM for 50 days in the absence (Vehicle) or presence of indicated inhibitor treatment. (**H**) Dissociated cells in Matrigel (500,000 cells/100 μL) were injected s.c. into NOD-SCID mice. In vivo tumor formed 45 days later (For those without tumor formation, observation was extended for up to 3 months, $n = 8$). ***, $p < 0.0001$ as determined with one-way ANOVA.

3.6. Lgr5$^+$ CSCs Are R-Spondin 1-Dependent

CSCs isolated from human intestinal tumor specimens express the Lgr5 crypt marker [34]. R-spondin is expressed by the intestinal stroma and is differentially upregulated during *Citrobacter rodentium*- and dextran sulfate sodium (DSS)-induced colitis in mice, which reflects human ulcerative colitis, a precancerous stage [35]. These data suggest a role for R-spondin 1 and its receptor Lgr5 in the maintenance of undifferentiated status and tumorigenesis of human colorectal CSCs. In contrast to tumorigenesis initiated by dedifferentiation (tumor generated from intestinal epithelial cell of *villin-creER^{T2}/APC $^{lox/lox}$/K-ras$^{G12D/+}$* mice) [4], where tumor cells were Lgr5$^-$ and generated independently of R-spondin 1, tumor cells in the current study were Lgr5$^+$ and generated dependently on R-spondin 1 (Figure 6), suggesting that tumor cells generated by DMBA treatment and PP2A deficiency in mouse Lgr5$^+$ cells were indeed intestinal CSCs and could serve as surrogates of human colorectal CSCs [34]. Previous reports using tumorigenesis models generated by genetic manipulation in differentiated villus cells, even with low efficiency in tumorigenesis [27], also demonstrated important roles for Lgr5+ cells in tumorigenesis/metastasis [7,8].

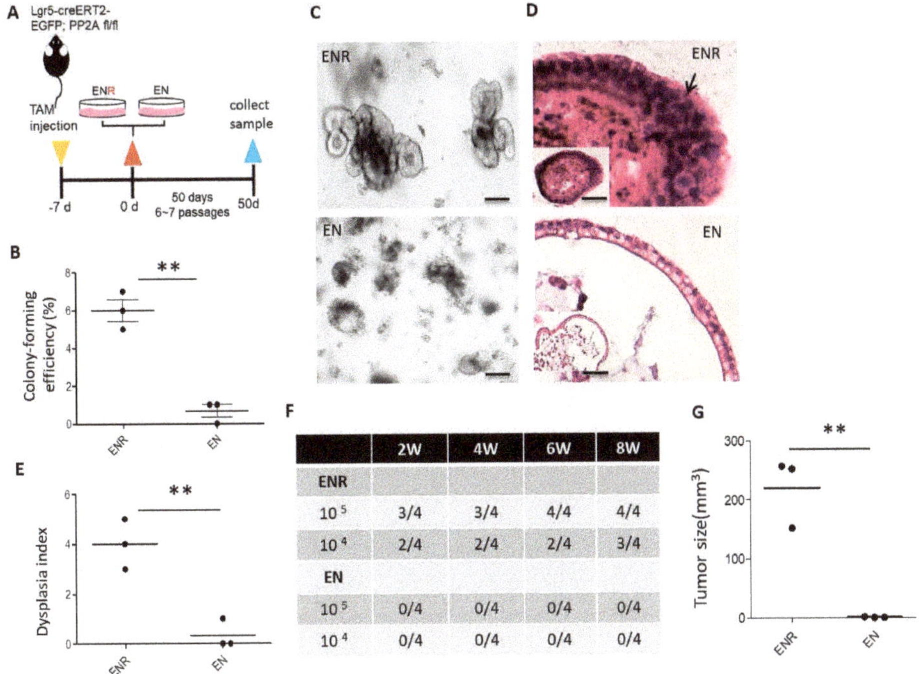

	2W	4W	6W	8W
ENR				
10^5	3/4	3/4	4/4	4/4
10^4	2/4	2/4	2/4	3/4
EN				
10^5	0/4	0/4	0/4	0/4
10^4	0/4	0/4	0/4	0/4

Figure 6. R-spondin 1 drives Wnt-dependent dysplasia and oncogenic transformation in the intestinal organoid culture. In vitro culture of *Lgr5-EGFP-CreERT2; Ppp2r1a^{flox/flox}* intestinal organoids (500 single cells/well) treated with DMBA and tamoxifen (TAM) in the presence of EGF, Noggin, and with (ENR) or without R-spondin-1 (EN). (**A**) Schematic illustration of the experimental design. (**B**) Colony (organoid)-forming efficiency at day 7. At least 100 organoids were counted in each group. (**C**) Bright-field images at day 50. Scale bar, 100 μm. (**D**) H&E staining and histologic characterization of cystic stratified epithelium with nuclear pleomorphism (arrow). Scale bar, 50 μm. (**E**) Dysplasia index at day 50 (experiment was repeated twice with *n* = 3 microscopic fields containing viable organoids). (**F**) Incidence of tumor formation at indicated time periods, and (**G**) tumor size in NOD-SCID mice after s.c. injection 45 days later of indicated cell numbers from 50-day organoid. **, *p* < 0.01 as determined with Student's *t*-test.

3.7. Combination of DMBA Treatment with PP2A Deficiency in Lgr5+ Also Induces Tumor Formation In Vivo

We investigated whether tumorigenesis generated by DMBA treatment and PP2A deficiency in mouse Lgr5+ cells reflects tumor formation in vivo. *Lgr5-EGFP-CreERT2; Ppp2r1a^{flox/flox}* mice treated with DMBA and tamoxifen formed adenocarcinoma in the small intestine and colorectal regions 36 days later. Histological analysis revealed that mice receiving DMBA and tamoxifen increased the incidence of multiple foci adenoma compared to other groups (Figure 7A). IHC analysis revealed that beta-catenin accumulated in the nuclei of tumor cells mainly in the crypt area (Figure 7B) and quantitative evaluation of nucleus beta-catenin accumulation [29] further showed that nucleus beta-catenin H-score was significantly greater in mice receiving DMBA and tamoxifen compared to other groups (Figure 7C,D). The efficiency and rapidity were much greater than tumorigenesis generated by *Apc* deficiency in either Lgr5+ crypt stem cell- [3] or differentiated villus cell-based models [4].

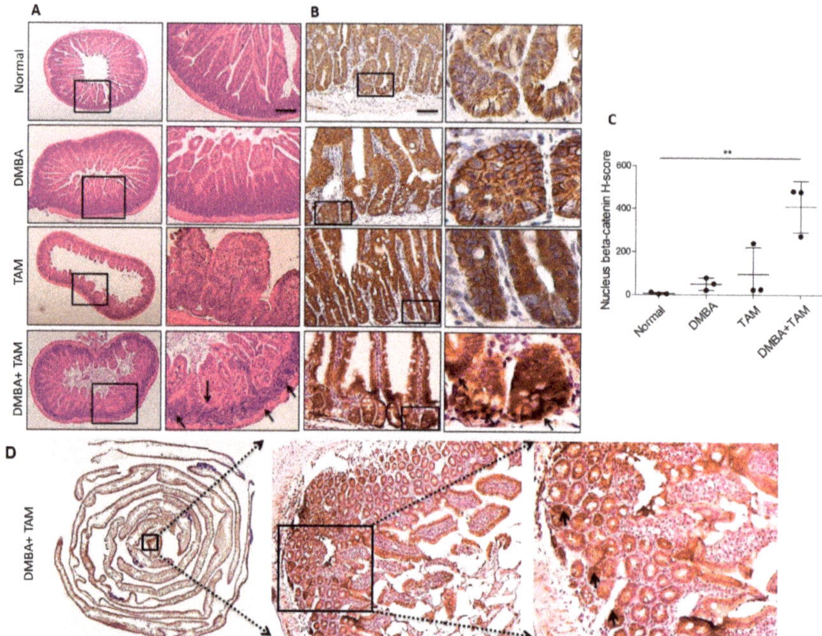

Figure 7. Combination of DMBA treatment with PP2A deficient in Lgr5$^+$ drives intestinal neoplasia in both the small intestine and colon. (**A**) H&E staining and (**B**) Beta-catenin IHC were performed on the serial sections. Multiple beta-catenin^{high} adenomas were observed throughout the colon 36 days after induction. (**C**) Quantitative analysis of nucleus beta-catenin H-score from (**B**). (**D**) High level expression of beta-catenin was apparent in the transformed stem cell compartment (Peyer's patches are stained blue). Multiple beta-catenin^{high} transformed cells were observed throughout the intestinal 36 day after induction (magnified at right panel, denoted by arrows). Original magnifications: (**A**) left, 4×; (**A**) right, 10×; (**B**) left, 20×; (**B**) right, 40×; (**D**) center, 20×; (**D**) right, 40×; Scale bar, 100 μm. **, $p < 0.01$ as determined with one-way ANOVA.

3.8. Not Only DMBA but Also Other Carcinogens Induce Tumors from Lgr5$^+$ Intestinal Stem Cells of PP2A Deficient Mice

To demonstrate carcinogen-induced tumor from Lgr5$^+$ intestinal stem cells of PP2A deficient mice was not limited to DMBA, *N*-methyl-*N*-nitrosourea (MNU) and 2-amino-1-methyl-6-phenylimidazo-[4,5-b]pyridine (PhIP) were also administered in combination with tamoxifen treatment in organoid cultures derived from *Lgr5-EGFP-CreERT2; Ppp2r1a$^{flox/flox}$* mice and from *Villin-Cre; Ppp2r1a$^{flox/flox}$* mice. Interestingly, increased early organoid forming efficiency, late organoid with irregular nuclei and prominent nucleoli morphology and dysplasia, and in vivo tumorigenicity were only observed in organoid cultures derived from *Lgr5-EGFP-CreERT2; Ppp2r1a$^{flox/flox}$* mice but not from *Villin-Cre; Ppp2r1a$^{flox/flox}$* mice (Figure S4A,B). More importantly, we found that tumorigenicity induced by combination of MUN or PhIP with PP2A deficiency also depended on the activation of PI3K, ERK, and Wnt signals (Figure S4C–F). Similarly, combination of MUN or PhIP treatment with PP2A deficiency in Lgr5$^+$ cells also induced tumor formation in vivo (Figure S4G).

3.9. Mutational Landscapes of Intestinal Tumors Derived from Lgr5$^+$ Cells Treated with Carcinogen and PP2A Deficiency

To gain insight of the genetic alterations that drove these pathways in the models, we performed whole-exome sequencing (WES) (Figure S5A). Somatic variants in each chemical induced tumor organoid sample were identified with a tumor-control paired strategy by removing the variants in their paired control samples and the variants affecting protein coding sequence were further filtrated (Figure S5B). The numbers of somatic mutations, including synonymous and nonsynonymous mutations, for organoid cultures with combined DMBA, MUN, or PhIP treatment with PP2A deficiency in Lgr5$^+$ cells were shown in Figure S5C. IPA analysis of 270 mutated genes shared by DMBA, MUN, and PhIP treatment with PP2A deficiency in Lgr5$^+$ cells (Figure S5D) revealed 10 top significantly enriched pathways (Figure S5E), including the intestinal tumor metastasis signaling (p-value = 2.12×10^{-57}) and Wnt/beta-catenin signaling (p-value = 1.40×10^{-55}). Furthermore, beta-catenin (*CTNNB1*) seemed to be a common downstream hub as identified by the Path Explorer function in IPA (Figure S5F). There were five core pathways related with human intestinal tumor found by TCGA, including p53, RAS-MAPK, PI3K, TGF-beta, and WNT pathways (Figure 8) [14]. We identified several recurrent mutations with FDR < 0.1, including *Braf*, *erbb2*, *kras*, *pten*, *Smad2*, *Smad4*, *Apc*, *DKK2*, *Wnt4*, *Wnt5a*, and *Wnt5b* (Figure 8). Based on these analyses, the mouse intestinal tumor organoid models based on combined use of chemical carcinogen and genetic modification can recapitulate the human colorectal cancer development process in response to complex interactions between the genotype and environmental factor.

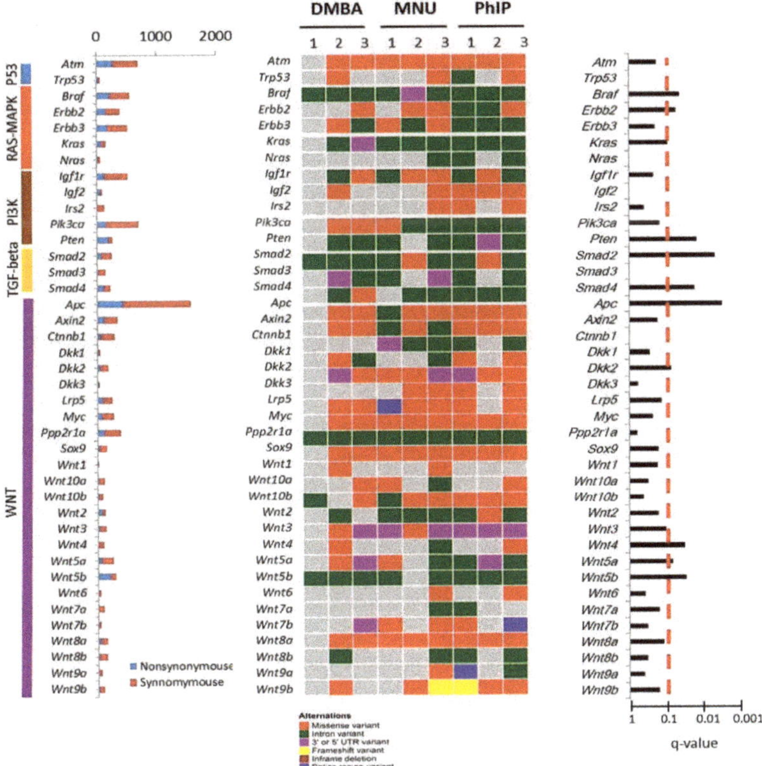

Figure 8. Landscape of somatic mutant genes in three carcinogen treated organoid cultures. Data matrix shows number of somatic mutant genes in each carcinogen treated organoid cultures and were classified according to their pathway. Somatic mutations are presented according to the type of mutation (missense variant, intron variant, 3' or 5' prime UTR variant, frameshift variant, inframe deletion, or splice region variant) On the left, the total number of mutations of each gene within all three groups is shown with a bar plot, while the q-value of each significantly mutated gene is shown on the right.

4. Discussion

To improve the efficiency of tumorigenesis in GEMMs, multiple gene mutations are necessary. Given that the frequency of point mutations varies from less than 0.1 to greater than 50 mutations per megabase [36], GEMMs are a greatly oversimplified view of the numbers and types of mutations found in human cancers [37]. The use of GEMMs in combination with carcinogenesis increased the tumor spectrum or speeded the tumor formation observed in some GEMMs, such as the $p53^{-/-}$ mouse model in combination with exposure to carcinogens or radiation [21,38]. In the current study, $Ppp2r1a^{-/-}$ mouse model in combination with exposure to different carcinogens induced tumorigenesis through similar signaling pathways, while GEMM or carcinogen exposure alone did not induce tumorigenesis. These data suggest that the combination of GEMMs with exposure to carcinogen or other environmental agents is a logical approach in studying tumorigenesis and cancer progression.

PP2A is a tumor suppressor that regulates many oncogenic pathways. In fact, decreased PP2A activity has been reported as a common event in colorectal cancer [39]. DMBA and MNU are important environmental carcinogens. PhIP, a food-borne carcinogen produced while cooking meat and fish, models human colon cancer in rodents [40]. Here, we demonstrated that DMBA, MNU, or PhIP each induced intestinal organoid transformation when combined with PP2A deficiency in $Lgr5^+$ intestinal stem cells but not in differentiated villus cells, suggesting that PP2A plays a role in suppressing colorectal tumorigenesis induced by chemical carcinogen exposure. Our results provide further experimental evidence to demonstrate that the cell of origin of intestinal tumor is crypt stem cells instead of differentiated villus cells. We did not focus on the molecular mechanism underlying the differential tumorigenicity pathways between stem cells and differentiated cells, which requires further investigation.

A recent study shows that unless the R-spondin and Wnt ligands are both present, the default fate of $Lgr5^+$ crypt stem cells is differentiation [41]. However, gain-of-function studies reveal that R-spondin and Wnt ligands have qualitatively distinct, non-interchangeable roles in crypt stem cells. Wnt proteins confer a basal competency unto the $Lgr5^+$ crypt stem cell by inducing R-spondin receptor expression that enables R-spondin-driven $Lgr5^+$ crypt stem cell self-renewal. In the current study, using a combination of carcinogen treatment and PP2A deficiency, intestinal organoids derived from Lgr5-cre; Ppp2r1aflox/flox, but not Villin-Cre; Ppp2r1aflox/flox, underwent oncogenic transformation and exhibited CSC phenotypes that were dependent on the presence of R-spondin 1 in the culture media. These studies together suggest the important roles of R-spondin signaling in stem cell self-renewal and preventing differentiation. The discrepancy of R-spondin 1 dependence between tumorigenicity induced by the current method and previous GEMM relied on *Apc* depletion or *CTNNB1* activation [4] is supported by the mutual exclusion of *R-spondin* fusion and *Apc* or *CTNNB1* mutation identified in human intestinal tumors [20]. The Lgr5/R-spondin 1 complex degrades Rnf43 and Znrf3, two transmembrane E3 ligases that remove Wnt receptors from the stem cell surface. Consistently, simultaneous deficient of these two E3 ligases in the intestinal epithelium induced the formation of unusual adenomas consisting entirely of $Lgr5^+$ stem cells and their niche [42].

Simultaneous carcinogen treatment and conditional deletion of PP2A in villus cells did not induce transformation or increased proliferation or dysplasia of intestine organoids. The current study did not aim to study the underlying mechanism that differentiates $Lgr5^+$ crypt stem cells from differentiated villus cells in terms of vulnerability to carcinogenesis. Based on our current results, both Wnt and Rspondin/Lgr5 signaling pathways play important roles in nuclear-beta catenin localization, which is important for stem cell self-renewal and may also initiate tumorigenesis once dysregulated.

5. Conclusions

In summary, we demonstrated that carcinogen-induced cancer arises from $Lgr5^+$ crypt stem cells in $Ppp2r1a^{-/-}$ mice. In addition, combing carcinogenesis with GEMM recapitulated the developmental process of environmentally induced human tumor, while increasing the rate and percentage of tumorigenesis. Interestingly, exposure to different carcinogens, such as DMBA, MNU, or PhIP, when combined with PP2A deficiency in Lgr5$^+$ intestinal stem cells induced tumorigenesis that was dependent on the activation of pathways including Wnt, PI3K, and RAS-MAPK signalings, the common altered pathways revealed by TCGA human intestinal tumor project [14]. Together, these data suggest that PP2A functions as a tumor suppressor in intestine carcinogenesis. This organoid platform provides experimental evidence as to its usefulness in detection of key oncogenes and suppressor genes as early molecular epidemiological biomarkers of carcinogenesis, and is useful in human cancer prevention practice as well.

Supplementary Materials: The following are available online at http://www.mdpi.com/2073-4409/9/1/90/s1, Figure S1: Screening strategy for phosphatase hub gene in TCGA-COAD data set, Figure S2: DMBA and PP2A deficiency did not induce dysplasia and oncogenic transformation in *Villin-Cre; Ppp2r1a$^{flox/flox}$* intestinal organoid culture, Figure S3: Serine/threonine phosphorylation protein array screening in individual organoid groups, Figure S4: Combination of MNU or PhIP and TAM induces dysplasia and oncogenic transformation in *Lgr5-EGFP-CreERT2; Ppp2r1a$^{flox/flox}$* intestinal organoid culture, Figure S5: Ingenuity pathway analysis (IPA) to identify significant canonical pathways and interactome in three carcinogens induced oncogenic transformation in *Lgr5-EGFP-CreERT2; Ppp2r1a$^{flox/flox}$* intestinal organoids.

Author Contributions: Y.-T.Y. designed and performed experiments, analyzed the data and wrote the paper. M.C. contributed to Lgr5-cre and Vilin-cre PP2A knockout mice analysis and helped tissue processing. Y.-C.L. performed the RNA-seq analysis. D.-P.C. performed the WES analysis. C.-M.C. reviewed and edited the paper. M.-C.H. reviewed and edited the paper. S.-C.H. wrote the proposal, designed and supervised the study, analyzed the data and wrote the paper. All authors have read and agreed to the published version of the manuscript.

Funding: Grants supported by Minister of Science and Technology (MOST 106-2321-B-039-003) and Integrative Stem Cell Center, China Medical University Hospital. This work was also financially supported by the "Drug Development Center, China Medical University" from The Featured Areas Research Center Program within the framework of the Higher Education Sprout Project by the Ministry of Education (MOE) in Taiwan. The funding sources had no involvement in study design, in the collection, analysis and interpretation of data, in the writing of the report, and in the decision to submit the article for publication.

Acknowledgments: We thank Bin Tean Teh (Cancer Science Institute of Singapore, National University of Singapore, Centre for Life Sciences, Singapore) for discussion of our data.

Conflicts of Interest: The authors declare no conflict of interest.

References

1. Poirier, M.C. Chemical-induced DNA damage and human cancer risk. *Nat. Rev. Cancer* **2004**, *4*, 630–637. [CrossRef]

2. Visvader, J.E. Cells of origin in cancer. *Nature* **2011**, *469*, 314–322. [CrossRef] [PubMed]

3. Barker, N.; Ridgway, R.A.; van Es, J.H.; van de Wetering, M.; Begthel, H.; van den Born, M.; Danenberg, E.; Clarke, A.R.; Sansom, O.J.; Clevers, H. Crypt stem cells as the cells-of-origin of intestinal cancer. *Nature* **2009**, *457*, 608–611. [CrossRef] [PubMed]

4. Schwitalla, S.; Fingerle, A.A.; Cammareri, P.; Nebelsiek, T.; Goktuna, S.I.; Ziegler, P.K.; Canli, O.; Heijmans, J.; Huels, D.J.; Moreaux, G.; et al. Intestinal tumorigenesis initiated by dedifferentiation and acquisition of stem-cell-like properties. *Cell* **2013**, *152*, 25–38. [CrossRef] [PubMed]

5. Tetteh, P.W.; Basak, O.; Farin, H.F.; Wiebrands, K.; Kretzschmar, K.; Begthel, H.; van den Born, M.; Korving, J.; de Sauvage, F.; van Es, J.H.; et al. Replacement of lost Lgr5−positive stem cells through plasticity of their enterocyte-lineage daughters. *Cell Stem Cell* **2016**, *18*, 203–213. [CrossRef]

6. Westphalen, C.B.; Asfaha, S.; Hayakawa, Y.; Takemoto, Y.; Lukin, D.J.; Nuber, A.H.; Brandtner, A.; Setlik, W.; Remotti, H.; Muley, A.; et al. Long-lived intestinal tuft cells serve as colon cancer-initiating cells. *J. Clin. Investig.* **2014**, *124*, 1283–1295. [CrossRef]

7. De Sousa e Melo, F.; Kurtova, A.V.; Harnoss, J.M.; Kljavin, N.; Hoeck, J.D.; Hung, J.; Anderson, J.E.; Storm, E.E.; Modrusan, Z.; Koeppen, H.; et al. A distinct role for Lgr5$^+$ stem cells in primary and metastatic colon cancer. *Nature* **2017**, *543*, 676–680. [CrossRef]

8. Shimokawa, M.; Ohta, Y.; Nishikori, S.; Matano, M.; Takano, A.; Fujii, M.; Date, S.; Sugimoto, S.; Kanai, T.; Sato, T. Visualization and targeting of LGR5(+) human colon cancer stem cells. *Nature* **2017**, *545*, 187–192. [CrossRef]

9.   Ruvolo, P.P. The broken "Off" switch in cancer signaling: PP2A as a regulator of tumorigenesis, drug resistance, and immune surveillance. *BBA Clin.* **2016**, *6*, 87–99. [CrossRef]

10. Bardelli, A.; Parsons, D.W.; Silliman, N.; Ptak, J.; Szabo, S.; Saha, S.; Markowitz, S.; Willson, J.K.; Parmigiani, G.; Kinzler, K.W.; et al. Mutational analysis of the tyrosine kinome in colorectal cancers. *Science* **2003**, *300*, 949. [CrossRef]

11. Wang, Z.; Shen, D.; Parsons, D.W.; Bardelli, A.; Sager, J.; Szabo, S.; Ptak, J.; Silliman, N.; Peters, B.A.; van der Heijden, M.S.; et al. Mutational analysis of the tyrosine phosphatome in colorectal cancers. *Science* **2004**, *304*, 1164–1166. [CrossRef]

12. Samuels, Y.; Wang, Z.; Bardelli, A.; Silliman, N.; Ptak, J.; Szabo, S.; Yan, H.; Gazdar, A.; Powell, S.M.; Riggins, G.J.; et al. High frequency of mutations of the PIK3CA gene in human cancers. *Science* **2004**, *304*, 554. [CrossRef] [PubMed]

13. Parsons, D.W.; Wang, T.L.; Samuels, Y.; Bardelli, A.; Cummins, J.M.; DeLong, L.; Silliman, N.; Ptak, J.; Szabo, S.; Willson, J.K.; et al. Colorectal cancer: Mutations in a signalling pathway. *Nature* **2005**, *436*, 792. [CrossRef] [PubMed]

14. Cancer Genome Atlas, N. Comprehensive molecular characterization of human colon and rectal cancer. *Nature* **2012**, *487*, 330–337. [CrossRef] [PubMed]

15. Westermarck, J.; Hahn, W.C. Multiple pathways regulated by the tumor suppressor PP2A in transformation. *Trends Mol. Med.* **2008**, *14*, 152–160. [CrossRef] [PubMed]

16. Lin, S.P.; Lee, Y.T.; Wang, J.Y.; Miller, S.A.; Chiou, S.H.; Hung, M.C.; Hung, S.C. Survival of cancer stem cells under hypoxia and serum depletion via decrease in PP2A activity and activation of p38-MAPKAPK2-Hsp27. *PLoS ONE* **2012**, *7*, e49605. [CrossRef] [PubMed]

17. Lin, S.P.; Lee, Y.T.; Yang, S.H.; Miller, S.A.; Chiou, S.H.; Hung, M.C.; Hung, S.C. Colon cancer stem cells resist antiangiogenesis therapy-induced apoptosis. *Cancer Lett.* **2013**, *328*, 226–234. [CrossRef] [PubMed]

18. Liu, C.C.; Lin, S.P.; Hsu, H.S.; Yang, S.H.; Lin, C.H.; Yang, M.H.; Hung, M.C.; Hung, S.C. Suspension survival mediated by PP2A-STAT3-Col XVII determines tumour initiation and metastasis in cancer stem cells. *Nat. Commun.* **2016**, *7*, 11798. [CrossRef] [PubMed]

19. Cheng, Y.; Ren, X.; Yuan, Y.; Shan, Y.; Li, L.; Chen, X.; Zhang, L.; Takahashi, Y.; Yang, J.W.; Han, B.; et al. eEF-2 kinase is a critical regulator of Warburg effect through controlling PP2A-A synthesis. *Oncogene* **2016**, *35*, 6293–6308. [CrossRef]

20. Seshagiri, S.; Stawiski, E.W.; Durinck, S.; Modrusan, Z.; Storm, E.E.; Conboy, C.B.; Chaudhuri, S.; Guan, Y.; Janakiraman, V.; Jaiswal, B.S.; et al. Recurrent R-spondin fusions in colon cancer. *Nature* **2012**, *488*, 660–664. [CrossRef]

21. Donehower, L.A.; Harvey, M.; Slagle, B.L.; McArthur, M.J.; Montgomery, C.A., Jr.; Butel, J.S.; Bradley, A. Mice deficient for p53 are developmentally normal but susceptible to spontaneous tumours. *Nature* **1992**, *356*, 215–221. [CrossRef] [PubMed]

22. Cheng, J.L.; Futakuchi, M.; Ogawa, K.; Iwata, T.; Kasai, M.; Tokudome, S.; Hirose, M.; Shirai, T. Dose response study of conjugated fatty acid derived from safflower oil on mammary and colon carcinogenesis pretreated with 7,12-dimethylbenz[a]anthracene (DMBA) and 1,2-dimethylhydrazine (DMH) in female Sprague-Dawley rats. *Cancer Lett.* **2003**, *196*, 161–168. [CrossRef]

23. Brown, K.; Buchmann, A.; Balmain, A. Carcinogen-induced mutations in the mouse c-Ha-ras gene provide evidence of multiple pathways for tumor progression. *Proc. Natl. Acad. Sci. USA* **1990**, *87*, 538–542. [CrossRef] [PubMed]

24. Kimura, K.; Satoh, K.; Kanno, A.; Hamada, S.; Hirota, M.; Endoh, M.; Masamune, A.; Shimosegawa, T. Activation of Notch signaling in tumorigenesis of experimental pancreatic cancer induced by dimethylbenzanthracene in mice. *Cancer Sci.* **2007**, *98*, 155–162. [CrossRef] [PubMed]

25. Reddy, B.S.; Narisawa, T.; Maronpot, R.; Weisburger, J.H.; Wynder, E.L. Animal models for the study of dietary factors and cancer of the large bowel. *Cancer Res.* **1975**, *35*, 3421–3426. [PubMed]

26. Wakabayashi, K.; Nagao, M.; Esumi, H.; Sugimura, T. Food-derived mutagens and carcinogens. *Cancer Res.* **1992**, *52*, 2092s–2098s.

27. Li, X.; Nadauld, L.; Ootani, A.; Corney, D.C.; Pai, R.K.; Gevaert, O.; Cantrell, M.A.; Rack, P.G.; Neal, J.T.; Chan, C.W.; et al. Oncogenic transformation of diverse gastrointestinal tissues in primary organoid culture. *Nat. Med.* **2014**, *20*, 769–777. [CrossRef]

28. Sato, T.; Vries, R.G.; Snippert, H.J.; van de Wetering, M.; Barker, N.; Stange, D.E.; van Es, J.H.; Abo, A.; Kujala, P.; Peters, P.J.; et al. Single Lgr5 stem cells build crypt-villus structures in vitro without a mesenchymal niche. *Nature* **2009**, *459*, 262–265. [CrossRef]

29. Bilalovic, N.; Sandstad, B.; Golouh, R.; Nesland, J.M.; Selak, I.; Torlakovic, E.E. CD10 protein expression in tumor and stromal cells of malignant melanoma is associated with tumor progression. *Mod. Pathol.* **2004**, *17*, 1251–1258. [CrossRef]

30. Mahe, M.M.; Aihara, E.; Schumacher, M.A.; Zavros, Y.; Montrose, M.H.; Helmrath, M.A.; Sato, T.; Shroyer, N.F. Establishment of gastrointestinal epithelial organoids. *Curr. Protoc. Mouse Biol.* **2013**, *3*, 217–240. [CrossRef]

31. Metsalu, T.; Vilo, J. ClustVis: A web tool for visualizing clustering of multivariate data using Principal Component Analysis and heatmap. *Nucleic Acids Res.* **2015**, *43*, W566–W570. [CrossRef] [PubMed]

32. Subramanian, A.; Tamayo, P.; Mootha, V.K.; Mukherjee, S.; Ebert, B.L.; Gillette, M.A.; Paulovich, A.; Pomeroy, S.L.; Golub, T.R.; Lander, E.S.; et al. Gene set enrichment analysis: A knowledge-based approach for interpreting genome-wide expression profiles. *Proc. Natl. Acad. Sci. USA* **2005**, *102*, 15545–15550. [CrossRef] [PubMed]

33. Madison, B.B.; Dunbar, L.; Qiao, X.T.; Braunstein, K.; Braunstein, E.; Gumucio, D.L. Cis elements of the villin gene control expression in restricted domains of the vertical (crypt) and horizontal (duodenum, cecum) axes of the intestine. *J. Biol. Chem.* **2002**, *277*, 33275–33283. [CrossRef] [PubMed]

34. Kemper, K.; Prasetyanti, P.R.; De Lau, W.; Rodermond, H.; Clevers, H.; Medema, J.P. Monoclonal antibodies against Lgr5 identify human colorectal cancer stem cells. *Stem Cells* **2012**, *30*, 2378–2386. [CrossRef]

35. Kang, E.; Yousefi, M.; Gruenheid, S. R-spondins are expressed by the intestinal stroma and are differentially regulated during citrobacter rodentium- and DSS-induced colitis in mice. *PLoS ONE* **2016**, *11*, e0152859. [CrossRef]

36. Vogelstein, B.; Papadopoulos, N.; Velculescu, V.E.; Zhou, S.; Diaz, L.A., Jr.; Kinzler, K.W. Cancer genome landscapes. *Science* **2013**, *339*, 1546–1558. [CrossRef]

37. Westcott, P.M.; Halliwill, K.D.; To, M.D.; Rashid, M.; Rust, A.G.; Keane, T.M.; Delrosario, R.; Jen, K.Y.; Gurley, K.E.; Kemp, C.J.; et al. The mutational landscapes of genetic and chemical models of Kras-driven lung cancer. *Nature* **2015**, *517*, 489–492. [CrossRef]

38. Lee, C.L.; Mowery, Y.M.; Daniel, A.R.; Zhang, D.; Sibley, A.B.; Delaney, J.R.; Wisdom, A.J.; Qin, X.; Wang, X.; Caraballo, I.; et al. Mutational landscape in genetically engineered, carcinogen-induced, and radiation-induced mouse sarcoma. *JCI Insight* **2019**, *4*. [CrossRef]

39. Cristobal, I.; Manso, R.; Rincon, R.; Carames, C.; Senin, C.; Borrero, A.; Martinez-Useros, J.; Rodriguez, M.; Zazo, S.; Aguilera, O.; et al. PP2A inhibition is a common event in colorectal cancer and its restoration using FTY720 shows promising therapeutic potential. *Mol. Cancer Ther.* **2014**, *13*, 938–947. [CrossRef]

40. Kakiuchi, H.; Watanabe, M.; Ushijima, T.; Toyota, M.; Imai, K.; Weisburger, J.H.; Sugimura, T.; Nagao, M. Specific 5′-GGGA-3′–>5′-GGA-3′ mutation of the Apc gene in rat colon tumors induced by 2-amino-1-methyl-6-phenylimidazo [4,5-b] pyridine. *Proc. Natl. Acad. Sci. USA* **1995**, *92*, 910–914. [CrossRef]

41. Yan, K.S.; Janda, C.Y.; Chang, J.; Zheng, G.X.Y.; Larkin, K.A.; Luca, V.C.; Chia, L.A.; Mah, A.T.; Han, A.; Terry, J.M.; et al. Non-equivalence of Wnt and R-spondin ligands during Lgr5(+) intestinal stem-cell self-renewal. *Nature* **2017**, *545*, 238–242. [CrossRef] [PubMed]

42. Koo, B.K.; Spit, M.; Jordens, I.; Low, T.Y.; Stange, D.E.; van de Wetering, M.; van Es, J.H.; Mohammed, S.; Heck, A.J.; Maurice, M.M.; et al. Tumour suppressor RNF43 is a stem-cell E3 ligase that induces endocytosis of Wnt receptors. *Nature* **2012**, *488*, 665–669. [CrossRef] [PubMed]

Article

Mouse Embryonic Stem Cell-Derived Ureteric Bud Progenitors Induce Nephrogenesis

Zenglai Tan [1,*], Aleksandra Rak-Raszewska [1], Ilya Skovorodkin [1] and Seppo J. Vainio [1,2,*]

1   Center for Cell Matrix Research, Faculty of Biochemistry and Molecular Medicine, Biocenter Oulu,
     Laboratory of Developmental Biology, Infotech Oulu, University of Oulu, Aapistie 5A, 90220 Oulu, Finland;
     Aleksandra.Rak-Raszewska@oulu.fi (A.R.-R.); ilya.skovorodkin@oulu.fi (I.S.)
2   Borealis Biobank of Northern Finland, Oulu Central Hospital, 90220 Oulu, Finland
*   Correspondence: zenglai.tan@oulu.fi (Z.T.); seppo.vainio@oulu.fi (S.J.V.)

Received: 18 December 2019; Accepted: 27 January 2020; Published: 31 January 2020

Abstract: Generation of kidney organoids from pluripotent stem cells (PSCs) is regarded as a potentially powerful way to study kidney development, disease, and regeneration. Direct differentiation of PSCs towards renal lineages is well studied; however, most of the studies relate to generation of nephron progenitor population from PSCs. Until now, differentiation of PSCs into ureteric bud (UB) progenitor cells has had limited success. Here, we describe a simple, efficient, and reproducible protocol to direct differentiation of mouse embryonic stem cells (mESCs) into UB progenitor cells. The mESC-derived UB cells were able to induce nephrogenesis when co-cultured with primary metanephric mesenchyme (pMM). In generated kidney organoids, the embryonic pMM developed nephron structures, and the mESC-derived UB cells formed numerous collecting ducts connected with the nephron tubules. Altogether, our study established an uncomplicated and reproducible platform to generate ureteric bud progenitors from mouse embryonic stem cells.

Keywords: mouse embryonic stem cell; differentiation protocol; ureteric bud progenitor cells; 3D kidney organoids

---

1. Introduction

Pluripotent stem cells (PSCs) possess great potential of differentiating into multiple cell types that are widely used for studies in developmental biology and regenerative medicine [1]. Kidney organoids derived from PSCs have been shown to be able to mimic the in vivo kidney structure development and function in vitro [2–4]. Renal organoids in a four-dimensional (4D) (3D plus time) culture system self-organize into highly complex tissue-specific morphology that is sufficient to model tissue development, disease, and injury [5–8]. A combination of genome editing and stem cell technologies allows for generation of personalized kidney organoids, which provide powerful tools for kidney disease treatment, drug toxicity screening, and tissue regeneration [9,10].

Recently, multiple protocols enabling generation of renal lineages from mouse and human PSCs have been published. We and several other groups reported induction of nephron progenitors, which have the potential to develop into epithelial nephron-like structures [2–4,8,11–18]. Other groups have shown the derivation of ureteric bud (UB) progenitors [19]. However, they did not show nephron progenitor induction and also lacked the connection between collecting ducts and nephrons [2,19]. A newly published study has shown generation of UB structures from PSCs, which possessed UB-like branching morphogenesis when aggregated with the primary metanephric mesenchyme (pMM) to form chimeric kidney organoids [8]. However, the protocol is technically complex, which limits its application. Analysis of these reports suggests that we still need further studies to develop simple, reproducible, and stable protocols for UB progenitor generation.

Here, we report a simple protocol to direct differentiation of mouse embryonic stem cells (mESCs) into UB progenitor cells. The newly generated UB progenitor cells have a potential to develop into ureteric bud structures and have a capacity to induce nephrogenesis when co-cultured with dissociated pMM. In reconstructed kidney organoids, the pMM developed into nephron structures and the UB progenitor cells formed collecting ducts, which also connected with the nephron tubules. The chimeric kidney organoids also display the presence of endothelial cells forming a vascular network. In conclusion, our study established an uncomplicated and reproducible method for generation of UB progenitors from PSCs that can be used for tubulogenesis induction.

2. Materials and Methods

Animal care and experimental procedures in this study were in accordance with Finnish national legislation on the use of laboratory animals, the European Convention for the protection of vertebrate animal used for experimental and other scientific purposes (ETS 123), and the EU Directive 86/609/EEC. Animal experimentation was also authorized by the Finnish National Animal Experiment Board (ELLA) as being compliant with the EU guidelines for animal research and welfare.

2.1. Mouse ES Cell Line Generation and Maintenance

mESC line Sv129S6 was obtained from the Biocenter Oulu Transgenic core facility. The Sv129S6 mESC line was derived from Taconic's W4/129S6 inbred mouse strain and has been tested to have a normal karyotype [19]. The mESC and mESC-GFP (Green Fluorescent Protein, pcDNA3.1 transfected) line was described previously [20]. It was maintained on mitotically inactivated mouse embryonic fibroblasts (MEFs) in mESC medium: Dulbecco's Modified Eagle's Medium (DMEM; Life Technologies, Waltham, MA, USA) supplemented with 10% fetal bovine serum (FBS, Gibco, Waltham, MA, USA), 1% (*v/v*) nonessential amino acids (Life Technologies), 100 mM 2-mercaptoethanol (Nacalai Tesque, Kyoto, Japan), and 1000 U/mL leukemia inhibitory factor (LIF, Millipore, Espoo, Finland).

2.2. Directed Differentiation of mESCs to UB Progenitors

Mouse ESCs were cultured as described above in MEF-coated 6-well plates in mESCs medium up to 70%–90% confluency. The cells were passaged on 1% geltrex-coated 24-well plates at 30,000 cells/cm^2 in 500 μL 50% ReproFF2 (ReproCELL, Glasgow, UK)—50% CM (conditioned MEF medium) supplied with 10 ng/mL fibroblast growth factor (FGF) 2 (PeproTech Nordic, Stockholm, Sweden) and 10 ng/mL Activin A (Cell Guidance Systems, Cambridge, UK). After 2 days, the cells reached 60–80% of confluency, and the medium was changed to differentiation medium: Advanced RPMI (Thermo Fisher Scientific, Waltham, MA, USA) supplemented with 8 μM CHIR99021 (Reagents Direct, Encinitas, CA, USA) and 4–8 ng/mL Noggin (R&D systems, Minneapolis, MN, USA) for 2 days, followed by treatment of the cells with Activin A (10 ng/mL) for 2 days and then 3 days with FGF9 (40 ng/mL) (PeproTech Nordic, Stockholm, Sweeden) in Advanced RPMI medium (ThermoFisher Scientific).

2.3. Gene Expression Analysis

A RNeasy kit (Qiagen, Germantown, MD, USA) was used according to the manufacturer's recommendations to extract total RNA. cDNA synthesis (First Strand cDNA Synthesis Kit, ThermoFisher Scientific, Waltham, MA, USA) was performed using standard protocols. qPCR was run using a CFX96 BioRad thermocycler. Brilliant III SYBR® Green QPCR Master Mix (Agilent Technologies, Santa Clara, CA, USA) was used according to the manufacturer's instructions. GAPDH probe served as a control to normalize the data. The data was analyzed using the $2^{-\Delta\Delta CT}$ method. Gene expression measurements were performed in triplicates on three independent experiments. The primer sequences are given in Supplementary Table S1.

2.4. Chimera 3D Kidney Organoids Formation

At day 9 of mESCs differentiation, which represents the UB progenitor cells stage, cells were dissociated into single cell suspension using TrypLE select (Life Technologies). After three washes in phosphate buffer saline (PBS), cells were reconstituted in organ culture medium (DMEM supplemented with 10% fetal bovine serum, 1% penicillin–streptomycin). For generation of kidney organoids, we mixed dissected and dissociated pMM cells (from CD-1 pregnant females, at E11.5), as described before [21], with differentiated UB progenitors at a ratio of 3:1, respectively. Briefly, kidney rudiments were dissected out of E11.5 embryos and treated with trypsin/pancreatin solution for 30–40 s to separate the UB from the pMM [22]. The MM cells were dissociated using 2.0 mg/mL Collagenase IV solution (Wortington, Lakewood, NJ, USA) in 0.1% bovine serum albumin (BSA) in 1×PBS. 10 min incubation was interrupted by pipetting and continued until cells were separated. The reaction was stopped by washing three times in complete organ culture medium [22]. The pMM- and mESC-derived UB cells were mixed in a 3:1 ratio and centrifuged for 4 min at 1380× g to form a pellet (5×10^4 cells) in Lo-Binding Eppendorf tubes. Following centrifugation, we carefully transferred the differentiated UB and pMM pellets to filter into Trowel culture to aggregate as an organoid. The organ culture medium was changed every 3–4 days.

For generation of whole kidney organoids, we dissected mouse kidney rudiments at E11.5 from CD-1 pregnant females. Kidney rudiments were dissociated into single cell suspension as described previously [19]. After dissociation, the embryonic kidney cells (7×10^4) were mixed with undifferentiated mESC or differentiated mESCs-derived UB progenitors (1×10^4) to make the pellet. We then continued the procedure as described above.

2.5. Whole-Mount Immunohistochemistry

Kidney organoids were washed 2 times with PBS and fixed with 100% cold methanol (–20 °C) for 30 min at room temperature (RT) or with 4% paraformaldehyde in PBS (organoid with GFP or dye) for 30 min at RT in the dark. After fixation, the organoids were washed at least three times in PBS and blocked in 0.1% Triton-X100 (Sigma, Lyon, France), 1% BSA, and 10% goat serum/0.02M glycine-PBS for 1–3 h at room temperature. Incubation of the organoids with primary antibodies was performed in a blocking buffer overnight at 4 °C. The samples were washed 6 times with PBS and incubated with secondary antibodies Alexa Fluor 405, 488, 568, 546, or 647 (1:1000, Life technologies) and fluorescein anti-LTL (Lotus Tetragonolobus Lectin, 1:350, #FL-1321, Vector Laboratories, Burlingame, CA, USA) overnight at 4 °C and counter-stained with Hoechst (Thermo Fisher Scientific). The primary antibodies used in stainings were: Wt1 (1:100, #05-753, Millipore), Pax2 (1:200, #PRB-276P, Covance, Cambridge, MA, USA), Troma1 (1:200, DSHB, Iowa City, IA, USA), Gata3 (1:20, #AF2605-SP, R&D Systems), E-cad (1:300, #610181, BD Biosciences, Franklin Lakes, NJ, USA), Synaptopodin (SYNPO) (1:4, #ABIN112223, antibody on line.com, Aachen, Germany), Umod (1:25, #LS-C150268, LSBio, Seattle, WA, USA), CD31 (1:100, #550274 BD Biosciences), Laminin (1:200, #L9393, Sigma), and Cleaved Caspase-3 (1:200, #9661s, Cell Signaling Technology, Leiden, Netherlands). Stained organoids were mounted with Shandon™ Immu-Mount™ (Thermo Scientific™). A Zeiss LSM780 microscope and Zeiss Axiolab (Zeiss, Oberkochen, Germany) were used for image capture and analysis.

2.6. Nephrotoxicity Assay

3D kidney organoids were cultured in organ culture medium supplemented with gentamicin at 5 mg/mL (#G1264, Sigma) for 48 h, or with cisplatin at 5, 20, or 50 µM (#P4394 Sigma) for 24 h after day 8 of organ culture. Organoids were then fixed with 100% cold methanol for 30 min for whole-mount immunohistochemistry. The Notch inhibitor, *N-S*-phenyl-glycine-*t*-butyl ester (DAPT, #D5942, Sigma), was used (10µM) to investigate toxicity towards proximal tubule development.

2.7. Statistical Evaluation

All data are presented as mean ± standard deviation (SD) and represent a minimum of three independent experiments. Student's two-tailed t-test was used for statistical evaluation. *p*-value < 0.05 was considered significant.

3. Results

3.1. Direct Differentiation of mESCs into UB Progenitor Cells

During development, both the nephron and ureteric bud progenitor cells are derived from the intermediate mesoderm (IM). To establish a protocol and direct differentiation of mESCs into UB lineage, we first differentiated mESCs into IM (Figure 1A). We treated the mESCs with FGF2 and activin A to differentiate mESCs into epiblast in monolayer culture (Supplementary Figure S1A). The differentiated cells showed expression of epiblast markers such as Fgf5 and T (Brachyury) (Supplementary Figure S1B). We then used glycogen synthase kinase-3β (GSK-3β) inhibitor, CHIR99021 (CHIR), together with a low concentration of bone morphogenetic protein (BMP) signaling inhibitor, Noggin (suppresses number of cells differentiated to lateral plate mesoderm) [2], to activate differentiation of mouse epiblast cells into primitive streak (PS). The Noggin-treated cells expressed PS markers Cdx2, T, Tbx6, and Mixl1 (Supplementary Figure S1B). Activin A has been previously used for specification of the mesodermal cells towards intermediate mesoderm [19]. Therefore, we followed for two days with activin A treatment, which differentiated the cells to the IM stage and expressed Osr1, Lhx1, and Pax2 (Supplementary Figure S1B).

Previous studies have demonstrated that FGF9 is able to induce renal lineage differentiation from the IM population [2]. Therefore, we treated these cells with a moderate concentration of FGF9 for an additional three days, directing them to differentiate into UB progenitor cells with expression of UB markers. These cells expressed UB tip markers: Ret, Wnt11, and Sox9, as well as other markers of UB: Lhx1, Ecad, Hnf1b, Wnt7b, Wnt9b, Calb1, Emx2, Gata3, Hoxb7, and Tacstd2 (Figure 1B and Supplementary Figure S1C). In addition, expression of stromal cell marker Foxd1 nephron progenitor cell markers, Six2 and Eya1 (Figure 1B), or other epithelial segment markers, were observed at day nine of differentiation (Supplementary Figure S1D). Immunofluorescence staining further revealed that the use of a moderate concentration of FGF9 induced the cells to express Pax2, E-cadherin (Ecad), and Gata3 (Figure 1C–F), which may suggest that these differentiated cells represent putative UB progenitor cells.

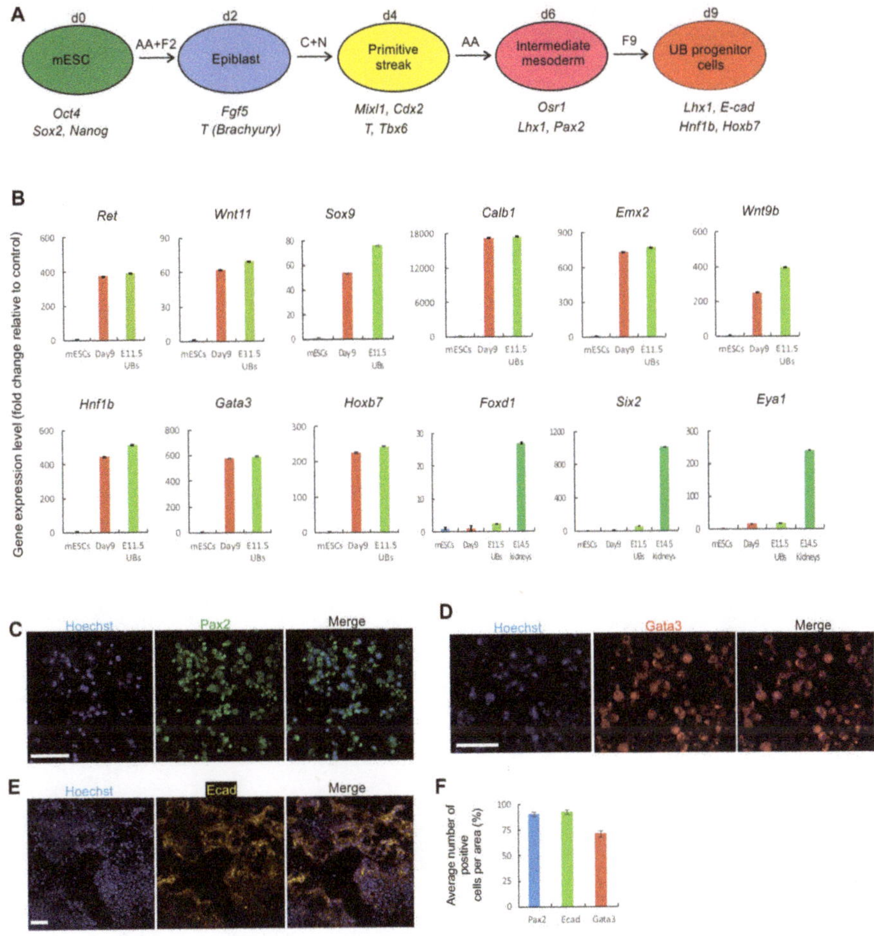

Figure 1. Differentiation of mouse embryonic stem cells (mESCs) to ureteric bud (UB) progenitor cells. (**A**) Schematic of the differentiation protocol of mESCs into UB progenitor cells. AA: Activin A; F2: FGF2; C: CHIR99021; N: Noggin; F9: FGF9. (**B**) Graphs of qPCR results showing the gene expression (fold change) of ureteric bud markers relative to mESC. No expression of stroma (Foxd1) and nephron progenitor cell markers (Six2 and Eya1) was observed at day 9 of differentiation. (*n* = 3). (**C–E**) Immunocytochemistry of Pax2, Ecad, and Gata3 in mESCs on day 9 of differentiation. Scale bars, 50 µm. (**F**) Quantification of the number of cells expressing Pax2, Ecad, and Gata3 at day 9 of differentiation. *n* = 3 samples per marker (3 randomly chosen areas in 3 independent experiments).

3.2. Generation of Kidney Organoids by mESC-Derived UB Progenitor Cells and Dissociated Primary MM Population

We and other groups previously reported that dissociation of mouse pMM into single cells maintains the nephron progenitor stemness. The dissociated MM population develops into nephrons when induced by the inducer such as the embryonic UB or spinal cord cells [8,21,23–27]. To establish the potential and function of the mESC-derived UB progenitor cells, we aggregated these cells with mouse E11.5-dissociated pMM cells to generate a kidney organoid. The cell aggregates were cultured for up to 11 days in a traditional Trowell organ culture system, during which they spontaneously formed kidney organoids with complex structures (Figure 2A,B). On day three, we observed Troma1+ UB structures and the formation of renal vesicles adjacent to the UB (Figure 2C–D).

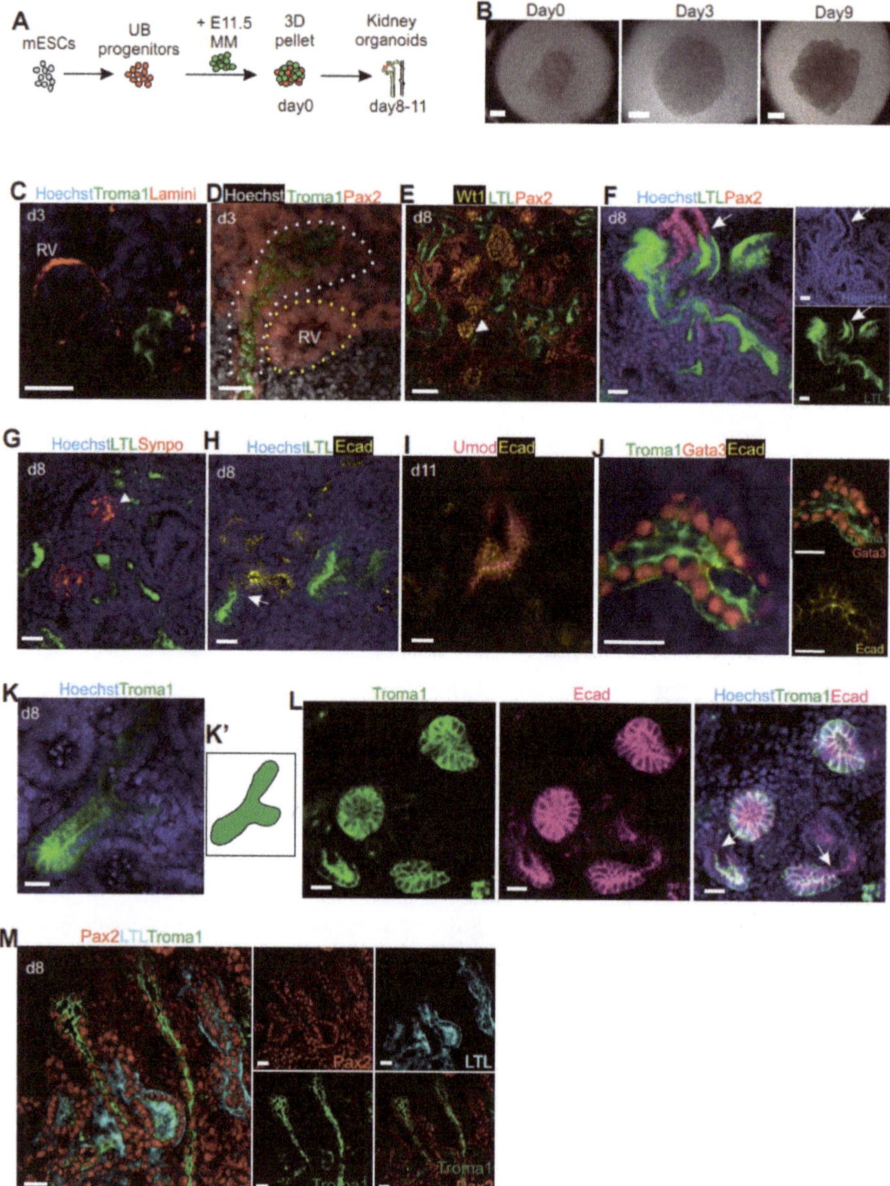

Figure 2. Generation of renal organoids by mESC-derived UB progenitors and primary metanephric mesenchyme (pMM) cells. (**A**) Schematic of generation of kidney organoids from mESC-derived UB progenitors with mouse E11.5-dissociated pMM. (**B**) Global bright field images of self-organizing kidney organoids in a Trowel organ culture system. Scale bars, 500 µm. (**C**) Immunofluorescence of

kidney organoids at day 3 show formation of a renal vesicle next to the Troma1+ structure generated by mESC-derived UB progenitor cells. **(D)** Confocal image at day 3 showing Pax2+ renal vesicle (yellow dotted line) surrounded by Troma+Pax2+ ureteric epithelium (white dotted line). **(E–M)** Immunofluorescence of kidney organoids at day 8 or 11. **(E)** Glomeruli are marked with Wt1, proximal tubules are marked with LTL, and distal tubules are marked with Pax2+LTL–. The arrowhead shows the connection location of glomeruli with proximal tubule. **(F)** Immunostaining of distal tubule marked with Pax2 and proximal tubule marked with LTL, with all nuclei stained with DAPI. The arrow shows the proximal tubules (Pax2+LTL+) connected with distal tubules (Pax2+LTL–). **(G)** Glomeruli (Synpo+) adjacent to proximal tubules (LTL+) (the arrowhead marks the place of connection). **(H)** Immunostaining of proximal tubule with LTL and distal tubule with Ecad. Confocal image shows proximal tubules (LTL+) connected with distal tubules (Ecad+LTL–) (marked with an arrow). **(I)** Confocal image of Loops of Henle marked by Umod and Ecad on day 11. **(J)** Confocal images of Troma1+Gata3+Ecad+ collecting duct structure. **(K-K′)** A "T" shaped UB structure in the kidney organoid. **(L)** mESC-derived UB progenitor cells-generated collecting ducts (Troma1+Ecad+) connecting with nephron's distal tubules (Troma1–Ecad+) (marked with an arrow). **(M)** Kidney organoids developed collecting duct trunk structures. Scale bars, **(B)** 500 μm, **(E)** 50 μm, **(C–D, F–M)** 20 μm.

Whole-mount immunostaining of day eight chimeric organoids showed development of nephrons with positive staining of glomerular markers Wilms tumor 1 (Wt1+) and Synaptopodin (Synpo+), proximal tubule marker Lotus Tetragonolobus Lectin (LTL+), and distal tubules markers Pax2+LTL– and Ecad+ (Figure 2E–H, Supplementary Figure S2A). Moreover, we found numerous Synpo+ and Wt1+ glomeruli adjacent to LTL+ proximal tubules, which connected with Pax2+LTL–/Ecad+ distal tubules (Figure 2E–H marked with arrowheads and arrows, Supplementary Figure S2A), indicating a proper nephron structure with glomerulus/proximal tubule/distal tubule organization. On day 11, the organoids also displayed Henle's loop with the expression of uromodulin (Umod+) and Ecad+ (Figure 2I).

We also tested whether the mESC-derived UB progenitor cells have in vivo UB capacity to form the collecting duct system. On day eight, we observed Troma1+Gata3+Ecad+Pax2+LTL– collecting duct structures in the kidney organoids by immunocytochemistry (Figure 2J–M, Supplementary Figure S2B). We also found some Troma1+ "T" shaped UB structures (Figure 2K, K′), indicating that the mESC-derived UB progenitors behave in a manner similar to UB cells in vivo. Importantly, we observed that the collecting ducts (Troma1+Ecad+) connected with the distal tubules (Ecad+Troma1–) of the nephron structure (Figure 2L, Supplementary Figure S2C), generating an interconnection between collecting ducts and nephrons, which is essential for urine drainage. The collecting ducts displayed the morphology of branches and long collecting duct trunks (Figure 2M). Altogether, the *in vitro* reconstructed organoids developed kidney structures that are similar to in vivo kidney, although it is unclear if the generated UB structures form a proper network.

We have also attempted to induce vascularization of developing glomeruli in these renal organoids. Previous studies demonstrated that rho-associated protein kinases (ROCK) are downstream effectors of vascular endothelial growth factor (VEGF), and negatively regulate the process of angiogenesis [28]. Therefore, we used the ROCK inhibitor to enhance angiogenesis in the renal organoids. However, this treatment did not increase the endothelial network area, and CD31+ cells could not be found in the developing glomerular tuft (Supplementary Figure S3).

3.3. Characterization of Kidney Organoids

To rule out the possibility that the differentiated cells (mESC-derived UB progenitors) could give rise to nephrons when induced by the embryonic UB, we aggregated day nine differentiated cells with E11.5-dissociated UB cells and grew them in organ culture. Since UB survival and development depend on the presence of metanephric mesenchyme, in the organ cultures, the purified UB failed to branch in the presence of UB-differentiated cells (Supplementary Figure S4A). Next, we wanted to verify that nephron structures in the organoids were generated via the MM population induced by the mESC-derived UB progenitors, and not by contamination of the MM with the primary UB cells. To assess this possibility, we cultured E11.5 MM tissue in isolation. The tissue already underwent apoptosis at the second day of culture and died at day three (Supplementary Figure S4B). This result also confirmed that without a suitable inducer, the MM cells will not undergo nephrogenesis and cannot survive for a long time under the organ culture conditions in vitro. These data suggest that we did differentiate mESC towards UB progenitors, and that they can further develop into collecting ducts when co-cultured with pMM.

To further confirm that the nephron structures and collecting ducts were derived from pMM- and mESC-derived UB progenitors respectively, we used pMM population isolated from mTmG mice (td-tomato+) to aggregate with differentiated mESCs—unlabeled (Figure 3A). The aggregates formed well-developed nephron structures (WT1+ glomeruli, LTL+ proximal tubules, Ecad+Troma1– distal tubules), which originated from MM cells (mTmG+) (Figure 3B) and Troma1-labeled collecting duct, which was mTmG– and therefore originated from the mESC-derived UB progenitors (Figure 3C–E). We also generated an mESC line with stable expression of the GFP. We differentiated these cells into UB progenitors and used them to induce pMM cells. Immunofluorescence analysis confirmed that the Troma1+Ecad+ collecting ducts were derived from GFP+ mESCs (Figure 3F, Supplementary Figure S4C). Altogether, these data demonstrate that the nephrogenesis occurring in generated kidney organoids is specifically derived from an interaction between the pMM and mESC-derived UB progenitors.

3.4. mESC-Derived UB Progenitor Cells Integrated into the 3D Ureteric Bud Structures

In order to identify the mESC-derived UB progenitors from wild-type UB cells in chimeric organoids, and to ensure that they will only generate UB cells, we generated an organoid with an mESC-GFP line. We dissociated a whole E11.5 embryonic kidney rudiment (MM and UB) and mixed it with undifferentiated mESC or mESC-derived UB progenitors generating chimeric kidney organoids [19,21,27] (Figure 4A).

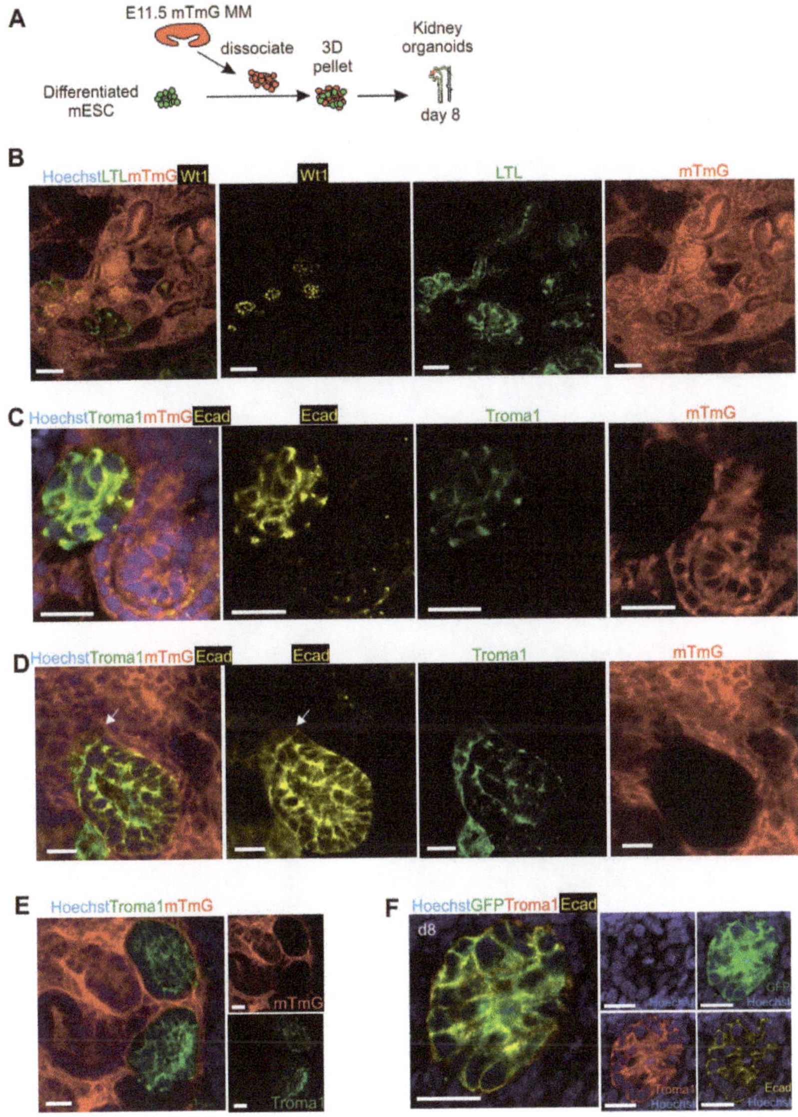

Figure 3. UB structures in kidney organoids are specifically derived from differentiated mESCs. (**A**) Schematic of generation of kidney organoids from mESC-derived UB progenitors with mouse E11.5-dissociated mTmG (td-tomato+) pMM. (**B**) Confocal images of three-dimensional (3D) kidney organoids derived from E11.5 mTmG+ pMM and mESC-derived UB progenitors. Nephron structures such as WT1+ glomeruli and LTL+ proximal tubules were derived from mTmG+ pMM. Scale bars, 50 μm. (**C**) Confocal images of kidney organoids showing ureteric bud structures (Ecad+Troma1+) being generated by mESC-derived UB progenitor cells. Scale bars, 20 μm. (**D**) Confocal images showed Troma1+Ecad+ collecting duct derived from mTmg− cells (mESC-derived UB progenitors) and connected with Troma1−Ecad+ renal tubules (connection marked with an arrow). Scare bars, 20 μm. (**E**) The Troma1+ UB structures derived from mTmG− cells (mESC-derived UB progenitors). Scale bar: 20 μm. (**F**) Troma1+Ecad+ collecting ducts are derived from GFP+ mESCs-derived UB progenitors. Scale bar: 20 μm.

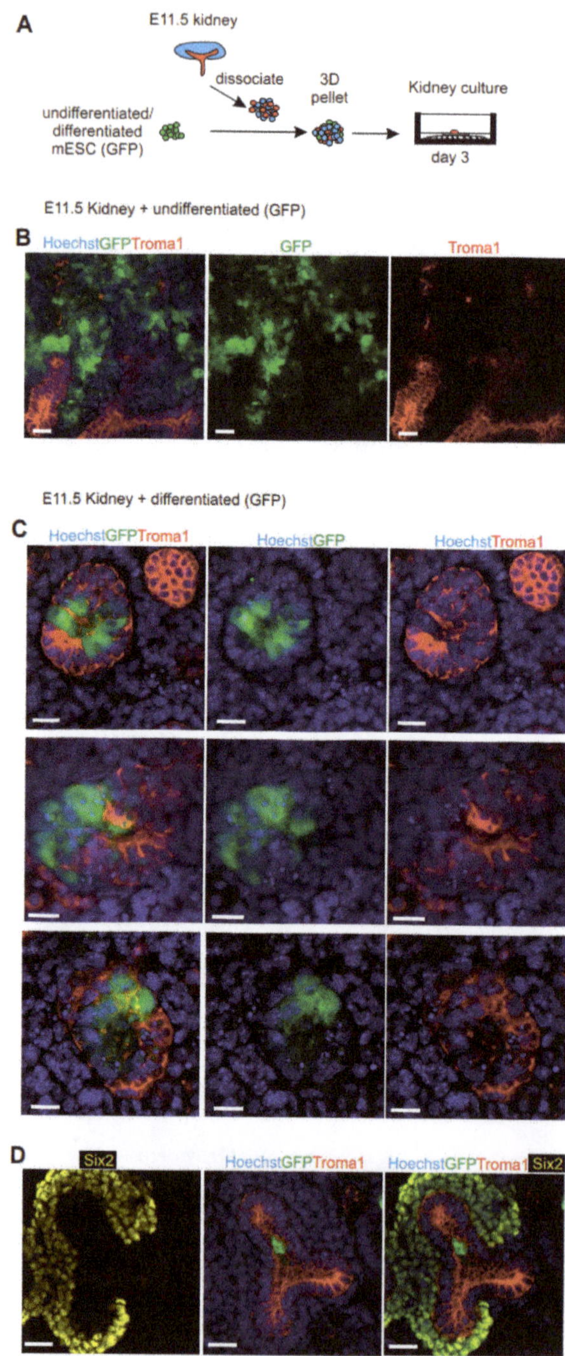

Figure 4. Mouse ESC-derived UB progenitor cells form anembryonic UB in 3D organ culture in vitro. (**A**) Schematic of kidney organoid generation from mESC-derived UB progenitors with mouse E11.5-dissociated kidney rudiments. (**B**) Immunofluorescence analysis demonstrating random

localization of undifferentiated mESCs in organ co-cultures. Scale bars, 20 μm. (**C**) Immunofluorescence analysis demonstrating mESC-derived UB progenitors integrated into the UB structures and enhanced chimeric ureteric bud formation. Scale bars, 20 μm. (**D**) Confocal image showing localization of mESC-derived UB progenitor cells (GFP+) in Troma1+ UB structures of the chimeric organoid and not in the Six2+ nephron progenitor's region. Scale bars: 20 μm.

The undifferentiated mESCs aggregated with dissociated embryonic kidney rudiments randomly localized within the organoids and had a negative effect on nephrogenesis. The undifferentiated mESC had formed groups within the renal structures; however, there were only a few structures, and very few Troma1+ ureteric buds could be observed (Figure 4B). When we aggregated the mESCs-derived UB progenitors with the dissociated embryonic kidney, we noticed much more robust tubulogenesis and numerous chimeric ureteric bud structures (Figure 4C). The UB progenitor cells did not interfere with renal development, but efficiently and specifically integrated into the mouse ureteric buds, as indicated by Troma1 and GFP+ staining (Figure 4C and D). Moreover, the immunostaining of day three organoids with nephron progenitor's marker, Six2, presented integration of mESC-derived UB progenitors (GFP+) into the UB structure (Figure 4D). Together, these data demonstrate that mESC-derived UB progenitors do integrate well into ureteric bud structures in chimeric renal organoids and are able not only to form UB de novo, but also generate chimeric structures.

3.5. Kidney Organoids as Models to Study Kidney Development and Drug Toxicity

After transfer of the 3D chimeric organoids to organ culture, they self-organized and generated a well-structured kidney organoid with proper nephron and collecting duct structures. As patterning of the nephron into its different segments begins at the renal vesicle stage during development [29,30], we postulated that developmental patterning could be changed by chemical modulation of these endogenous signals. A previous report revealed that Notch signaling is required for proximal tubule fate acquisition in a mammalian nephron [31]. We therefore treated the organoids (pMM cells mixed with mESC-derived UB progenitor cells in a ratio of 3:1, Figure 2A) with a Notch signaling inhibitor, *N-S*-phenyl-glycine-*t*-butyl ester (DAPT). DAPT leads to suppression of proximal tubules' formation in the human nephron organoid culture [4,31,32]. We added DAPT to the culture medium from day 3 to 11 of 3D organoid culture. Immunofluorescence analysis of the cells demonstrated that with the DAPT treatment, formation of glomerulus and ureteric bud ducts was normal, but development of proximal tubules was severely suppressed (Supplementary Figure S5A).

In order to use stem cell-derived kidney organoids for disease modelling and drug screening, they need to present functional maturation of the nephrons within the organoids. To test whether these organoids could be used to study kidney injury and toxicity in vitro, we focused on drug-induced nephrotoxicity, which has been shown as an important cause of acute kidney injury in hospitalized patients [33]. We treated the chimeric kidney organoids with gentamicin, a commonly used antibiotic with well-established proximal tubular toxicity, after nine days of organ culture for 48 h [34]. We also treated organoids with another nephrotoxicant, cisplatin, from day 10 for 24 h. Cisplatin induces caspase-mediated acute apoptosis of proximal tubular cells in the kidney [35]. Whole-mount immunostaining of control organoids with caspase 3 showed random apoptotic interstitial cells; however, both gentamicin and cisplatin induced acute apoptosis in LTL+ proximal tubules (Supplementary Figure S5B). The percentage of apoptotic proximal tubular cells induced by cisplatin increased in a dose-dependent manner: the 5 and 20 μM cisplatin doses mainly affected the proximal tubule compartment, 32% and 62% respectively, while 50 μM cisplatin led to apoptosis of almost all of the proximal tubule cells (≈96%), but was also toxic to other cell types presenting a global type of nephrotoxicity at this concentration (Supplementary Figure S5C).

To summarize, our work shows that mESC-derived UB progenitors induce nephrogenesis in pMM cells, and furthermore, chimeric renal organoids generated from these progenitors show an expected response to toxic chemicals and drugs. We also demonstrated that the kidney organoid system can be used to test nephrotoxicity of drugs and other chemicals in vitro.

4. Discussion

Kidney development starts from an interaction between two precursor tissues of the kidney, UB and MM. A major part of the MM cell population comprises the nephron progenitor cells (NPCs) which will differentiate into nephrons, and the ureteric bud will form the collecting duct tree. Protocols for differentiation of mouse and human pluripotent stem cells to renal progenitor cells, and further to self-organized kidney organoids containing nephrons, have been well established [2–4,13,36], but current methods of differentiation of pluripotent stem cells, specifically to ureteric bud progenitor cells, need further development [8,19].

Previous studies on generation of ESC-derived UB have shown derivation of a UB-like population by selective induction with metanephric mesenchyme cells [19] or through CHIR99021 treatment for nephron differentiation of both ureteric bud and nephron structures [2]. However, the ESC-derived UB-like cells did not show nephron progenitor induction [19], and therefore, an inter-nephron connection with collecting ducts was lacking [2]. Another group recently published a protocol for derivation of UB structures from PSCs, and generation of kidney organoid composed of mESCs-derived UB aggregated with pMM, or mESCs-derived UB combined with mESC-derived NPC and primary stromal progenitor cells (SPs). Although successful, their protocol requires a knock-in of markers in the PSCs, which later involves sorting of cells for the specific marker [8]. More recently, Mae and colleagues reported a protocol for generation of branching ureteric bud tissues from human pluripotent stem cells (hPSC) with a series of growth factors, but there was no evidence of nephrogenesis [37] (Supplementary Table S2). While these published protocols produce functional kidney organoids, they are technically complex. Here, we reported an establishment of a simple (directed with growth factors), efficient (>90% of Pax2+, Ecad+ and >70% Gata3+ cells), and reproducible differentiation protocol of mESC to ureteric bud progenitor cells. These mESC-derived UB progenitor cells induced pMM cells to undergo nephrogenesis leading to development of well-structured nephrons. These nephrons consisted of glomeruli, proximal tubules, loops of Henle, and distal tubules, and were connected with collecting duct structures generated by mESC-derived UB cells. Moreover, these mESC-derived UB progenitor cells formed a UB de novo when combined with pMM cells, and they generated chimeric structures when combined with kidney rudiment cells (pUB and pMM) (Supplementary Table S2).

While our culturing conditions produced well-functioning kidney organoids, further studies are needed to fine-tune the culturing conditions. Previous studies showed that ROCK kinases are VEGF downstream effectors, which negatively regulate the process of angiogenesis [28]. We added ROCK inhibitor (Y27) to culture expecting an increase in angiogenesis. However, we failed to see a difference between control and Y27-treated samples. Better results could be obtained by supplementing the medium with VEGF, which was already published by Freedman and co-workers, although they still did not observe endothelial cells entering the glomerular tuft in their system [9]. Similarly, development of a vascular glomeruli in organotypic kidney cultures and renal organoids was reported earlier by our research team [38]. It seems that in the absence of the blood flow, the endothelial cells are not able to properly interact with developing nephrons [39] and formation of glomerular vasculature does not proceed further than migration of endothelial cells into the vascular cleft region of an S-shaped stage nephron. Sufficient vascularization of the organoids may be achieved by treatment with angiogenic factors, co-culture with blood vessel organoids [40], by providing flow to the system [15], or with a combination of the aforementioned treatments.

Even though organoids generated by mESC-derived UB progenitor cells did not have a proper vascular network, the developed nephrons did respond to toxicological tests as expected. We and others [4] have shown that the use of gentamicin and cisplatin induces apoptosis in proximal tubular cells. Thus, these organoids present a functional platform to test drug-induced nephrotoxicity.

In summary, we have developed an easy and reproducible protocol for generation of UB progenitors from mESCs. This work generates a strong foundation for in vitro kidney studies, including disease modelling and drug discovery approaches, which are difficult to perform, and require animal models and/or primary cells which may not faithfully recapitulate all features of developmental or disease processes. Given the rapid progress in the field, we hope that in the near future, researchers will be able to generate fully functional nephrons in kidney organoids where the UB and MM parts are derived from PSC. Using these cells will enable generation of not only well-structured nephrons, but also the collecting duct tree. This is the first step for generating high-throughput gene discovery models and advancing tissue engineering for producing organs for transplanting. However, these organoids need to be successfully vascularized and grown to appropriate size. The studies presented here produce new insights into renal pathophysiology and open new avenues for developing new treatment options.

Supplementary Materials: The following are available online at http://www.mdpi.com/2073-4409/9/2/329/s1. Table S1: Primers (5′-3′) used in the study. Table S2: Summary of available protocols of differentiation of mouse and human PSC to UB linages. Figure S1: Differentiations of mESC into ureteric bud progenitor cells. Figure S2: Differentiated mESC induce embryonic MM to nephrogenesis. Figure S3: Vascularization of the kidney organoids. Figure S4: Characterization of kidney organoids. Figure S5: Kidney organoids model kidney development and injury.

Author Contributions: Conceptualization, Z.T., A.R.-R., I.S., and S.J.V.; Funding acquisition, S.J.V.; Investigation, Z.T.; Visualization, Z.T.; Writing—original draft, Z.T.; Writing—review and editing, A.R.-R. All authors have read and agreed to the published version of the manuscript.

Funding: This work was supported financially by H2020 Marie Skłodowska-Curie Actions Innovative Training Network "RENALTRACT" (ID 642937), Academy of Finland (ID 315030, Centre of Excellence, ID 251314), Sigrid Jusélius Foundation, Cancer Research Foundation, and Finnish Cultural Foundation (personal grant to A.R.-R).

Acknowledgments: We thank Paula Haipus, Hannele Härkman, Johanna Kekolahti-Liias, and Sanna Kauppinen for technical assistance; Biocenter Oulu Transgenic core facility for the mouse ESCs; Dr. Susanna Kosamo for critical reading of the manuscript and English language corrections; Dr. Jingdong Shan, Dr. Florence Naillat, MSc. Abhishek Sharma, Dr. Fariba Jian Motamedi, and Prof. Andreas Schedl for the discussion.

Conflicts of Interest: The authors declare no conflict of interest.

References

1. Robinton, D.A.; Daley, G.Q. The promise of induced pluripotent stem cells in research and therapy. *Nature* **2012**, *481*, 295–305. [CrossRef]
2. Takasato, M.; Er, P.X.; Chiu, H.S.; Maier, B.; Baillie, G.J.; Ferguson, C.; Parton, R.G.; Wolvetang, E.J.; Roost, M.S.; Chuva de Sousa Lopes, S.M.; et al. Kidney organoids from human iPS cells contain multiple lineages and model human nephrogenesis. *Nature* **2015**, *526*, 564–568. [CrossRef]
3. Freedman, B.S.; Brooks, C.R.; Lam, A.Q.; Fu, H.; Morizane, R.; Agrawal, V.; Saad, A.F.; Li, M.K.; Hughes, M.R.; Werff, R.V.; et al. Modelling kidney disease with CRISPR-mutant kidney organoids derived from human pluripotent epiblast spheroids. *Nat. Commun.* **2015**, *6*, 8715. [CrossRef]
4. Morizane, R.; Lam, A.Q.; Freedman, B.S.; Kishi, S.; Valerius, M.T.; Bonventre, J.V. Nephron organoids derived from human pluripotent stem cells model kidney development and injury. *Nat. Biotechnol.* **2015**, *33*, 1193–1200. [CrossRef]
5. Boreström, C.; Jonebring, A.; Guo, J.; Palmgren, H.; Cederblad, L.; Forslöw, A.; Svensson, A.; Söderberg, M.; Reznichenko, A.; Nyström, J.; et al. A CRISP(e)R view on kidney organoids allows generation of an induced pluripotent stem cell–derived kidney model for drug discovery. *Kidney Int.* **2018**, *94*, 1099–1110. [CrossRef] [PubMed]
6. Kim, Y.K.; Nam, S.A.; Yang, C.W. Applications of kidney organoids derived from human pluripotent stem cells. *Korean J. Intern. Med.* **2018**, *33*, 649–659. [CrossRef] [PubMed]
7. Hale, L.J.; Howden, S.E.; Phipson, B.; Lonsdale, A.; Er, P.X.; Ghobrial, I.; Hosawi, S.; Wilson, S.; Lawlor, K.T.; Khan, S.; et al. 3D organoid-derived human glomeruli for personalised podocyte disease modelling and drug screening. *Nat. Commun.* **2018**, *9*, 1–17. [CrossRef]
8. Taguchi, A.; Nishinakamura, R. Higher-Order Kidney Organogenesis from Pluripotent Stem Cells. *Cell Stem Cell* **2017**, *21*, 730–746. [CrossRef] [PubMed]

9. Czerniecki, S.M.; Cruz, N.M.; Harder, J.L.; Menon, R.; Annis, J.; Otto, E.A.; Gulieva, R.E.; Islas, L.V.; Kim, Y.K.; Tran, L.M.; et al. High-Throughput Screening Enhances Kidney Organoid Differentiation from Human Pluripotent Stem Cells and Enables Automated Multidimensional Phenotyping. *Cell Stem Cell* **2018**, *22*, 929–940. [CrossRef] [PubMed]

10. Cruz, N.M.; Song, X.; Czerniecki, S.M.; Gulieva, R.E.; Churchill, A.J.; Kim, Y.K.; Winston, K.; Tran, L.M.; Diaz, M.A.; Fu, H.; et al. Organoid cystogenesis reveals a critical role of microenvironment in human polycystic kidney disease. *Nat. Mater.* **2017**, *16*, 1112–1119. [CrossRef]

11. Araoka, T.; Mae, S.; Kurose, Y.; Uesugi, M.; Ohta, A.; Yamanaka, S.; Osafune, K. Efficient and Rapid Induction of Human iPSCs/ESCs into Nephrogenic Intermediate Mesoderm Using Small Molecule-Based Differentiation Methods. *PLoS ONE* **2014**, *9*, e84881. [CrossRef]

12. Takasato, M.; Er, P.X.; Becroft, M.; Vanslambrouck, J.M.; Stanley, E.G.; Elefanty, A.G.; Little, M.H. Directing human embryonic stem cell differentiation towards a renal lineage generates a self-organizing kidney. *Nat. Cell Biol.* **2014**, *16*, 118–126. [CrossRef] [PubMed]

13. Taguchi, A.; Kaku, Y.; Ohmori, T.; Sharmin, S.; Ogawa, M.; Sasaki, H.; Nishinakamura, R. Redefining the in vivo origin of metanephric nephron progenitors enables generation of complex kidney structures from pluripotent stem cells. *Cell Stem Cell* **2014**, *14*, 53–67. [CrossRef] [PubMed]

14. Garreta, E.; Prado, P.; Tarantino, C.; Oria, R.; Fanlo, L.; Martí, E.; Zalvidea, D.; Trepat, X.; Roca-Cusachs, P.; Gavaldà-Navarro, A.; et al. Fine tuning the extracellular environment accelerates the derivation of kidney organoids from human pluripotent stem cells. *Nat. Mater.* **2019**, *18*, 397–405. [CrossRef] [PubMed]

15. Homan, K.A.; Gupta, N.; Kroll, K.T.; Kolesky, D.B.; Skylar-Scott, M.; Miyoshi, T.; Mau, D.; Valerius, M.T.; Ferrante, T.; Bonventre, J.V.; et al. Flow-enhanced vascularization and maturation of kidney organoids in vitro. *Nat. Methods* **2019**, *16*, 255–262. [CrossRef]

16. Allison, S.J. Fluid flow enhances vascularization and maturation of kidney organoids. *Nat. Reviews. Nephrol.* **2019**, *15*, 254. [CrossRef] [PubMed]

17. Low, J.H.; Li, P.; Chew, E.G.Y.; Zhou, B.; Suzuki, K.; Zhang, T.; Lian, M.M.; Liu, M.; Aizawa, E.; Rodriguez Esteban, C.; et al. Generation of Human PSC-Derived Kidney Organoids with Patterned Nephron Segments and a De Novo Vascular Network. *Cell Stem Cell* **2019**, *25*, 373–387. [CrossRef] [PubMed]

18. Przepiorski, A.; Sander, V.; Tran, T.; Hollywood, J.A.; Sorrenson, B.; Shih, J.; Wolvetang, E.J.; McMahon, A.P.; Holm, T.M.; Davidson, A.J. A Simple Bioreactor-Based Method to Generate Kidney Organoids from Pluripotent Stem Cells. *Stem Cell Rep.* **2018**, *11*, 470–484. [CrossRef] [PubMed]

19. Xia, Y.; Nivet, E.; Sancho-Martinez, I.; Gallegos, T.; Suzuki, K.; Okamura, D.; Wu, M.; Dubova, I.; Esteban, C.R.; Montserrat, N.; et al. Directed differentiation of human pluripotent cells to ureteric bud kidney progenitor-like cells. *Nat. Cell Biol.* **2013**, *15*, 1507–1515. [CrossRef]

20. Aro, E.; Salo, A.M.; Khatri, R.; Finnilä, M.; Miinalainen, I.; Sormunen, R.; Pakkanen, O.; Holster, T.; Soininen, R.; Prein, C.; et al. Severe Extracellular Matrix Abnormalities and Chondrodysplasia in Mice Lacking Collagen Prolyl 4-Hydroxylase Isoenzyme II in Combination with a Reduced Amount of Isoenzyme I. *J. Biol. Chem.* **2015**, *290*, 16964–16978. [CrossRef]

21. Junttila, S.; Saarela, U.; Halt, K.; Manninen, A.; Pärssinen, H.; Lecca, M.R.; Brändli, A.W.; Sims-Lucas, S.; Skovorodkin, I.; Vainio, S.J. Functional genetic targeting of embryonic kidney progenitor cells ex vivo. *J. Am. Soc. Nephrol.* **2015**, *26*, 1126–1137. [CrossRef]

22. Rak-Raszewska, A. Experimental Tubulogenesis Induction Model in the Mouse. In *Kidney Organogenesis*; Methods in Molecular Biology; Vainio, S., Ed.; Humana Press: New York, NY, USA, 2019; Volume 1926, pp. 39–51.

23. Ganeva, V.; Unbekandt, M.; Davies, J.A. An improved kidney dissociation and reaggregation culture system results in nephrons arranged organotypically around a single collecting duct system. *Organogenesis* **2011**, *7*, 83–87. [CrossRef] [PubMed]

24. Hendry, C.E.; Vanslambrouck, J.M.; Ineson, J.; Suhaimi, N.; Takasato, M.; Rae, F.; Little, M.H. Direct Transcriptional Reprogramming of Adult Cells to Embryonic Nephron Progenitors. *J. Am. Soc. Nephrol. Jasn.* **2013**, *24*, 1424–1434. [CrossRef]

25. Lusis, M.; Li, J.; Ineson, J.; Christensen, M.E.; Rice, A.; Little, M.H. Isolation of clonogenic, long-term self renewing embryonic renal stem cells. *Stem Cell Res.* **2010**, *5*, 23–39. [CrossRef] [PubMed]

26. Saxén, L.; Sariola, H. Early organogenesis of the kidney. *Pediatr. Nephrol.* **1987**, *1*, 385–392. [CrossRef] [PubMed]

27. Unbekandt, M.; Davies, J.A. Dissociation of embryonic kidneys followed by reaggregation allows the formation of renal tissues. *Kidney Int.* **2010**, *77*, 407–416. [CrossRef] [PubMed]

28. Kroll, J.; Epting, D.; Kern, K.; Dietz, C.T.; Feng, Y.; Hammes, H.P.; Augustin, H.G. Inhibition of Rho-dependent kinases ROCK I/II activates VEGF-driven retinal neovascularization and sprouting angiogenesis. *Am. J. Physiol.—Heart Circ. Physiol.* **2009**, *296*, H893–H899. [CrossRef]

29. Georgas, K.; Rumballe, B.; Valerius, M.T.; Chiu, H.S.; Thiagarajan, R.D.; Lesieur, E.; Aronow, B.J.; Brunskill, E.W.; Combes, A.N.; Tang, D.; et al. Analysis of early nephron patterning reveals a role for distal RV proliferation in fusion to the ureteric tip via a cap mesenchyme-derived connecting segment. *Dev. Biol.* **2009**, *332*, 273–286. [CrossRef]

30. Little, M.; Georgas, K.; Pennisi, D.; Wilkinson, L. Kidney development: Two tales of tubulogenesis. *Curr. Top. Dev. Biol.* **2010**, *90*, 193.

31. Cheng, H.T.; Kim, M.; Valerius, M.T.; Surendran, K.; Schuster-Gossler, K.; Gossler, A.; McMahon, A.P.; Kopan, R. Notch2, but not Notch1, is required for proximal fate acquisition in the mammalian nephron. *Development* **2007**, *134*, 801–811. [CrossRef]

32. Cheng, H.; Miner, J.H.; Lin, M.; Tansey, M.G.; Roth, K.; Kopan, R. Gamma-secretase activity is dispensable for mesenchyme-to-epithelium transition but required for podocyte and proximal tubule formation in developing mouse kidney. *Development* **2003**, *130*, 5031–5042. [CrossRef]

33. Uchino, S.; Kellum, J.A.; Bellomo, R.; Doig, G.S.; Morimatsu, H.; Morgera, S.; Schetz, M.; Tan, I.; Bouman, C.; Macedo, E.; et al. Beginning and Ending Supportive Therapy for the Kidney (BEST Kidney) Investigators, for the Acute Renal Failure in Critically Ill Patients: A Multinational, Multicenter Study. *JAMA* **2005**, *294*, 813–818. [CrossRef]

34. Whiting, P.H.; Brown, P.A.J. The Relationship Between Enzymuria and Kidney Enzyme Activities in Experimental Gentamicin Nephrotoxicity. *Ren. Fail.* **1996**, *18*, 899–909. [CrossRef] [PubMed]

35. Mese, H.; Sasaki, A.; Nakayama, S.; Alcalde, R.E.; Matsumura, T. The role of caspase family protease, caspase-3 on cisplatin-induced apoptosis in cisplatin-resistant A431 cell line. *Cancer Chemother. Pharm.* **2000**, *46*, 241–245. [CrossRef] [PubMed]

36. Takasato, M.; Little, M.H. A strategy for generating kidney organoids: Recapitulating the development in human pluripotent stem cells. *Dev. Biol.* **2016**, *420*, 210–220. [CrossRef] [PubMed]

37. Mae, S.; Ryosaka, M.; Toyoda, T.; Matsuse, K.; Oshima, Y.; Tsujimoto, H.; Okumura, S.; Shibasaki, A.; Osafune, K. Generation of branching ureteric bud tissues from human pluripotent stem cells. *Biochem. Biophys. Res. Commun.* **2018**, *495*, 954–961. [CrossRef]

38. Halt, K.J.; Pärssinen, H.E.; Junttila, S.M.; Saarela, U.; Sims-Lucas, S.; Koivunen, P.; Myllyharju, J.; Quaggin, S.; Skovorodkin, I.N.; Vainio, S.J. CD146+ cells are essential for kidney vasculature development. *Kidney Int.* **2016**, *90*, 311–324. [CrossRef] [PubMed]

39. Serluca, F.C.; Drummond, I.A.; Fishman, M.C. Endothelial Signaling in Kidney Morphogenesis: A Role for Hemodynamic Forces. *Curr. Biol.* **2002**, *12*, 492–497. [CrossRef]

40. Wimmer, R.A.; Leopoldi, A.; Aichinger, M.; Wick, N.; Hantusch, B.; Novatchkova, M.; Taubenschmid, J.; Hämmerle, M.; Esk, C.; Bagley, J.A.; et al. Human blood vessel organoids as a model of diabetic vasculopathy. *Nature* **2019**, *565*, 505–510. [CrossRef]

Review

Traditional and Advanced Cell Cultures in Hematopoietic Stem Cell Studies

Antonio Carlos Ribeiro-Filho [1], Débora Levy [2], Jorge Luis Maria Ruiz [3], Marluce da Cunha Mantovani [1] and Sérgio Paulo Bydlowski [1,2,4,*]

[1] Organoid Development Team, Center of Innovation and Translational Medicine (CIMTRA), University of São Paulo School of Medicine, Sao Paulo 05360-130, Brazil; antonio_biomed@hotmail.com (A.C.R.-F.); marlucem@gmail.com (M.d.C.M.)

[2] Lipids, Oxidation and Cell Biology Team, Laboratory of Immunology (LIM19), Heart Institute (InCor), University of São Paulo School of Medicine, Sao Paulo 05403-900, Brazil; d.levy@hc.fm.usp.br

[3] Life and Nature Science Institute, Federal University of Latin American Integration-UNILA, Foz de Iguaçú, PR 858570-901, Brazil; jorge.ruiz@unila.edu.br

[4] National Institute of Science and Technology in Regenerative Medicine (INCT-Regenera), CNPq, Rio de Janeiro 21941-902, Brazil

* Correspondence: spbydlow@usp.br

Received: 13 August 2019; Accepted: 4 December 2019; Published: 12 December 2019

Abstract: Hematopoiesis is the main function of bone marrow. Human hematopoietic stem and progenitor cells reside in the bone marrow microenvironment, making it a hotspot for the development of hematopoietic diseases. Numerous alterations that correspond to disease progression have been identified in the bone marrow stem cell niche. Complex interactions between the bone marrow microenvironment and hematopoietic stem cells determine the balance between the proliferation, differentiation and homeostasis of the stem cell compartment. Changes in this tightly regulated network can provoke malignant transformation. However, our understanding of human hematopoiesis and the associated niche biology remains limited due to accessibility to human material and the limits of in vitro culture models. Traditional culture systems for human hematopoietic studies lack microenvironment niches, spatial marrow gradients, and dense cellularity, rendering them incapable of effectively translating marrow physiology ex vivo. This review will discuss the importance of 2D and 3D culture as a physiologically relevant system for understanding normal and abnormal hematopoiesis.

Keywords: hematopoiesis; hematopoietic stem cells; stem cell culture; 2D culture; 3D culture

---

1. Introduction

Blood is a connective tissue made up of approximately 34% cells and 66% plasma, transporting nutrients, gases and molecules in general to the whole body. Hematopoiesis is the process by which blood cells are formed, replenishing the blood system over the life of an individual. The hematopoietic process is a highly hierarchical phenomenon, in which hematopoietic stem cell (HSCs) differentiation and proliferation are of vital importance. Each cell within the hematopoietic hierarchy can be distinguished based on specific surface markers, which contain epitopes that are recognized by antibodies [1–3]. Figure 1 shows the main markers in human hematopoietic hierarchy. The hematopoietic process in humans starts in the yolk sac (mesoblastic phase). Then, it is transferred to the liver and spleen. Finally, the bone marrow becomes the main organ responsible for hematopoiesis. In the bone marrow, HSCs have the capacity of unlimited self-renewal, producing progeny that is the same as the original cell. They are generally in the G0 phase of the cell cycle and have the capacity to differentiate into specialized cells.

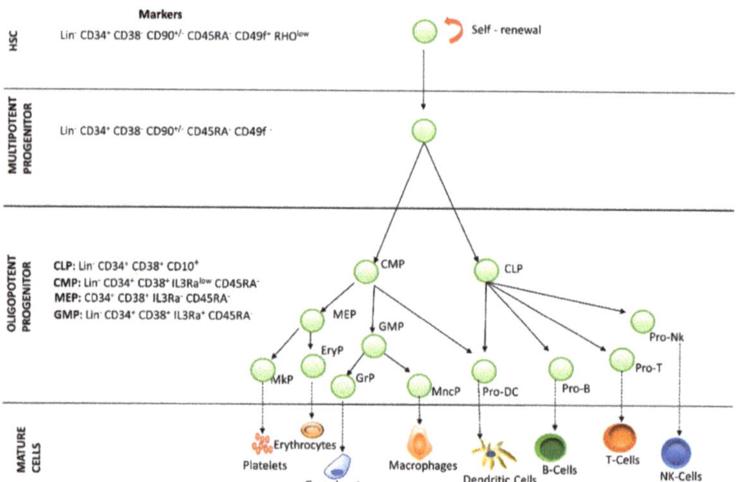

Figure 1. Hierarchy of human hematopoiesis. LT-HSC: Long Term-Hematopoietic Stem Cell; ST-HSC: Short Term-Hematopoietic Stem Cell; MPP: Multipotent Progenitor; OPP: Oligopotent Progenitor; LRP: Lineage Restricted Progenitor; MEC: Mature Effector Cell. The markers of the most important lineages are listed: Common Lymphoid Progenitor (CLP); Common Myeloid Progenitor (CMP); Megakaryocyte-Erythrocyte Progenitor (MEP); Granulocyte-Macrophage Progenitor (GMP). Restricted lineage progenitor cells: Megakaryocyte Progenitor (MkP); Erythrocytic Progenitor (EryP); Granulocytic Progenitor (GrP); Monocyte Progenitor (MncP); Dendritic Progenitor Cell (Pro DC); Progenitor Cell-T (Pro-T); Progenitor Cell-B (Pro-B); Progenitor Cell-Nk (Pro-Nk).

2. Hematopoietic Stem Cells

The medullary microenvironment participates in the quiescence, self-renewal, proliferation, maturation and apoptosis of HSCs and contains several cells (i.e., mesenchymal stem cells, endothelial cells, fibroblasts, osteoblasts, reticular cells, adipocytes). These cells are sources of cytokines, growth factors, glycoproteins and glycosaminoglycans, among other regulators. Different combinations of these molecules lead to the formation of specific microenvironments within the medullary cavity, known as niches [4]. Histologically defined microenvironments are subdivided into four regions: endosteal, subendosteal, central, and perisinusoidal. Granulocytes and monocytes are found in all regions of the bone marrow, whereas erythroblasts proliferate preferentially in the central region [5]. Concerning the dynamics of the lymphoid lineage, B lymphocyte precursors are found in the subendosteal region, gradually decreasing toward the central region, whereas mature B cells are found throughout the bone marrow [6]. HSCs are located in the endosteal region, also known as the osteoblastic niche, but studies suggest that HSCs may migrate to the perisinusoidal region or vascular niche and remain quiescent or differentiate depending on the needs of the organism [2,5,7–9]. In fact, studies with new markers for HSCs and niche cells, new image techniques, including labeling protocols, have shown that most HSCs reside adjacent to sinusoidal vessels, leading to the proposed existence of a perivascular niche for HSCs [10]. It is assumed that in the bone marrow, there are at least two different niches: the endosteal niche, which would harbor quiescent HSCs, and the perivascular niche, which would harbor cycling HSCs [11]. Although most studies have been done on non-humans, researchers suppose that the data reflect what happens in humans. It has previously been proposed that HSCs are maintained in the endosteal (osteoblastic) niche; however, the available evidence does not seem to support this model. Nevertheless, the endosteal niche seems to support the maintenance of a subset of lymphoid progenitors [10]. Approximately 80% of dividing and non-dividing HSCs have been described to be associated with sinusoidal vessels, with another 10% of HSCs being adjacent to arterioles, and almost another 10% in transition zone vessels [10]. A small percentage of HSCs are located in the endosteum. In fact, a quantitative model of cellular components that could define these microenvironments, and the preferential location of HSCs in the bone marrow are still lacking [12]. Obstacles to recognizing the HSCs in the bone marrow include the low frequencies at which HSCs are found in the bone marrow and the cellular complexity of the bone marrow microenvironment.

Cell signaling in the HSC niche is a complex process and passes through an extensive signaling network balancing self-renewal and differentiation [13]. This signaling involves several substances, such as growth factors and cytokines that are secreted by both the medullary stroma cells and hematopoietic stem cells and are important signaling factors for hematopoiesis, proliferation, and differentiation [14].

The Jak/STAT, Ras/Erk, and PI3K/Akt signaling pathways have been described as important inducers of erythropoiesis transduction by activation of the erythropoietin receptor, and these intracellular pathways are responsible for the survival, proliferation, and differentiation of normal erythropoietic progenitors [15].

Other intracellular signaling pathways that are important for the control of hematopoiesis have been described, such as the Notch, Wingless-type (Wnt), and Hedgehog pathways [16], which have been associated with self-renewal and maintenance of HSCs. Notch proteins are highly conserved receptors on the surface of the cell membrane that regulate the development of stem cells, and mutations in this receptor may cause leukemia [17] and breast cancer [18]. Activation of the Notch pathway is necessary to keep HSCs undifferentiated. This pathway is more active in HSCs and less active in differentiated cells. Inhibition of the Notch pathway potentiates the differentiation of HSCs and loss of the bone marrow reconstitution capacity of sub-lethally irradiated animals; thus, Notch has been used as a marker of undifferentiated HSCs [19]. Wnt protein regulates several phenomena during fetal development, and this protein has been related to the self-renewal of stem cells [20]. Hedgehog (Hh) protein has been described as regulating embryonic and adult stem cell activity. In mammals, three genes are known to be responsible for this protein—Sonic Hedgehog (SHh), Indian Hedgehog (IHh), and Desert Hedgehog (DHh) [21,22].

Soluble factors are also closely associated with the maintenance and regulation of the undifferentiated state of HSCs in the bone marrow of adults, in addition to regulating the proliferation and differentiation of this population. Stromal-cell-derived factor-1 (SDF-1/CXCL12) and its CXCR4 linker are activated to recruit endothelial progenitor cells (EPCs) and regulate HSCs [2,13,23]. Other soluble factors act to promote the maintenance of HSCs in their niche; for example, the stem cell factor (SCF/Kit-ligand) and its c-Kit receptor (CD117) are both required by HSCs for their maintenance. SCF is an important soluble cytokine for hematopoiesis, and THE c-Kit receptor is expressed on the HSCs surface; altered forms of this receptor have been associated with several types of cancer [24,25]. Thrombopoietin (TPO) and its MPL ligand are also important soluble factors necessary for the maintenance of HSCs in their niche. TPO is a primary physiological regulator responsible for the stimulation of platelet production, a primary dominant factor and megakaryocytopoiesis stimulator. In addition, recent in vitro studies have shown that TPO alone or in combination with growth factors, such as a c-Kit ligand, IL-3, or even Flt-3, stimulates the proliferation of hematopoietic progenitor cells [26,27]. Many other factors also modulate the function of HSCs but are not necessarily required, such as angiogenin, angiopoietin-1, G-CSF, IL-6, and TGFβ, among others [25].

3. Stem Cell Culture Methods

The complex microenvironment and cellular interactions are difficult to reproduce in vitro. Some improved techniques can help researchers mimic the bone marrow architecture for hematopoiesis studies and cell production in regenerative medicine applications.

Cell culture is the growth of cells from animal or plant in a favorable, artificial, and controlled environment. Today, cell culture is the basis of biology techniques and essential for regenerative medicine procedures [28]. These cells can originate directly from the tissue or after enzymatic or mechanically dissociated tissue or can be derived from an already established cell line.

Traditional cell culture, also known as 2D cell culture, is a very well-established method and easy-to-use culture model (Figure 2A,C). This technique depends on adherence to a flat surface, typically a culture flask, dish or polystyrene glass, to provide mechanical support for its growth in monolayers and access to nutrients and growth factors present in the culture media. Cells can also grow in suspension, as those derived from blood or bone marrow [29]. This method may present some advantages, such as low cost and performance of the functional assays, but this type of culture may also present some disadvantages, such as not mimicking the natural structures of the tissue and not being able to efficiently mimic cell–cell and cell–environment extracellular interactions [29].

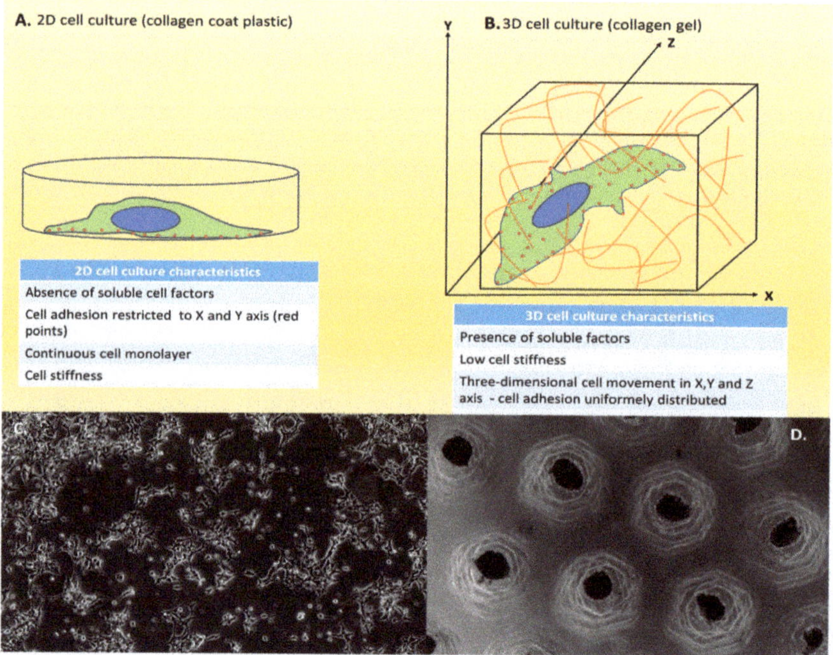

Figure 2. Two-dimensional (2D) and three-dimensional (3D) cell culture. (**A**) Schematic model of 2D cell culture. Standard model of 2D cell culture. Cells are cultured as a single layer in a culture flask. (**B**) Schematic model of 3D cell culture. 3D model of cell culture giving the notion of height, width and depth; cells are surrounded by the medium. (**C**) Example of 2D culture. HEK 293 cells in 2D cell culture; the adherence to a flat surface provides mechanical support for growth in monolayers. Scale bar 400 μm. (**D**) Example of 3D culture. HEK 293 cells 3D cell culture in agarose allowing the cells to grow or interact with their surroundings in all three dimensions. Scale bar 1000 μm.

Primary cell culture has been one of the greatest tools of cellular biology for evaluating cellular aspects such as chemoresistance, karyotypes, cellular parameters, metabolism and in vitro modeling of physiological and pathological models [30]. The isolated cells resemble their tissue of origin. Primary cell culture is complex and requires specific care, including storage, thawing-freezing procedures, or choice of enzymatic treatment, but the main problem is how to keep the primary cell in culture long enough to be established for experimental tests [31–33]. Primary culture can be representative of the cell types of the tissue from which they were isolated; although these cells are difficult to maintain, they best mimic a pathological condition or physiological function.

Immortalized cells or established cell lineages are cells that can grow in vitro indefinitely due to natural or induced transformation (e.g., embryonic stem cells or viral transformation) [34]. Immortalized cells have several advantages over primary cells, such as profitability, easy manipulation, unlimited supply of material, and lack of ethical concerns associated with the use of animal and human tissues [35]. Immortalized cells also provide a uniform and homogeneous population of cells [36]. Immortalized cell lines emerged as a solution to some problems that appeared with primary cell cultures, such as misidentification of the cell line used and genotypic and phenotypic instability. Care should also be taken to avoid the use of aged cell lines that do not maintain the original physiological characteristics. The immortalized strains constitute a simple and representative model system for functional studies and therapeutic targets. As each cell line may have unique characteristics, specific studies should take these characteristics into account [37].

The co-culture technique is used for numerous applications, including the study of natural or synthetic interactions between distinct populations of cells. Co-culture methods are of great importance in research, as they are used to observe cell–cell interactions, how cells are organized, how they participate in the development of diseases such as cancers, in which different types of cells are involved, including the microenvironment [38].

Both self-renewal and differentiation are required abilities in any cell of the hematopoietic lineage [39]. However, the expansion of these cells in vitro has been challenging for the scientific community, as regulation of these cells depends on several mechanisms of intercellular communication resulting from the secretion of local- and systemic-acting factors [40]. HSCs, which can also be obtained and expanded from umbilical cord blood, should respond to and integrate events in their microenvironment to ensure sustained production of all hematopoietic lineages [41].

Although HSCs have been extensively analyzed and characterized, their ex vivo expansion remains a problem [42–44]. It has been shown that the cell culture of HSCs is viable for at least 169 days but very much depends on the quality of the HSCs. There are currently several approaches to ex vivo expansion. Most first require the isolation of $CD34^+$ or $CD133^+$ cells from frozen or fresh hematopoietic tissue and incubation in culture medium supplemented with cytokines, granulocyte colony stimulating factor (G-CSF), stem cell factor (SCF), and thrombopoietin (TPO) [45].

Therefore, the cell culture of HSCs must take into account the microenvironment in which these cells are inserted so that it can reproduce the whole framework of the hematopoietic structure. The hematopoietic niche contains several types of cells, and mimicking this microenvironment in vitro requires a stromal layer, which controls multiple cellular parameters, including quiescence, self-renewal, differentiation, apoptosis and migration. Under the artificial culture conditions, HSCs undergo differentiation and apoptosis [46]. A support layer is required so HSCs can survive and proliferate.

Co-culture of HSCs is one of the most frequently used models for understanding how the highly specific bone marrow niche interacts ex vivo with hematopoietic cells, promoting their differentiation and expansion. The technique can focus on the importance of non-contact culture systems on the successful maintenance of hematopoietic cells [47] and on the use of the stroma as a cell layer to provide support for HSC culturing, as mentioned above. Several studies have shown that contact between HSCs and stromal cells is important for maintaining HSC function [48,49]. The stromal cells from bone marrow include osteoblasts, macrophages, endothelial cells, and mesenchymal cells [50,51]. Mesenchymal cells are multipotent progenitor cells that can differentiate into mesenchymal cells, such as osteoblasts, adipocytes, and chondrocytes, and also have the potential for differentiation into cells such as neurons and lung cells [52]. These multipotent cells have interactions with hematopoietic cells forming the framework of the HSC niche, supporting the development of hematopoiesis and acting as immunological regulators [53,54]. In addition, recent studies have indicated that other molecules, such as N6-methyladenosine (M6A), by modulating the expression of a group of YTHDF2 genes at the mRNA level, are important regulators of HSC self-renewal. Some authors have suggested that although extensive efforts have led to multiple methods for in vivo expansion of HSCs, it is not possible for some single molecules or pathways to be manipulated simultaneously due to a large number of essential targets for self-renewal of stem cells [55].

In this regard, there are still important challenges to overcome, including the development of more efficient methods for the maintenance of HSCs in vitro, and methods of ex vivo expansion for therapeutic development in regenerative medicine; then, it will be possible to have a better understanding of the hematopoietic niche and its intrinsic and extrinsic regulators from a physiological and pathophysiologic point of view. The ability to genetically reprogram HSCs for clinical therapeutic use [56] still needs to be improved.

4. 3D Hematopoietic Stem Cell Culture

Cell culture in 2D format is simple and provides excellent material for studying homogeneous populations [57]. However, it does not consider several other important parameters of cell physiology, such as cell–cell communication or communication between the cell and microenvironment or adjacent molecules [58]. Therefore, the major problem with 2D cell culture is its limits [59]. One of the main criticisms of this study format is that animal physiology cannot be mimicked using only one layer of cells, which certainly does not correspond to the original tissue considering the complexity of the cellular microenvironment in the tissue of origin [60].

As an alternative to the technical limitations of 2D cell culture, 3D cell culture allows a better simulation of the in vivo structural complexity, replicating several characteristics present in tissues (Figure 2B,D), not only the interaction of cells with their microenvironment, but also morphology, differentiation, polarity, proliferation rate, gene expression profiles, and cell heterogeneity [61–64]. In fact, 3D culture has proven to be a realer model for translating research results into in vivo applications.

Current 3D cell culture models and methods include spheroids, organoids, microcarrier cultures, organs-on-chips, and 3D bioprinting [65,66]. Organotypic explant culture methods are used mostly when a technical/specific requirement of the tissue is to be studied and mimicked [67]. However, although the complexity of the 3D system is evident, some criteria must be considered and cannot be disregarded, such as the choice of material for the scaffold and hydrogels and cell type and culture methods, which vary considerably according to the tissue studied [66].

Organoid culture is most commonly used to describe constructs derived from pluripotent stem cells (embryonic or induced cells) or adult stem cells from various organs, including the hematopoietic tissue [68]. Organoid culture is classified as either tissue organoids (i.e., organotypic) or stem cells depending on how the cell layers are formed [69]. Organotypic tissue refers to free stromal cells without parenchyma, and its application is mainly with epithelial cells because of their high intrinsic ability to self-organize. With this method, it is possible to study HSCs, and Christopher et al. [70] were able to produce mature T cells from stem cells and hematopoietic progenitors in a thymus organoid culture.

Spheroid culture models simulate the microenvironment conditions of a living cell. Compared to the classic 2D model, the spheroid culture model emphasizes the interactions between cells and their relationship to the extracellular matrix (ECM) [71,72]. Tissues are not composed of a homogeneous population of cells; they are complex structures formed by several different components, with an intricate relationship such as vessels, nerves, and stroma, which should be considered in any kind of tissue engineering. This complex interaction between cells and the extracellular compartment requires a 3D environment to best represent these interactions [73,74].

Spheroid cell culture can be used in scaffold and scaffold-free models. The scaffold methods comprise hydrogel support, decellularized extracellular matrix, and resistant polymeric material support. Hydrogel consists of polymeric groups with a water-swollen hydrophilic structure [75]. Depending on their nature, these compounds are classified into different categories, including ECM protein–containing hydrogels (ECMPs), natural hydrogels, or synthetic hydrogels, each one with its own properties [76,77]. The decellularization technique consists of removing cells from the native tissue by chemical treatment, preserving the ECM. The tissue can then be replenished with cultured cells and grafted in vivo or used for ex vivo cell differentiation. In contrast to artificial scaffolds, decellularized scaffolds preserve the vascular structure, tridimensional niches, and chemical composition of bone marrow. Extracellular matrix components are directly related to several important factors for cell survival support, including cell behavior, signaling for survival, and proliferation, among others. Resistant polymeric material supports are structures similar to fibers or sponges. Cells maintained in this type of support exhibit a physiological behavior close to that of native tissue (such as those of cartilaginous tissue). Regarding the polymeric support, the most commonly used material is polystyrene, but biodegradable supports such as polycaprolactone have also been employed.

The development of a bioprinting system for the bone marrow microenvironment is important since it cannot be mimicked by methods such as organoids. Advanced 3D bone marrow models could serve for several different studies, including hematopoietic regulation, BM, MSCs, HSC interaction and expansion, interactions of hematological cancer cells, and evolution of several hematological diseases. In addition, these models could be used as a platform for expanding HSCs for transplantation. However, it is difficult to obtain a reliable BM model because of several critical technical challenges, from currently available 3D printing techniques to the possibility of precisely mimicking the different BM niches [78].

Relatively few studies have been done using hematopoietic stem cells in these methods. Decellularization of cartilaginous tissues has been tested for ex vivo culturing of hematopoietic cells [79]. A co-culture model of hematopoietic stem/progenitor cell (HSPC) spheroids using polydimethylsiloxane (PDMS) has shown that the effectiveness of three-dimensional culture for HSPC expansion for clinical use is still a strategy that needs further improvement [80].

A three-dimensional collagen-based culture using HSPC, bone-marrow-derived MSCs or the umbilical cord (UC) to mimic the main compartments of the bone marrow hematopoietic niche has been proposed [81]. Data analysis generated the following compartments: (I) HSPC in suspension above collagen and (II) migratory HSPC in collagen fiber matrix. The different sites were representative of the distinct microenvironments that make up the bone marrow and have a significant impact on the fate of HSPC. The authors suggested that this 3D culture system using collagen and BM-MSC allowed HSPC expansion and provided a potential platform for advanced study of niches and hematopoiesis and their regulatory mechanisms.

A bone marrow composed of two compartments, solid and liquid, that act harmoniously, has been proposed [82]. A bone marrow structure was created using a macroporous PEG hydrogel that resembled the macroporous 3D architecture of the trabecular bone, site of the red bone marrow, and, therefore, where the HSPC niches are located. This bone marrow analog was found to be suitable for HSPC culture and for enhancing HSPC expansion compared to conventional 2D cell culture. The developed model of a perfused 3D bone marrow analogue mimicked the HSCs niche under steady-state or activated cell conditions that favor the maintenance or differentiation, respectively, of HSCs and allowed drug testing [82] It was concluded that the system reflected the behavior of HSPC in the niche under physiological conditions.

Some studies have shown that the interaction of bone marrow stromal cells with leukemic cells increases the resistance of these leukemic cells. Numerous scaffolds have been created to provide a minimal structure based on the 3D leukemic microenvironment [73]. As 3D culture could result in resistance to drugs, it can be a good screening tool for drug evaluation prior to the administration of chemotherapic drugs [74,83]. A classic example of scaffolds used to recreate the leukemic niche is polycaprolactone, which is an aliphatic and biodegradable polyester with good biocompatibility [84,85].

Recently, a scaffold of degradable zwitterionic hydrogel was tested for human HSPC expansion [86]. A 73-fold increase of long-term hematopoietic stem cells was observed (LT-HSC). The viscoelasticity and smoothness of the highly hydrated zwitterionic hydrogels seems to be important for the creation of cell niches, by their unique mechanical and antifouling properties.

The organs-on-chips technique was developed to study the mechanical and physiological response of a tissue, combining concepts of tissue engineering and microfluidics [87]. A bioelectronic device based on a conductive polymer scaffold was integrated with an electrochemical transistor configuration that allows 3D cell growth and the real-time monitoring of cell adhesion and growth.

This technique consists of manipulating small amounts (10^{-9} to 10^{-18} L) of fluids using small openings with micrometer dimensions. This methodology has been widely reported because it offers several advantages, such as using small amounts of samples and reagents, performing separations and detections with high resolution, and a short time for analysis [61]. Such techniques allow the precise control of fluids and particles in a given cell culture. This culture method enables the control of such nanoliter-scale fluids as described above and further enables and facilitates the simultaneous manipulation of cultured cells from a single cell [62]. Through precise manipulation of the components of the microfluidic culture medium, it is possible to transport nutrients, hormones, and oxygen growth factors to facilitate homeostasis and recreate mechanical signals that are absent in traditional culture [88].

This cell culture system offers several advantages for basic research applications, including the precision of micromanufacturing, which allows the presentation of a controllable and reproducible microenvironment [64]. Another advantage of this model is that it provides complete control over the conditions of the cell culture, including dynamic cell control, nutrient addition, removal of metabolites, stimulation with drugs and proteins, and simultaneous image and chip format [62,89,90].

The organs-on-chips technique was described in cancer studies (cancer-on-chip), where it can replicate the microenvironment to achieve robust and reliable results [91,92]. However, up to now, it has been difficult to culture HSCs in an organs-on-chips model. A device with a central cavity that successfully mimics the bone and bone marrow has been described. This device is made from polydimethylsiloxane (PDMS) with bone inducers inserted. The central chamber of this device is composed of porous PDMS membranes, and cytokines are added in the microfluids [93]. Although several methods have been used, results using HSCs are generally poor [94].

5. Final Remarks

Bone marrow HSCs are the stem cell most used in medical practice. Finding effective ways to mimic the bone marrow HSCs and microenvironment, the hematopoietic niche, in vitro is a challenge. In addition to HSCs, several other bone marrow cell types, including megakaryocytes, macrophages, monocytes and endothelial cells, directly or indirectly regulate HSCs and niche function. Two-dimensional cell cultures have been widely used due to their low maintenance cost and easy learning. However, 2D culture has several limitations. As a result, 3D cell culture techniques have been developed. Although the current cost is relatively high, the fact that a 3D culture has characteristics much closer to the tissue being studied is a great advantage. There are several methods currently employed or in development, such as spheroids, organoids, 3D bioprinting and organs-on-chip, and these are facing the challenge of culturing HSCs while maintaining their properties. This achievement will allow ex vivo HSC expansion, bringing new medical perspectives for HSCs transplantation, drug testing, and personalized treatment.

Author Contributions: Conceptualization, A.C.R.-F.; D.L. and S.P.B.; writing—original draft preparation, A.C.R.-F.; D.L.; J.L.M.R. and M.d.C.M.; writing—review and editing, A.C.R.-F.; D.L.; J.L.M.R.; M.d.C.M. and S.P.B.; supervision, S.P.B.; funding acquisition, S.P.B.

Funding: This research was funded by Conselho Nacional de Desenvolvimento Científico e Tecnológico (CNPq), Coordenação de Aperfeiçoamento de Pessoal de Nível Superior (CAPES), Instituto Nacional de Ciência e Tecnologia–Fluidos Complexos (INCT-FCx), Instituto Nacional de Ciência e Tecnologia—Medicina Regenerativa (INCT-Regenera), all from Brazil.

Conflicts of Interest: The authors declare no conflict of interest.

References

1. Ng, A.P.; Alexander, W.S. Haematopoietic stem cells: Past, present and future. *Cell Death Discov.* **2017**, *3*, 17002. [CrossRef]
2. Chotinantakul, K.; Leeanansaksiri, W. Hematopoietic stem cell development, niches, and signaling pathways. *Bone Marrow Res.* **2012**, *2012*, 270425. [CrossRef]
3. Wasnik, S.; Tiwari, A.; Kirkland, M.A.; Pande, G. Osteohematopoietic stem cell niches in bone marrow. *Int. Rev. Cell Mol. Biol.* **2012**, *298*, 95–133.
4. Bianco, P.; Robey, P.G. Stem cells in tissue engineering. *Nature* **2001**, *414*, 118–121. [CrossRef]
5. Calvi, L.M.; Link, D.C. Cellular complexity of the bone marrow hematopoietic stem cell niche. *Calcif. Tissue Int.* **2014**, *94*, 112–124. [CrossRef]
6. Nilsson, S.K.; Johnston, H.M.; Coverdale, J.A. Spatial localization of transplanted hemopoietic stem cells: Inferences for the localization of stem cell niches. *Blood* **2001**, *97*, 2293–2299. [CrossRef]
7. Kiel, M.J.; Morrison, S.J. Maintaining hematopoietic stem cells in the vascular niche. *Immunity* **2006**, *25*, 862–864. [CrossRef]
8. Orkin, S.H.; Zon, L.I. Hematopoiesis: An evolving paradigm for stem cell biology. *Cell* **2008**, *132*, 631–644. [CrossRef]
9. Calvi, L.M.; Adams, G.B.; Weibrecht, K.W.; Weber, J.M.; Olson, D.P.; Knight, M.C.; Martin, R.P.; Schipani, E.; Divieti, P.; Bringhurst, F.R.; et al. Osteoblastic cells regulate the haematopoietic stem cell niche. *Nature* **2003**, *425*, 841–846. [CrossRef]
10. Oh, I.H.; Kwon, K.R. Concise review: Multiple niches for hematopoietic stem cell regulations. *Stem Cells* **2010**, *28*, 1243–1249.
11. Crane, G.M.; Jeffery, E.; Morrison, S.J. Adult haematopoietic stem cell niches. *Nat. Rev. Immunol.* **2017**, *17*, 573–590. [CrossRef]
12. Nies, C.; Gottwald, E. Advances in Tissue Engineering and Regenerative Medicine Artificial Hematopoietic Stem Cell Niches-Dimensionality Matters. *Adv. Tissue Eng. Regen. Med.* **2017**, *2*, 42.
13. Gomariz, A.; Isringhausen, S.; Helbling, P.M.; Nombela-Arrieta, C. Imaging and spatial analysis of hematopoietic stem cell niches. *Ann. N. Y. Acad. Sci.* **2019**. [CrossRef]
14. Blank, U.; Karlsson, G.; Karlsson, S. Signaling pathways governing stem-cell fate. *Blood* **2008**, *111*, 492–503. [CrossRef]
15. Majka, M.; Janowska-Wieczorek, A.; Ratajczak, J.; Ehrenman, K.; Pietrzkowski, Z.; Kowalska, M.A.; Gewirtz, A.M.; Emerson, S.G.; Ratajczak, M.Z. Numerous growth factors, cytokines, and chemokines are secreted by human CD34(+) cells, myeloblasts, erythroblasts, and megakaryoblasts and regulate normal hematopoiesis in an autocrine/paracrine manner. *Blood* **2001**, *97*, 3075–3085. [CrossRef]
16. Jafari, M.; Ghadami, E.; Dadkhah, T.; Akhavan-Niaki, H. PI3k/AKT signaling pathway: Erythropoiesis and beyond. *J. Cell. Physiol.* **2019**, *234*, 2373–2385. [CrossRef]
17. Cerdan, C.; Bhatia, M. Novel roles for Notch, Wnt and Hedgehog in hematopoesis derived from human pluripotent stem cells. *Int. J. Dev. Biol.* **2010**, *54*, 955–963. [CrossRef]
18. Aster, J.C.; Pear, W.S.; Blacklow, S.C. Notch signaling in leukemia. *Annu. Rev. Pathol.* **2008**, *3*, 587–613. [CrossRef]
19. Tekmal, R.R.; Keshava, N. Role of MMTV integration locus cellular genes in breast cancer. *Front. Biosci.* **1997**, *2*, 519–526.
20. Duncan, A.W.; Rattis, F.M.; DiMascio, L.N.; Congdon, K.L.; Pazianos, G.; Zhao, C.; Pazianos, G.; Zhao, C.; Yoon, K.; Cook, J.M.; et al. Integration of Notch and Wnt signaling in hematopoietic stem cell maintenance. *Nat Immunol.* **2005**, *6*, 314–322. [CrossRef]
21. Gordon, M.D.; Nusse, R. Wnt signaling: Multiple pathways, multiple receptors, and multiple transcription factors. *J. Biol. Chem.* **2006**, *281*, 22429–22433. [CrossRef] [PubMed]
22. Dyer, M.A.; Farrington, S.M.; Mohn, D.; Munday, J.R.; Baron, M.H. Indian hedgehog activates hematopoiesis and vasculogenesis and can respecify prospective neurectodermal cell fate in the mouse embryo. *Development* **2001**, *128*, 1717–1730.
23. Shi, X.; Wei, S.; Simms, K.J.; Cumpston, D.N.; Ewing, T.J.; Zhang, P. Sonic Hedgehog Signaling Regulates Hematopoietic Stem/Progenitor Cell Activation during the Granulopoietic Response to Systemic Bacterial Infection. *Front. Immunol.* **2018**, *9*, 349. [CrossRef] [PubMed]

24. Mendez-Ferrer, S.; Michurina, T.V.; Ferraro, F.; Mazloom, A.R.; Macarthur, B.D.; Lira, S.A.; Scadden, D.T.; Ma'ayan, A.; Enikolopov, G.N.; Frenette, P.S. Mesenchymal and haematopoietic stem cells form a unique bone marrow niche. *Nature* **2010**, *466*, 829–834. [CrossRef]

25. McNiece, I.K.; Briddell, R.A. Stem cell factor. *J. Leukoc. Biol.* **1995**, *58*, 14–22. [CrossRef]

26. Edling, C.E.; Hallberg, B. c-Kit–a hematopoietic cell essential receptor tyrosine kinase. *Int. J. Biochem. Cell Biol.* **2007**, *39*, 1995–1998. [CrossRef]

27. Nakamura-Ishizu, A.; Takizawa, H.; Suda, T. The analysis, roles and regulation of quiescence in hematopoietic stem cells. *Development* **2014**, *141*, 4656–4666. [CrossRef]

28. Solar, G.P.; Kerr, W.G.; Zeigler, F.C.; Hess, D.; Donahue, C.; de Sauvage, F.J.; Eaton, D.L. Role of c-mpl in early hematopoiesis. *Blood* **1998**, *92*, 4–10. [CrossRef]

29. Mao, A.S.; Mooney, D.J. Regenerative medicine: Current therapies and future directions. *Proc. Natl. Acad. Sci. USA* **2015**, *112*, 14452–14459. [CrossRef]

30. Duval, K.; Grover, H.; Han, L.H.; Mou, Y.; Pegoraro, A.F.; Fredberg, J.; Chen, Z. Modeling Physiological Events in 2D vs. 3D Cell Culture. *Physiology (Bethesda)* **2017**, *32*, 266–277.

31. Lerescu, L.; Tucureanu, C.; Caras, I.; Neagu, S.; Melinceanu, L.; Salageanu, A. Primary cell culture of human adenocarcinomas–practical considerations. *Roum. Arch. Microbiol. Immunol.* **2008**, *67*, 55–66.

32. Lopes, A.A.; Peranovich, T.M.; Maeda, N.Y.; Bydlowski, S.P. Differential effects of enzymatic treatments on the storage and secretion of von Willebrand factor by human endothelial cells. *Thromb. Res.* **2001**, *101*, 291–297. [CrossRef]

33. Janz Fde, L.; Debes Ade, A.; Cavaglieri Rde, C.; Duarte, S.A.; Romao, C.M.; Moron, A.F.; Zugaib, M.; Bydlowski, S.P. Evaluation of distinct freezing methods and cryoprotectants for human amniotic fluid stem cells cryopreservation. *J. Biomed. Biotechnol.* **2012**, *2012*, 649353.

34. Ruiz, J.L.; Fernandes, L.R.; Levy, D.; Bydlowski, S.P. Interrelationship between ATP-binding cassette transporters and oxysterols. *Biochem. Pharmacol.* **2013**, *86*, 80–88. [CrossRef] [PubMed]

35. Kirchhoff, C.; Araki, Y.; Huhtaniemi, I.; Matusik, R.J.; Osterhoff, C.; Poutanen, M.; Samalecos, A.; Sipilä, P.; Suzuki, K.; Orgebin-Criste, M.C. Immortalization by large T-antigen of the adult epididymal duct epithelium. *Mol. Cell. Endocrinol.* **2004**, *216*, 83–94. [CrossRef] [PubMed]

36. Kaur, G.; Dufour, J.M. Cell lines: Valuable tools or useless artifacts. *Spermatogenesis* **2012**, *2*, 1–5. [CrossRef]

37. Rosa Fernandes, L.; Stern, A.C.; Cavaglieri, R.C.; Nogueira, F.C.; Domont, G.; Palmisano, G.; Bydlowskia, S.P. 7-Ketocholesterol overcomes drug resistance in chronic myeloid leukemia cell lines beyond MDR1 mechanism. *J. Proteomics* **2017**, *151*, 12–23. [CrossRef] [PubMed]

38. Romano, P.; Manniello, A.; Aresu, O.; Armento, M.; Cesaro, M.; Parodi, B. Cell Line Data Base: Structure and recent improvements towards molecular authentication of human cell lines. *Nucleic Acids Res.* **2009**, *37*, D925–D932. [CrossRef]

39. Miki, Y.; Ono, K.; Hata, S.; Suzuki, T.; Kumamoto, H.; Sasano, H. The advantages of co-culture over mono cell culture in simulating in vivo environment. *J. Steroid Biochem. Mol. Biol.* **2012**, *131*, 68–75. [CrossRef]

40. Eridani, S.; Sgaramella, V.; Cova, L. Stem cells: From embryology to cellular therapy? An appraisal of the present state of art. *Cytotechnology* **2004**, *44*, 125–141. [CrossRef]

41. Csaszar, E.; Kirouac, D.C.; Yu, M.; Wang, W.; Qiao, W.; Cooke, M.P.; Boitano, A.E.; Ito, C.; Zandstra, P. Rapid expansion of human hematopoietic stem cells by automated control of inhibitory feedback signaling. *Cell Stem Cell* **2012**, *10*, 218–229. [CrossRef] [PubMed]

42. Le Blanc, K.; Frassoni, F.; Ball, L.; Locatelli, F.; Roelofs, H.; Lewis, I.; Lanino, E.; Sundberg, B.; Bernardo, B.M.; Remberger, M.; et al. Mesenchymal stem cells for treatment of steroid-resistant, severe, acute graft-versus-host disease: A phase II study. *Lancet* **2008**, *371*, 1579–1586. [CrossRef]

43. Zon, L.I. Intrinsic and extrinsic control of haematopoietic stem-cell self-renewal. *Nature* **2008**, *453*, 306–313. [CrossRef] [PubMed]

44. Heike, T.; Nakahata, T. Ex vivo expansion of hematopoietic stem cells by cytokines. *Biochim. Biophys. Acta* **2002**, *1592*, 313–321. [CrossRef]

45. Ando, K.; Yahata, T.; Sato, T.; Miyatake, H.; Matsuzawa, H.; Oki, M.; Miyoshi, H.; Tsuji, T.; Kato, S.; Hotta, T. Direct evidence for ex vivo expansion of human hematopoietic stem cells. *Blood* **2006**, *107*, 3371–3377. [CrossRef] [PubMed]

46. McNiece, I.K.; Stoney, G.B.; Kern, B.P.; Briddell, R.A. CD34+ cell selection from frozen cord blood products using the Isolex 300i and CliniMACS CD34 selection devices. *J. Hematother.* **1998**, *7*, 457–461. [CrossRef] [PubMed]

47. Xie, J.; Zhang, C. Ex vivo expansion of hematopoietic stem cells. *Sci. China Life Sci.* **2015**, *58*, 839–853. [CrossRef] [PubMed]

48. Vaidya, A.; Kale, V. Hematopoietic Stem Cells, Their Niche, and the Concept of Co-Culture Systems: A Critical Review. *J. Stem Cells* **2015**, *10*, 13–31.

49. Breems, D.A.; Blokland, E.A.; Siebel, K.E.; Mayen, A.E.; Engels, L.J.; Ploemacher, R.E. Stroma-contact prevents loss of hematopoietic stem cell quality during ex vivo expansion of CD34+ mobilized peripheral blood stem cells. *Blood* **1998**, *91*, 111–117. [CrossRef]

50. Walenda, T.; Bork, S.; Horn, P.; Wein, F.; Saffrich, R.; Diehlmann, A.; Eckstein, V.; Ho, A.D.; Wagner, W. Co-culture with mesenchymal stromal cells increases proliferation and maintenance of haematopoietic progenitor cells. *J. Cell Mol. Med.* **2010**, *14*, 337–350.

51. Crisan, M.; Dzierzak, E. The many faces of hematopoietic stem cell heterogeneity. *Development* **2016**, *143*, 4571–4581. [CrossRef] [PubMed]

52. Boulais, P.E.; Frenette, P.S. Making sense of hematopoietic stem cell niches. *Blood.* **2015**, *125*, 2621–2629. [CrossRef] [PubMed]

53. Takano, T.; Li, Y.J.; Kukita, A.; Yamaza, T.; Ayukawa, Y.; Moriyama, K.; Uehara, N.; Nomiyama, H.; Koyano, K.; Kukita, T. Mesenchymal stem cells markedly suppress inflammatory bone destruction in rats with adjuvant-induced arthritis. *Lab. Invest.* **2014**, *94*, 286–296. [CrossRef] [PubMed]

54. Haynesworth, S.E.; Baber, M.A.; Caplan, A.I. Cytokine expression by human marrow-derived mesenchymal progenitor cells in vitro: Effects of dexamethasone and IL-1 alpha. *J. Cell Physiol.* **1996**, *166*, 585–592. [CrossRef]

55. Devine, S.M.; Hoffman, R. Role of mesenchymal stem cells in hematopoietic stem cell transplantation. *Curr. Opin. Hematol.* **2000**, *7*, 358–363. [CrossRef] [PubMed]

56. Li, Z.; Qian, P.; Shao, W.; Shi, H.; He, X.C.; Gogol, M.; Yu, Z.; Wang, Y.; Qi, M.; Zhu, Y.; et al. Suppression of m(6)A reader Ythdf2 promotes hematopoietic stem cell expansion. *Cell Res.* **2018**, *28*, 904–917. [CrossRef]

57. Ravi, M.; Paramesh, V.; Kaviya, S.R.; Anuradha, E.; Solomon, F.D. 3D cell culture systems: Advantages and applications. *J. Cell Physiol.* **2015**, *230*, 16–26. [CrossRef]

58. Murray-Dunning, C.; McArthur, S.L.; Sun, T.; McKean, R.; Ryan, A.J.; Haycock, J.W. Three-dimensional alignment of schwann cells using hydrolysable microfiber scaffolds: Strategies for peripheral nerve repair. *Methods Mol. Biol.* **2011**, *695*, 155–166.

59. Tee, D.E.H. Culture of Animal Cells: A Manual of Basic Technique. *J. R. Soc. Med.* **1984**, *77*, 902–903.

60. Melissinaki, V.; Gill, A.A.; Ortega, I.; Vamvakaki, M.; Ranella, A.; Haycock, J.W.; Fotakis, C.; Farsari, M.; Claeyssens, F. Direct laser writing of 3D scaffolds for neural tissue engineering applications. *Biofabrication* **2011**, *3*, 045005. [CrossRef] [PubMed]

61. Whitesides, G.M. The origins and the future of microfluidics. *Nature* **2006**, *442*, 368–373. [CrossRef] [PubMed]

62. Mehling, M.; Tay, S. Microfluidic cell culture. *Curr. Opin. Biotechnol.* **2014**, *25*, 95–102. [CrossRef] [PubMed]

63. El-Ali, J.; Sorger, P.K.; Jensen, K.F. Cells on chips. *Nature* **2006**, *442*, 403–411. [CrossRef] [PubMed]

64. Dittrich, P.S.; Manz, A. Lab-on-a-chip: Microfluidics in drug discovery. *Nat. Rev. Drug Discov.* **2006**, *5*, 210–218. [CrossRef] [PubMed]

65. Pampaloni, F.; Reynaud, E.G.; Stelzer, E.H. The third dimension bridges the gap between cell culture and live tissue. *Nat. Rev. Mol. Cell Biol.* **2007**, *8*, 839–845. [CrossRef] [PubMed]

66. Haycock, J.W. 3D cell culture: A review of current approaches and techniques. *Methods Mol. Biol.* **2011**, *695*, 1–15.

67. Toda, S.; Watanabe, K.; Yokoi, F.; Matsumura, S.; Suzuki, K.; Ootani, A.; Aoki, S.; Koike, N.; Sugihara, H. A new organotypic culture of thyroid tissue maintains three-dimensional follicles with C cells for a long term. *Biochem. Biophys. Res. Commun.* **2002**, *294*, 906–911. [CrossRef]

68. Lancaster, M.A.; Huch, M. Disease modelling in human organoids. *Dis. Model Mech.* **2019**, *12*. [CrossRef]

69. Lancaster, M.A.; Knoblich, J.A. Organogenesis in a dish: Modeling development and disease using organoid technologies. *Science* **2014**, *345*, 1247125. [CrossRef]

70. Seet, C.S.; He, C.; Bethune, M.T.; Li, S.; Chick, B.; Gschweng, E.H.; Zhu, Y.; Kim, K.; Kohn, D.B.; Baltimore, D.; et al. Generation of mature T cells from human hematopoietic stem and progenitor cells in artificial thymic organoids. *Nat. Methods* **2017**, *14*, 521–530. [CrossRef]

71. Yuhas, J.M.; Li, A.P.; Martinez, A.O.; Ladman, A.J. A simplified method for production and growth of multicellular tumor spheroids. *Cancer Res.* **1977**, *37*, 3639–3643.

72. Fennema, E.; Rivron, N.; Rouwkema, J.; van Blitterswijk, C.; de Boer, J. Spheroid culture as a tool for creating 3D complex tissues. *Trends Biotechnol.* **2013**, *31*, 108–115. [CrossRef] [PubMed]

73. Tanner, K.; Gottesman, M.M. Beyond 3D culture models of cancer. *Sci. Transl. Med.* **2015**, *7*, 283ps9. [CrossRef] [PubMed]

74. Gillet, J.P.; Calcagno, A.M.; Varma, S.; Marino, M.; Green, L.J.; Vora, M.I.; Patel, C.; Orina, J.N.; Eliseeva, T.A.; Singal, V.; et al. Redefining the relevance of established cancer cell lines to the study of mechanisms of clinical anti-cancer drug resistance. *Proc. Natl. Acad. Sci. USA* **2011**, *108*, 18708–18713. [CrossRef] [PubMed]

75. Li, Y.; Huang, G.; Li, M.; Wang, L.; Elson, E.L.; Jian Lu, T.; Genin, G.M.; Xu, F. An approach to quantifying 3D responses of cells to extreme strain. *Sci. Rep.* **2016**, *6*, 19550. [CrossRef] [PubMed]

76. Polymeric Scaffolds in Tissue Engineering Application: A Review. *Int. J. Polymer Sci.* **2011**, *2011*. [CrossRef]

77. Kleinman, H.K.; Philp, D.; Hoffman, M.P. Role of the extracellular matrix in morphogenesis. *Curr. Opin. Biotechnol.* **2003**, *14*, 526–532. [CrossRef]

78. Raic, A.; Naolou, T.; Mohra, A.; Chatterjee, C.; Lee-Thedieck, C. 3D models of the bone marrow in health and disease: Yesterday, today and tomorrow. *MRS Commun.* **2019**, *9*, 37–52. [CrossRef]

79. Llames, S.; Garcia, E.; Otero Hernandez, J.; Meana, A. Tissue bioengineering and artificial organs. *Adv. Exp. Med. Biol.* **2012**, *741*, 314–336.

80. Futrega, K.; Atkinson, K.; Lott, W.B.; Doran, M.R. Spheroid Coculture of Hematopoietic Stem/Progenitor Cells and Monolayer Expanded Mesenchymal Stem/Stromal Cells in Polydimethylsiloxane Microwells Modestly Improves In Vitro Hematopoietic Stem/Progenitor Cell Expansion. *Tissue Eng. Part C Methods.* **2017**, *23*, 200–218. [CrossRef]

81. Leisten, I.; Kramann, R.; Ventura Ferreira, M.S.; Bovi, M.; Neuss, S.; Ziegler, P.; Wagner, W.; Knüchel, R.; Schneider, R.K. 3D co-culture of hematopoietic stem and progenitor cells and mesenchymal stem cells in collagen scaffolds as a model of the hematopoietic niche. *Biomaterials* **2012**, *33*, 1736–1747. [CrossRef] [PubMed]

82. Rodling, L.; Schwedhelm, I.; Kraus, S.; Bieback, K.; Hansmann, J.; Lee-Thedieck, C. 3D models of the hematopoietic stem cell niche under steady-state and active conditions. *Sci. Rep.* **2017**, *7*, 4625. [CrossRef] [PubMed]

83. Reticker-Flynn, N.E.; Malta, D.F.; Winslow, M.M.; Lamar, J.M.; Xu, M.J.; Underhill, G.H.; Hynes, R.O.; Jacks, T.E.; Bhatia, S.N. A combinatorial extracellular matrix platform identifies cell-extracellular matrix interactions that correlate with metastasis. *Nat. Commun.* **2012**, *3*, 1122. [CrossRef] [PubMed]

84. Chen, Y.; Gao, D.; Wang, Y.; Lin, S.; Jiang, Y. A novel 3D breast-cancer-on-chip platform for therapeutic evaluation of drug delivery systems. *Anal. Chim. Acta* **2018**, *1036*, 97–106. [CrossRef] [PubMed]

85. Song, K.; Wang, Z.; Liu, R.; Chen, G.; Liu, L. Microfabrication-Based Three-Dimensional (3-D) Extracellular Matrix Microenvironments for Cancer and Other Diseases. *Int. J. Mol. Sci.* **2018**, *19*, 935. [CrossRef] [PubMed]

86. Wu, J.; Xiao, Y.; Chen, A.; He, H.; He, C.; Shuai, X.; Li, X.; Chen, S.; Ren, B.; Zheng, J.; et al. Sulfated zwitterionic poly(sulfobetaine methacrylate) hydrogels promote complete skin regeneration. *Acta Biomater.* **2018**, *71*, 293–305. [CrossRef] [PubMed]

87. Yi, H.G.; Lee, H.; Cho, D.W. 3D Printing of Organs-On-Chips. *Bioengineering* **2017**, *4*, 10. [CrossRef]

88. Ozbolat, I.T. Bioprinting scale-up tissue and organ constructs for transplantation. *Trends Biotechnol.* **2015**, *33*, 395–400. [CrossRef]

89. Novo, P.; Volpetti, F.; Chu, V.; Conde, J.P. Control of sequential fluid delivery in a fully autonomous capillary microfluidic device. *Lab. Chip.* **2013**, *13*, 641–645. [CrossRef]

90. Ong, S.M.; Zhang, C.; Toh, Y.C.; Kim, S.H.; Foo, H.L.; Tan, C.H.; van Noort, D.; Park, S.; Yu, H. A gel-free 3D microfluidic cell culture system. *Biomaterials* **2008**, *29*, 3237–3244. [CrossRef]

91. Zhang, Y.S.; Duchamp, M.; Oklu, R.; Ellisen, l.W.; Langer, R.; Khademhosseini, A. Bioprinting the Cancer Microenvironment. *ACS Biomater Sci. Eng.* **2016**, *2*, 1710–1721. [CrossRef] [PubMed]

92. Shafiee, A.; Atala, A. Printing Technologies for Medical Applications. *Trends Mol. Med.* **2016**, *22*, 254–265. [CrossRef] [PubMed]

93. Torisawa, Y.S.; Spina, C.S.; Mammoto, T.; Mammoto, A.; Weaver, J.C.; Tat, T.; Collins, J.J.; Ingber, D.E. Bone marrow-on-a-chip replicates hematopoietic niche physiology in vitro. *Nat. Method.* **2014**, *11*, 663–669. [CrossRef] [PubMed]

94. Di Maggio, N.; Piccinini, E.; Jaworski, M.; Trumpp, A.; Wendt, D.J.; Martin, I. Toward modeling the bone marrow niche using scaffold-based 3D culture systems. *Biomaterials* **2011**, *32*, 321–329. [CrossRef]

MDPI

St. Alban-Anlage 66

4052 Basel

Switzerland

Tel. +41 61 683 77 34

Fax +41 61 302 89 18

www.mdpi.com

Cells Editorial Office

E-mail: cells@mdpi.com

www.mdpi.com/journal/cells